# LANDSCAPE AND SETTLEMENT IN ROMANOV RUSSIA

# LANDSCAPE AND SETTLEMENT IN ROMANOV RUSSIA 1613–1917

JUDITH PALLOT
DENIS J. B. SHAW

CLARENDON PRESS · OXFORD
1990

Oxford University Press, Walton Street, Oxford OX2 6DP

Oxford New York Toronto
Delhi Bombay Calcutta Madras Karachi
Petaling Jaya Singapore Hong Kong Tokyo
Nairobi Dar es Salaam Cape Town
Melbourne Auckland

and associated companies in
Berlin Ibadan

Oxford is a trade mark of Oxford University Press

Published in the United States
by Oxford University Press, New York

British Library Cataloguing in Publication Data
Pallot, Judith
Landscape and settlement in Romanov Russia 1613–1917.
1. Russia. Economic conditions, 1613–1917.
Geographical aspects
I. Title II. Shaw, Denis J. B.
300.947′046
ISBN 0–19–823246–2

Library of Congress Cataloging in Publication Data
Pallot, Judith.
Landscape and settlement in Romanov Russia, 1613–1917/Judith Pallot, Denis J. B. Shaw
Bibliography: p.    Includes index.
1. Russian S.F.S.R.—Economic conditions.   2. Russian S.F.S.R.—Social conditions.
I. Shaw, Denis J. B.   II. Title.
HC337.R85P35   1990   306′.0947—dc20   89–34127
ISBN 0–19–823246–2

Printed in Great Britain by
Bookcraft, Midsomer Norton, Bath

# Preface

THIS book is a product of our long-standing interest in the historical geography of Russia, an interest which dates back to our postgraduate days. While relatively few geographers have worked on Russian historico-geographical problems, the contribution of historians has been considerable, and especially of those with some geographical training. We believe that a geographical approach has much to contribute to an understanding of Russian history and that a historical approach can in turn assist the development of geographical theory. If our book serves to stimulate interest in the way the two disciplines can contribute to each other, we shall be well satisfied.

A work of this kind would have been impossible without the help, co-operation, and advice of more people than we could possibly name individually. We wish, however, to express our particular thanks to our former teacher, Dr R. A. French of University College London, whose interest, cheerfulness, and optimism have never failed to spur us on, and whose advice on matters both academic and practical has been invaluable. We also wish to record our sincere gratitude to Professor S. A. Kovalev of the Faculty of Geography, Moscow State University, whose kindness, guidance, and support have been a great encouragement to us both over the years. Colleagues at the Faculty of Geography, Moscow State University, and the Institute of History of the USSR Academy of Sciences must also be thanked for their hospitality and assistance. We are indebted to colleagues and friends in Britain, the Soviet Union, and other countries who have helped and advised us in many different ways.

Inevitably our work was dependent upon the willing labours of librarians and archivists in Britain, the USSR, and the USA. We would especially like to mention librarians at the Lenin Library, Moscow, the Gorkii Library, Moscow State University, the Saltykov–Shchedrin Public Library, Leningrad, the British Library, the Library of the University of Birmingham (including the Alexander Baykov Library), the Bodleian Library, Oxford, the Map Room of the Royal Geographical Society, the New York Public Library, and the Library of Congress, Washington DC. We owe a particular debt to the Soviet archivists who provided so many essential materials from archives in Moscow and Leningrad. The maps were efficiently drawn by Jean Dowling of the

School of Georgraphy, University of Birmingham and by Angela Newman and Sarah Rhodes of the School of Geography, University of Oxford. Parts of the manuscript were typed by Margaret Smith and Lynn Ford of the School of Geography, University of Birmingham. In thanking all these many people, we reserve to ourselves the credit for such shortcomings as remain in this book.

Denis Shaw wishes to express his gratitude for the financial support for his research provided by the British Council, the British Academy's Personal Research Fund, and the University of Birmingham. Judith Pallot likewise wishes to record her gratitude to Christ Church, Oxford, the University of Oxford, the British Academy, and the British Council.

Finally, we should like to express our appreciation of the encouragement given by Andrew Schuller of Oxford University Press while this book was prepared, and not least for his exemplary patience during the period before it finally saw the light of day.

Apart from the Introduction and Conclusion, which we wrote jointly, we are individually responsible for our separate chapters. Chapters one, two, three, eight, and ten were written by Denis Shaw and chapters four, five, six, seven, and chapter nine by Judith Pallot.

D. J. B. S.
J. P.

*University of Birmingham*
*Christ Church, University of Oxford*

# Contents

# List of Figures

# List of Tables

# Note on Territorial-Administrative Subdivisions

DURING the three-hundred-year Romanov period, Russian territorial-administrative divisions underwent numerous changes. Some discussion of these changes down to the nineteenth century will be found in Chapter Ten below, and the meanings of terms are given in the

Glossary. However, in order to minimize confusion, a brief initial synopsis is given here.

In the seventeenth century the basic territorial-administrative unit was the district (*uezd*). Districts were often subdivided into *stany* while the *volost* (which we have translated 'rural district') was an administrative area beneath the *uezd* for non-private peasants. In the early eighteenth century Peter the Great divided Russia into provinces (*gubernii*) which were in turn subdivided into sub-provinces (*provintsii*) and the sub-provinces into districts (*uezdy*). With certain changes, this system lasted until the 1770s when Catherine II redivided Russia into provinces (at first generally known as 'vice royalties' or *namestnichestva* but later as *gubernii*), which were in turn subdivided into districts (*uezdy*). This basic two-tier system lasted until the 1917 revolution.

The 1861 Emancipation of the Serfs ushered in a period of change in the pattern of administration for peasants. The rural district (*volost*) now became a general unit of peasant administration below the level of the *uezd*. The *volost* was in turn subdivided into rural societies (*selskie obshchestva*). The latter might or might not correspond with the peasant commune (*obshchina*). In 1864 an elected agency of rural self-government was introduced, known as the *zemstvo*. The *zemstvo* existed at provincial and district levels. Urban administration is discussed in Chapter Ten.

# Note on Transliteration and Dates

TRANSLITERATION of Russian terms and names is based on the Library of Congress system with minor adaptations. Thus the letters *ia*, *iu*, and initial *e* are rendered as *ya*, *yu*, and *ye*, diacritical marks are omitted, and the soft and hard signs have also been omitted throughout.

Dates are given according to the Julian calendar, which was used in Russia before the 1917 revolution. This was ten days behind the Gregorian calendar in the seventeenth century, eleven days in the eighteenth century, twelve days in the nineteenth century, and thirteen days in the twentieth century.

# Abbreviations

| | |
|---|---|
| d. | *delo* (file) |
| *des.* | *desyatina, -y* |
| f. | *fond* (collection) |
| l. | *list* (folios) |
| ob. | *oborot* (verso) |
| op. | *opis* (inventory) |
| *PSZ* | *Polnoe sobranie zakonov Rossiiskoi imperii* (1830– ) |
| *S.* | *Sazhen* (*i*) |
| TsGADA | Central State Archive of Ancient Acts, Moscow |
| TsGIA | Central State Historical Archive, Leningrad |
| TsGVIA | Central State Military History Archive, Moscow |
| yed. khr. | *yedinitsa khraneniya* (unit *or* file) |

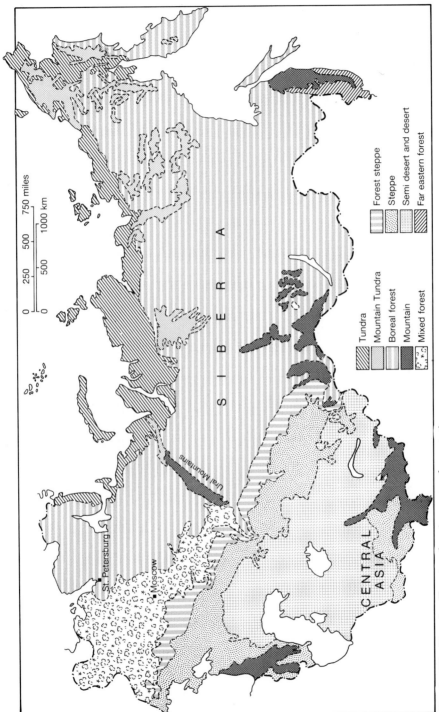

Fig 1. The Russian Empire showing natural vegetation

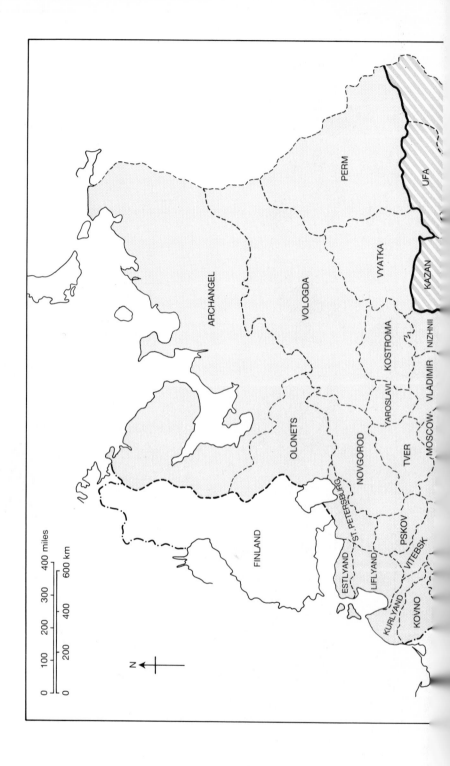

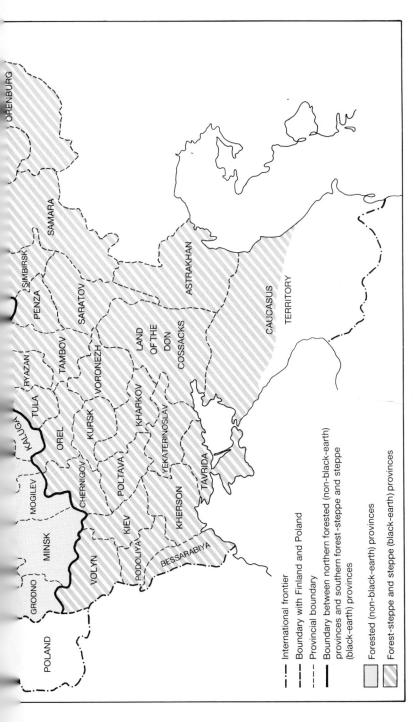

Fig 2.  European Russian provinces in the late nineteenth century

ORENBURG

SAMARA

SIMBIRSK

PENZA

SARATOV

ASTRAKHAN

RYAZAN

TAMBOV

VORONEZH

LAND OF THE DON COSSACKS

CAUCASUS TERRITORY

TULA

KALUGA

OREL

KURSK

KHARKOV

MOGILEV

CHERNIGOV

POLTAVA

YEKATERINOSLAV

MINSK

KIEV

TAVRIDA

GRODNO

VOLYN

PODOLIYA

KHERSON

POLAND

BESSARABIYA

- · · - International frontier
- — — Boundary with Finland and Poland
- - - - Provincial boundary
- —— Boundary between northern forested (non-black-earth) provinces and southern forest-steppe and steppe (black-earth) provinces

Forested (non-black-earth) provinces

Forest-steppe and steppe (black-earth) provinces

# Introduction

FOR the nineteenth-century historian V. O. Klyuchevskii, 1613 marked the beginning of modern Russian history: 'a new dynasty, new and enlarged territorial boundaries, a new class structure with a new ruling class at its head, and new economic developments'.[1] For the present-day historian or historical geographer, by contrast, the break with the past is less clear-cut. The unification of the Russian state had been achieved a century or so earlier. The great advance to the east had begun in the middle of the sixteenth century, and some of the profound social and political developments that were to affect urbanism, landholding, and settlement were already in train. Nevertheless, it is clear that many of these developments were strengthened and enhanced under the Romanov dynasty, while others were soon to make their mark. Thus, the choice of the year 1613 as the starting-point for the present work seems a logical one. Until its ultimate collapse in 1917, to be replaced by a very different social and political system, the Romanov dynasty allowed sufficient continuity in Russian development for its period to provide a convenient and useful framework for study.

The changes which affected Russia during the three centuries of Romanov rule were profound indeed. At the beginning of the era Russia's political structure was exceedingly frail, subject to both internal strife and external challenge, suffering from an array of social and economic distresses, peripheral to European developments and influences in almost every way. By the end of it, and in spite of many internal and external troubles, Russia had become a major world power, a serious contender in the race for economic supremacy, and a great military presence in Europe and Asia. The processes of geographical significance accompanying such developments were many. Territorially, Russia had expanded to incorporate a considerable section of the Baltic coast, including Finland, much of Poland, the Ukraine, and the Black Sea coast, the Transcaucasus, the deserts and mountains of Central Asia, the whole of Siberia and the Far East as far as the Bering Strait, and even, for a time, Alaska and adjacent parts of the American Pacific coast. Many countries on Russia's borders, such as China, India, Persia, and Turkey, felt threatened by such expansion, which was accompanied by migration, agricultural settlement, the foundation of towns, and much commercial development.

Territorial expansion was one factor which helped bring about change in Russian society and landscapes. Despite Klyuchevskii's statement that 1613 marked the beginning of modern Russian history, Russia could only by the wildest flight of imagination be termed 'modern' at that time. In the grip of a strengthening serfdom, characterized by xenophobia and by superstitious and conservative attitudes then under serious challenge in the rest of Europe, her economy and social structure were more medieval than modern. Yet, already in the sixteenth and seventeenth centuries new ideas and winds of change began to make themselves felt, especially as more and more Russians became aware of Western influences and technology. Such developments went much further under Peter the Great and his successors. By the late eighteenth century population growth and commercial expansion marked a veritable economic renaissance with inevitable consequences for agriculture, industry, towns, and handicrafts. Significant as such processes were, they were of minor consequence by comparison with the nineteenth century, when the forces of industrialization were finally unleashed, towns and cities grew as never before, railways were built, and the lives of many Russians were transformed. The social and psychological consequences of such events were momentous. And yet, in the context of Europe as a whole, Russia still appeared conservative and backward in 1917. It is the tension and interaction between 'traditional' and 'modern', between resistance and change, which forms the context for much of this book.

It is beyond the capabilities of a single book to encompass all the processes of interest to the historical geographer which affected Russia in the Romanov era. In the case of certain countries and regions of the world this has indeed been attempted.[2] For Russia the task would be barely possible. Despite the achievements of pre-revolutionary scholars, historical geography has largely atrophied in the USSR since 1917.[3] Sterling work has been done by Soviet historians, particularly in such areas as demographic history, certain aspects of urban history, and the history of rural society and agriculture, but the coverage is partial, selective, and far too uneven. As J. H. Bater and R. A. French have written, 'Given the presence of so many lacunae in historical geographical research on Russia, it would scarcely be possible to write an all-embracing book which gave adequate scholarly cover to all the multifarious geographical aspects of over one seventh of the world over more than a millennium.'[4] Much the same could be said of the period covered by this book.

Rather than striving for a comprehensive treatment, therefore, our aim is to present an analysis of selected facets of change in settlement, landscape, and the geographical manifestations of economic and social life as examples of the complex processes at work in Russia over the long period of Romanov rule. Our selection has been conditioned partly by our own research and acquaintance with original sources—hence those chapters which focus on relatively small areas or specific problems—and partly by the need we see to consider some of the principal changes taking place in Russia during the Romanov period. One of the most serious charges that can be brought against the 'comprehensive' approach to historical geography (or the attempt to describe entire, changing 'geographies') is that such synthetic treatments are rarely successful in capturing the bewildering variety of forces at work in moulding settlement and the landscape, or in giving simultaneous recognition to the fact of variation through space as well as through time. Selection is therefore inescapable. Our selection and analysis are informed as much by historical as by geographical considerations. In other words, we are interested not only in landscape and settlement change for their own sake, but also in their significance for Russia's development. At the same time, since we are geographers, the tangible landscape is never far from our view, nor the necessity of relating general processes to particular places in that landscape.

Our interest lies primarily in European Russia, particularly in those regions colonized by the Russians themselves. Since the Russians were both culturally and demographically dominant in the multiracial Russian state, this is perhaps no undue bias. Most Russians lived in European Russia, as indeed did most of the population of the Russian Empire. We are aware, however, that what was true of European Russians was not necessarily true of their Siberian cousins, and probably a good deal less true of the Ukrainians, the Georgians, or the Kazakhs. We have also focused our attention more upon rural than upon urban society. Even in 1914 only about 17 per cent of the Russian population was urban and, although towns were certainly not unimportant to Russia, it was probably in the vast rural regions that her modernizing rulers faced their greatest challenges. There are, moreover, a growing number of excellent texts dealing with industrial and urban topics, texts which we could not hope to rival in a single volume.[5]

## Theories and Themes

A fairly recent feature of Western, and particularly British, historical geography has been the attempt to move away from what A. H. Baker has termed the 'experimental and pragmatic scholarship' of H. C. Darby and his followers in the 'quest for a new problematic' in the sub-discipline.[6] This move, he contends, has been in response to new developments in contemporary human geography and within social history and the social sciences in general. Thus, there have been critical examinations of 'positivistic' modes of explanation, particularly in the guise of spatial analysis, and there have been pleas that historical geographers should seek to reconstruct their approaches according to behavioural, phenomenological, idealist, Marxian, or structuralist canons. Yet, in spite of the spirited and often polemical tone of many of these discussions, their theoretical harvest has as yet been meagre and their application in empirical research limited. No single theoretical viewpoint is currently dominant in historical geography and none has been developed which seems capable of incorporating the highly complex and variable interrelationships between human societies and their environments as they change through time. Perhaps what is needed is a broader appreciation of the necessity for a multiplicity of approaches and theoretical standpoints as different questions are addressed, or as similar but complex problems are tackled from different angles. In this way new and fruitful insights might be forthcoming even if the varied theories and hypotheses which give rise to them are hard or even impossible to reconcile.

The present study eschews any overt commitment to a single theoretical perspective in the belief that, for the present authors at least, such a commitment would be premature. Problems and issues which appear to be important and interesting on theoretical and empirical grounds have been chosen and analysed with sensitivity to different theoretical positions. Whether or not this pragmatic approach has been successful is left to the reader to judge.

Issues which have long been of concern to historical geographers form the substance of the volume: the ways in which human societies and their settlements are adjusted to and affected by the natural environment, questions of regional specialization and integration, patterns of landholding, land use, and agriculture. The justification for historical geography lies, in our view, in the observation that the forces which shape human society invariably have different consequences in different places, and that the experience of these places in turn affects the evolution of society at large. 'Place' is a central concept to the geographer and for the historical geographer it is past places which matter. It is perhaps nowhere more necessary to understand the importance of place than in the case of Russia, especially European Russia. In a state so vast and displaying apparently few great contrasts in topography or culture, and in a polity so obviously centralized, it is easy to be carried away by seeming uniformities and to overlook the local scale. Much of Western and Soviet research into the history of Russia has been concerned with large-scale and general processes of change.

The chapters that follow certainly consider such processes at a macro-level, but it is our belief that they cannot be properly comprehended without reference to other developments, political, economic, and social, taking place locally.

## THE NATURAL MILIEUX

European Russia, which forms the context for the local studies in this volume, is physically varied, crossed from east to west by belts of differing soil, climatic, and vegetation types. Nineteenth-century Russian geographers, such as V. V. Dokuchaev (1846–1903), identified particular combinations of these as 'natural regions'. Like their counterparts in Western Europe at the time, the Russians were convinced of the power of the natural environment to shape human society. The varied environments of European Russia did, indeed, evoke different responses from the people who came to settle in them. Wherever and whenever they settled, Russians had to use locally available resources to meet their basic requirements of food, shelter, and clothing, and later to meet their needs in trade goods. The most visible imprint of the natural environment was in the materials people used to build their dwellings, to make their utensils and their clothes, and in what they grew in their fields. But the influence of the environment went deeper than this since systems of production that were developed to utilize a particular combination of resources helped to create a 'local society' which was shaped by values and beliefs, and was characterized by structures of power and social relations, to some extent different from those in other localities. In anthropological and geographical literature such adaptations of peasant communities in pre-industrial societies to particular sets of resources available in the physical environment have been referred to as 'peasant ecotypes'.[7] As the chapters that follow will show, the Russian plain was made up of a mosaic of such human ecotypes.

During most of the period dealt with in this book, the most important physical divide for the Slavic peoples was between the forest and the steppe. Forests occupied the greater part of European Russia north of a line running some two hundred miles south of Moscow and they petered out into tundra only in the most northern parts of Karelia and the lower reaches of the Pechora River. In the south of the region forests were mixed, stands of pines, spruce, and birch interspersed with broad-leaved deciduous species such as lime, oak, elm, and maple and underlain by moderately leached podzols and grey forest soils. The deciduous stands have been greatly diminished in European Russia today. In the north the forests consisted of continuous stands of pine and spruce which constituted the western extension of the vast Siberian boreal forest. They were underlain by highly acidic soils, some with iron-pan. Throughout the forest zone short, but hot, summers alternated with long cold winters, the severity of which increased to the north and east. Abundant precipitation combined in these northern regions with poor drainage and low evaporation rates to create many marshes and swamps.

To the south of the southern margin of the mixed forests, the natural vegetation of European Russia changes into steppe through a transitional belt of forest steppe. Today, the natural vegetation of the steppe has largely been replaced by crops, but at the beginning of the Romanov period it still grew with high steppe grasses, sedges, legumes, and flowering herbs. The soils of the steppe are black earths, or chernozems, formed by the decay of the close-matted roots of the grasses. In their virgin state they were extremely fertile. In the south-east of the steppe, towards the Caspian Sea, the black earths grade into chestnut soils and vegetation becomes sparse and less varied. The climatic regime of the steppe is continental, as in the forests to the north, but the summers are hotter and much drier and aridity is a problem for human occupance. The exception to this rule is the Black Sea littoral, where the ameliorating influences of the sea reduce continentality.

At the beginning of the Romanov period most of the Russian population lived in the mixed forest zone. Within this region, and over many centuries, the Russians had learnt to adapt their agricultural practices to the afforested environment, to clear woodland, and to cultivate the podzols and grey forest soils, at first with extensive methods but gradually with more intensive farming systems. They also raised livestock on meadow and pasturelands, and learnt to exploit forests for their timber, fuel, game, and wild plants: berries, mushrooms, fruits, medicinal grasses, and herbs. The settlers in the mixed forest belt were able to create for themselves a genuinely mixed economy that was based on the exploitation of a range of farm and non-farm resources. They were woodland peasants, who met the challenge of a difficult environment by becoming jacks of all trades. They preserved this characteristic throughout the Romanov period, and by the end of the eighteenth century it is possible that a majority of households were multi-occupational, engaged in a variety of domestic and factory-based industries alongside their arable and livestock husbandry.

The experience which the mixed forest dwellers gained in using the resources of the forest proved to be valuable for the settlement of the coniferous forests. The Russian conquest of the taiga began several centuries before the start of the Romanov period and involved the gradual settlement of lands stretching north from Muscovy to the coast of the White and Barents Seas and ultimately eastwards towards the Urals and Siberia. The people who settled in the European north had to demonstrate the same ingenuity in using the resources around them as their southern neighbours, but the range available to them was different. The dense forests, swamps, and acidic soils that covered the region meant that possibilities for arable cultivation were limited except in the more favoured areas, particularly those towards the south, where shrub and bush vegetation in the fertile river basins could be easily cleared, or where clay loams had inhibited podzolization. The principal agricultural resource of the region was natural meadows, which were used for pasture and hay-making, and supported livestock husbandry. In parts of the north, and especially on the Kola peninsula (where it persisted into the twentieth century) transhumance was practised and herds were moved to upland summer pastures and returned to the lowland valleys and coasts in the winter. Elsewhere in the northern interior, the forest provided summer grazing for livestock, and along the coasts dunes and flats were also used for grazing. There were relatively few places where livestock husbandry, combined where possible with the cultivation of hardy cereals and roots, was the sole source of the peasants' livelihood. For many the most important economic activity was non-agricultural. On the coasts, lakesides, and along the river courses fishing provided the basis of peasants' subsistence and commerce, and it supported whole communities of peasant-fishermen. Freshwater bodies and the sea were extremely rich in stocks of fish, which included salmon, sturgeon, pike, cod, herring, sole, and some less familiar varieties. Fur-bearing forest animals and game supported other communities. The European taiga was an important source of furs for Russian merchants and remained so in the Romanov period even though it was overtaken by Siberia in the seventeenth century. Sable, marten, fox, hare, ermine, beaver, squirrel, and other animals were hunted by the Russians, or expropriated from native peoples. Elk, reindeer, roebuck, and bear were also hunted for their skins and meat, as were forest fowl and walruses, seals, and whales. Like the woodland peasants of the mixed forest belt, communities of peasant-fishermen, hunter-peasants and transhumant peasants in the north used the forests to supplement their diets, collecting wild plants and fruits, and as a source of building materials for their dwellings, boats, implements, and utensils.

The colonization of the steppe to the south of Muscovy took place between the sixteenth and nineteenth centuries. By comparison with the settlement of the coniferous forests, progress across the steppe was slow as the Russians encountered resistance from Tatar horsemen who roamed the area. It was further hindered by the need for Russian settlers to adapt to an environment very different from the one to which they were used in the forests. Those who made the most dramatic transition were the cossacks, escaped peasants who fled south and succeeded in the steppe by emulating the freebooting life-style of the nomads. They set up fortified villages and lived by hunting, fishing, raiding, and, only gradually, herding. The southern cossacks remained contemptuous of arable farming until population pressure and the erosion of their traditional way of life made it unavoidable. However, the majority of settlers in the steppe developed their livelihood around the region's principal resource: the black earths. By the eighteenth century the steppe had become Russia's principal 'bread-basket' and, reflecting this, the economy of serf and free settler alike was largely based upon arable husbandry.

The first settlements established in the steppe tended to be situated in the river valleys, and in the arid south this was a permanent feature. The valleys provided water, fish from the rivers, timber and fuel from the woods, meadows and open sites for agriculture, and often defensive locations for settlement. Beyond the valleys the natural steppe grasses could be used for livestock grazing. The ploughing up of the steppe for arable cultivation demanded special tools, such as the heavy iron-shod plough (*plug*) drawn by several oxen, much favoured by the Ukrainian settlers. The more successful communities also developed other tools for coping with the abundant weeds that infested the fields, and the cracking and drying of the soil that took place in the summer heat. Everywhere they farmed the settlers faced unfamiliar problems with the environment. Aridity was, of course, the principal of these, and it was exacerbated by the scorching *sukhovei*, hot, dry winds that had a devastating effect on crops and vegetation. Lack of winter snow in some years resulted in shortages of spring ground water, while very severe temperatures could kill the winter crop in the ground. Deforestation of the wooded steppe and the cultivation of fragile loess-based soils in some areas led to soil erosion, which had become a serious problem by the nineteenth century. The felling of trees also caused shortages of fuel and of timber for building. Despite all these difficulties, the rewards in favourable years were good and the centre of gravity of Russian settlement continued to shift southwards throughout the Romanov period.

Whether they relied upon the exploitation of a diverse range of natural resources, as in the forests, or were more narrowly dependent upon a limited range, rural dwellers were influenced by the growth of towns and the expansion of markets which took place in fits and starts during the Romanov period, and which accelerated towards the end of the nineteenth century. Urbanization and industrialization were to undermine some of the old ways of survival, but at the same time promoted the further development of those branches of the traditional economies for which particular places were well suited. On the national scale, particularly from the late eighteenth century, there emerged a geographical division of labour in Russia between the forest and steppe. The forest provided the steppe dwellers with manufactured goods (many produced by peasant craftsmen in their own homes), building materials, and fuel, and the steppe provided the forest dwellers with grain. The result was that differences in the life-styles of the inhabitants of the two realms were further entrenched, rather than diminished. Similar processes of increasing differentiation occurred between places located within the forest realm, and also between places located in the steppe.

Industrialization in the nineteenth century brought other changes to rural dwellers. Factories attracted people away from the land and competed for certain resources such as wood, but they also made use of new resources which had scarcely been needed before. By the end of Romanov rule, the divide between forest and steppe began to be intersected by other dividing-lines, such as those between mineral-rich and mineral-poor regions and between town and countryside. Such processes signified the dawn of a new era in the history of landscape and settlement, an era whose full impact was felt only after the 1917 revolution.

## The Political and Social Milieux

The many social and economic changes which influenced Russia during the Romanov period have already been commented upon, but something must also be said about the complex structure of Russian society and the government policy which helped to mould it. The Romanov dynasty came to power in 1613, at a time when Russia was in the throes of a dynastic crisis and subject to foreign invasion. The aim of the new dynasty was to restore national unity and social order, and to defend the land from outsiders. In this, the Romanovs were able to build on the work of their predecessors, who had created a unified state based on Moscow, assimilated the hitherto separate principalities, and developed a centralized political system. Seventeenth-century Russia was thus ruled by an absolute monarchy increasingly able to free itself from the restraints imposed by such institutions as the boyar council and *zemskii sobor* or 'Assembly of the Land'. The monarchy was sustained by a small group of aristocratic advisers and a developing administration, but the nobility did not constitute a cohesive class except in so far as they all owed service to the sovereign (those owing military service to the tsar being known as 'servitors'). One of the successes of sixteenth-century tsardom was the imposition of the service principle even upon hereditary estates. Service tenure finally disappeared in the eighteenth century, but not before the nobility had been able to secure from the tsar added powers over their peasants.

The peasants constituted the majority of the Russian population throughout the Romanov era. They were divided into a number of separate categories, each subject to different rights and obligations until the mid-nineteenth century, when uniformity was imposed. The seventeenth-century categories were as follows: slaves, becoming fewer during the course of the century; court peasants, who belonged to the tsar and the court; church and monastic peasants; 'black' or 'black-ploughing' peasants (*chernososhnye krestyane*),[8] who lived on land not formally held by any lord but which belonged to the tsar in his capacity as sovereign; proprietal peasants, who belonged to secular lords. In 1649 proprietal peasants lost their right to move from one estate to another, which effectively completed their enserfment. From this period the other categories of peasant also lost many of their former freedoms.

A third, although relatively minor, category within seventeenth-century society should also be mentioned—the merchants and towns-people (*posadskie lyudi*). As most of the towns were under the control of the sovereign, these people were regarded, like the 'black' peasants, as owing their duties and taxes directly to the tsar. The 1649 Law Code forbade their migration from town to town, and thus most of them had a social status hardly better than that enjoyed by the peasants. A few of the merchants were extremely wealthy and enjoyed the special favour of the tsar. Yet another category within seventeenth-century society was the lower class of servitors, or 'contract' servitors. They were found mostly in Moscow and in towns on the periphery of the expanding state, and were essentially military personnel. They enjoyed few of the rights and privileges of the nobility, many sharing the life-style of peasants or townspeople. In the eighteenth century most lower servitors were absorbed into these groups.

During the course of the seventeenth and eighteenth centuries, Russian society became increasingly regimented and divided into a set of discrete estates. An important stage in the process was reached in the reign of Peter the Great (1682–1725) who was one of Russia's most important modernizers. With Peter's Table of Ranks, promulgated in 1722, a final demarcation was made between the nobility, who from now on were to enjoy a virtual monopoly of serf- and landownership in exchange for their state service, and the rest of the population, whose obligations were to their lords and to the state. A further feature of Peter's policies was the imposition of a poll tax on all males, excluding the nobility and clergy. In order to facilitate this policy a series of national censuses, or 'revisions' as they were called, was implemented. The first revision was completed in the 1720s.

One result of these policies was the emergence of a new coherence and self-confidence among the landowning nobility (who still, however, varied greatly both in status and wealth).[9] The members of the nobility finally liberated themselves from the obligations to service in 1762, and obtained even greater rights and privileges from the Charter to the Nobility issued by Catherine II in 1785. The peasants, on the other hand, were now more than ever subjected to the whims and wishes of their lords, liable to be conscripted into the army or into factories, to be beaten, exiled to Siberia, or worked to death in their lord's fields. Those peasants who belonged to no lord were perhaps only slightly better off, being subject to considerable exactions by the state. In 1764 their numbers were swelled by the peasants from newly secularized church and monastic estates, these people now being known as 'economic peasants'. In the late eighteenth and nineteenth centuries the various groups of non-proprietal peasants gradually came to be regarded as forming a single category of 'state peasants'.

During the nineteenth century the pace of economic development began to break down the existing social structures. Thus, although Russian society continued to be divided into the old formal estates of peasants, merchants, townspeople, and nobility, by the end of the century these titles quite often no longer corresponded to their owners' status or occupation. For example, people categorized by virtue of their birth as peasants were to be found among the emerging class of industrial entrepreneurs or, at the other end of the scale, among the urban proletariat; if they remained on the land, they could be private landholders or agricultural wage-labourers. The same lack of corres-pondence between formal status and actual social and economic position was to be found among other classes in society. Barriers to social mobility continued to exist, however, and some of the new classes, such as industrial workers, were given no effective political or legal recognition. The tensions caused between the old and new social structures by these developments were a contributory factor in the revolutions at the beginning of the twentieth century.

The most important political landmark in the nineteenth century was the Emancipation Statute of 1861, which formally ended serfdom for proprietal peasants. It was followed by further statutes which changed the position of other peasants. Under the terms of the emancipation settlement, peasants in Russia were granted their freedom and their land. Land was not given to the peasants free of charge, however; they had to pay for its redemption and were given a forty-nine-year period in which to do this. Until the redemption payments were made on their land, peasants were bound to remain members of their communes, the institutions which historically had played a key role in the lives of the peasants. Such institutions continued to govern many aspects of peasant life, in particular their mobility, throughout the second half of the nineteenth century and into the twentieth. Despite this, large numbers of peasants migrated, albeit often temporarily, to the towns in search of work in industry and service, or to the south of Russia to work on large estates, or they embarked upon the long journey east to Siberia to take up farming in the virgin steppe. The reason for these migrations was the increasing inability of peasant households to meet their needs and wants from the land. Land shortage and rising expectations were powerful motives forcing peasants to leave their villages in search of new opportunities elsewhere.

The era of liberal reforms begun in 1861 was not sustained, and in the period of reaction that followed the traditions of autocracy quickly resurfaced and remained the dominant factor shaping state–society relations for the rest of the century.

# 1

# The Frontier Experience in Romanov Russia

*The Settlement of the Central Black Earth Region in the Seventeenth Century*

'THE American frontier', proclaimed the celebrated historian Frederick Jackson Turner in his much-quoted essay of 1893, 'is sharply distinguished from the European frontier—a fortified boundary line running through dense populations. The most significant thing about the American frontier is, that it lies at the hither edge of free land.'[1] This was a huge misconception of the European frontier, at least as far as Russia was concerned, but it was one which Turner later modified when he recommended that his method of analysis be adopted in the study of such diverse regions as Russia, Germany, Canada, Australia, and Africa.[2] He thus eventually came to recognize that the American frontier experience might have certain parallels elsewhere, including parts of Europe, though he remained convinced that in crucial respects the American record was unique.

In Russia the long and often difficult experience of the open frontier, which began when the eastern Slavs migrated on to the plains of European Russia in the first millennium AD, could hardly have escaped the notice of her later historians. Indeed V. O. Klyuchevskii, possibly the most famous of them all, argued that migration and colonization constituted the 'fundamental factor' in Russian history and attempted to use the different stages in the colonization process as a framework for ordering his greatest work.[3] Neither did the parallels with the American experience evade their attention. P. N. Milyukov, a disciple of Klyuchevskii, compared the two countries and noted numerous similarities, while nineteenth-century historians of the south European frontier, such as N. O. Vtorov and M. Skiada, felt moved to refer to a 'Russian America'.[4] Few such perceptive scholars could have encountered the name of Turner, and they would perhaps have been surprised by the suggestion that the frontier experience had uniquely moulded any one nation. Their evident interest in America anticipates more recent scholarly concern with cross-national comparisons and attempts to see historical events in broader explanatory frameworks. In frontier studies Turner's thesis has provided both the most influential and also the most provocative framework. It therefore proves a most appropriate one against which to examine the Russian experience.[5]

Russia's most dramatic frontier movement in the seventeenth century was her seizure and occupation of the vast regions lying to the east of the Urals. This eastward expansion, which eventually gave her the status of a Pacific power, was remarkable both for its speed and for the extent of the territories it embraced. By contrast, lying to the south of European Russia across the forest-steppe and steppe was a much more difficult and dangerous frontier zone, which was to prove particularly trouble-some to the early Romanovs. The steppe grasslands were occupied, as they had been for centuries, by nomadic bands of horsemen whose warlike behaviour constituted a continuing menace to the Russian state and a significant drain on its resources. Not only did the bands of Crimean and Nogai Tatars hinder Russia's long-standing communica-tions with the Black Sea and threaten her agricultural communities, but they also provided a weapon in the hands of one of her most determined adversaries, the Ottoman Turks. The southern frontier, therefore, played a particularly vital role in Russia's seventeenth-century develop-ment and was an important arena for the furnishing of her particular frontier experience.

In the early seventeenth century Russian settlement was but lightly scattered over the territories that were eventually to form the Central Black Earth Region, as shown in Fig. 1.1. To Turner, of course, the open frontier formed 'the hither edge of free land', although his use of the word 'free' in that context can be questioned. Russia's claims certainly extended far to the south over the mainly unsettled regions of the forest-steppe and steppe,[6] and these claims were to some extent bolstered by her network of patrols and guards, who kept watch over Tatar wanderings and burnt the steppe grasslands in order to deny forage to the Tatar horses.[7] Individual Russians and organized groups, or institutions such as monasteries, frequently exploited the southern resources of fish, insects, and game, sometimes paying the tsar for the privilege. But such activities with their implied claims to territory were by no means recognized by the Tatar bands, who regarded the entire region as their historic possession. Neither the Muscovite tsar nor the Crimean khan would admit that these unsettled lands were, in Bagalei's words, 'res nullius'.[8]

At the beginning of the Romanov era only the northernmost part of

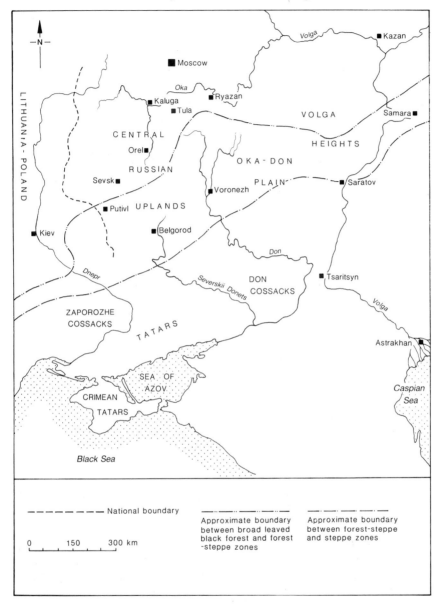

Fig. 1.1. South European Russia in the early seventeenth century

the Central Black Earth Region was occupied by an established network of fortified towns and supporting agricultural settlement, as indicated in Fig. 1.2. Many of these towns, grouped around such centres as Ryazan,

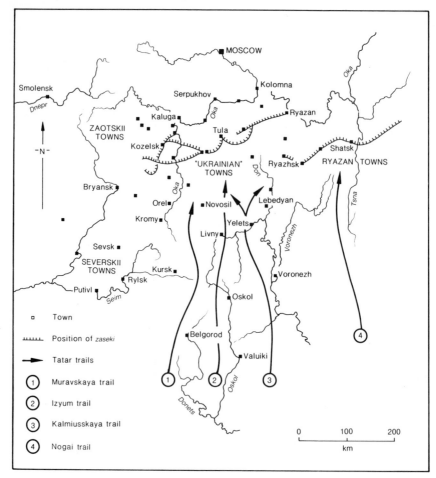

FIG. 1.2. Russia's southern defensive frontier at the beginning of the
seventeenth century

Tula, Sevsk, and Putivl, dated from the first part of the sixteenth century
or even earlier. Eventually they were linked together by a sophisticated
system of fortified lines or *zaseki*, cutting across the principal Tatar
invasion routes (*shlyakhi*) and helping to prevent incursions in the
direction of Moscow. However, after the sack of Moscow by the Tatars
in 1571 the Russian government was spurred to decisive action. Already,
in February of that year, the decision had been taken to reorganize the
system of patrols and guard-posts on a more systematic basis. This
action undoubtedly improved the method whereby early warning was
given of the approach of Tatar raiders, but it could do relatively little to

ensure their repulsion. Matters improved somewhat in the 1580s and 1590s with the building of a further series of military towns in the valleys of the Severskii Donets, Oskol, and Don. The last and most southerly of these, Tsarevborisov, was constructed close to the confluence of the Severskii Donets and Oskol rivers in 1599. However, in the ensuing Time of Troubles (1604–13) the defensive system suffered a severe reverse and Tsarevborisov disappeared entirely.

The establishment of the Romanov dynasty in 1613 heralded a gradual restoration of the southern defences. From the 1630s there followed a positive and determined effort to stem and even to challenge the Tatar menace. The completion of the Belgorod Line in the 1650s, as shown in Fig. 1.3, and further modernization of the military forces, proved a decisive stage in the occupation of the Central Black Earth territories. By the end of the century the continuing influx of Russian and Ukrainian settlers encouraged the rulers of Russia to think in terms of an aggressive policy in the south. Their attitude was encouraged by evidence of the growing inability of the Tatars to resist, despite the support of their Turkish backers.[9]

Russia's successes in the seventeenth century were only partially the consequence of action by her government. Acquisition of the southern territories was aided by, indeed to a considerable degree dependent upon, the southward migration of the peasant populace, frequently occurring against the declared wishes of the authorities. This process has been investigated for the first half of the seventeenth century by A. A. Novoselskii.[10] During the previous century the flow of refugees escaping the economic and social deprivations of the centre had been adding to the numerical strength of the cossacks (*kazaki*), who lived a freebooting existence in various parts of the forest-steppe and steppe. By the early seventeenth century Tatar control of the steppe was being seriously challenged by these wild marauders, who were now beginning to form a series of organized forces, or 'hosts', on the lower Don and Dnepr.

While the southern environs into which the Russian settlers penetrated were endowed with numerous attractions, they also presented the newcomers with many difficulties. The tough steppe grasses in particular were very different from the predominantly forested environment to which the Russians had been used further north. As I. N. Miklashevskii wrote of the Russian colonizer, 'It was not the forest but the steppe which arrested his progress',[11] and he was not alluding to the Tatar raiders alone. Much more conducive to settlement were the broad-leaved forests, especially the oak woodlands of the river valleys and their associated meadow grass, grey forest, and alluvial soils. Such an environment not only provided the settler with much-needed resources, but also with secure hiding places against the sudden raid,

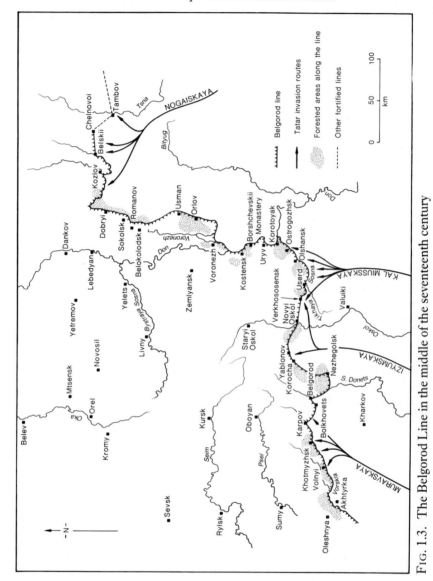

Fig. 1.3. The Belgorod Line in the middle of the seventeenth century

such places being frequently mentioned in the cadastres of the day.[12] In locating their settlements the Russians took advantage of every strategic opportunity the environment had to offer. The protection provided by high river-bank, deep stream, dense forest or thicket, or commanding viewpoint was well utilized in the construction of village, town, or fortified line. Where nature itself afforded inadequate defence human artifice substituted for its shortcomings in the construction of watchtowers, defensive ditches, earth banks, and pallisades. In this way the freedom the Tatars had enjoyed to wander and pounce at will was gradually limited, and their nomadic habits constrained by the organized power of the Russian state.

## 'Savagery and Civilization'

Seventeenth-century Russians would certainly have endorsed Turner's Eurocentric description of the frontier as 'the meeting point between savagery and civilization'.[13] Upholders of Orthodoxy and deeply suspicious of outsiders, Russians long held themselves aloof from untoward foreign influence. The prevailing attitude towards the frontier is well summarized in the later words of Klyuchevskii, who described it as 'the very edge of the world of Christian culture . . . the historical scourge of ancient Rus'.[14] When in the 1660s the Croat Yurii Krizhanich denounced the enemy infidel's incursions into the Russian lands and called upon Moscow to aid its suffering co-religionists, his words were well understood by the Russians.[15] Russian documents of the period are full of references to the Tatar 'foe' and to their many raids on people and livestock, raids which reached almost unprecedented levels of ferocity in the 1630s, the 1640s, and again after 1670. An official Russian account, written in 1681, relates how earlier in the century the southern towns and their districts were repeatedly attacked by the 'Crimean and Azov thieves and Nogai nomadic Tatars' because the towns were 'underpopulated and the distance from town to town was not small, and the ramparts and fortifications were few'.[16] Subsequently, however, according to this same account, the region was triumphantly occupied by the tsar and his peoples. Yet such triumphs could hardly eradicate the memories of those who had suffered personally at the hands of the Tatars. The Voronezh junior boyar Martin Sentsov, for example, explained in a petition to the tsar how he had been travelling from his estate to Voronezh when he had been captured by a band of Crimean Tatars. He had subsequently been sold as a galley-slave to the Turks, among whom he had suffered for fifteen years.[17] Bitter memories such as these fuelled the emnity between Russian and Tatar; the Russian attitude towards the steppe adversary can hardly have differed from the common hatred felt by American settlers towards the Plains Indians some two centuries later.[18]

As in the case of the American West, however, it would be easy to overgeneralize about relations between the settlers and the nomads. There were many exceptions to the frequently antagonistic state of affairs, and this is hardly surprising given the fact that the pastoral economy of the Tatars was to some extent complementary to the agriculture of the Russians, and the fact that the various Tatar bands often acted independently. Thus, trade was by no means uncommon, and Tatars were sometimes employed by the Russians to act as guides or in some other capacity. On some parts of the frontier, especially in Tambov and more easterly areas, they were often recruited as military servitors.[19] Relations between settler and nomad were also complicated by the existence of other steppe peoples, such as the cossacks and the Kalmyks.[20]

'At the frontier', wrote Turner, 'the environment is at first too strong for the man'; 'the wilderness masters the colonist'.[21] In the West, he believed, the pioneer had to come to terms with his environment; the sophisticated European became an American. Stripped of the hyperbole, Turner's observations contain a germ of truth which can be applied to all frontiers. That truth is that the pioneer, whatever his background, cannot afford to remain aloof from his surroundings; he must adapt to and learn from his new milieu. This was certainly the case with the seventeenth-century Russians. The Tatar economy, based on the herding of horses, cattle, and sheep, a little agriculture in places, and on raiding and plunder, had evolved over centuries and was finely attuned to the rhythms and changes of the steppe environment. The Russians who most quickly and successfully adapted to that environment were the cossacks, who accomplished this feat by emulating the Tatars. Within a short time the most adventuresome among the cossacks had, like the Tatars, become masters in the use of the horse, able warriors and raiders, fishermen, hunters, herders, and traders. Only gradually did they take to agriculture—an activity which was long despised. As a military force the cossacks proved a fierce adversary to both Tatar and Russian alike, but the rulers of Russia were well placed to make use of the military skills of their cossack co-religionists, bribing them, flattering them, and occasionally compelling them to take their side in their numerous campaigns against the Tatars and Turks.[22] Many of the less independent cossacks were recruited directly by the Russians, granted land near one of the military towns, and employed on cavalry duties in defence of the frontier.[23]

The cossacks, therefore, constituted 'the outer edge of the wave',[24] the nearest Russian equivalent to Turner's hardy pioneer. Other frontiersmen, such as the hunters, beekeepers, and fishermen who ventured south in search of natural wealth, were also forced to adapt to the demands and dangers of the untamed wilderness, as indeed were the servitors and other settlers around the frontier towns. The apparent immediacy of the Russian centralized state and of its military institutions constitutes an important difference from those parts of the American West where movement was freer, but the extent of this difference can be exaggerated. Even in southern Russia there appear to have been many individuals eager to evade the organizing arms of authority by moving with the frontier. Not all of these individuals joined the cossacks. Evidence of high turnover among the frontier servitors and townsmen suggests the existence of a frontier 'type', migrating with the frontier and seizing the opportunities and advantages it offered.[25] Attempts by the government to control such migration seem to have been only partially successful.

### SEQUENCES OF FRONTIERS

One of Turner's most vivid metaphors concerns the succession of frontiers which, he maintained, had swept across the United States from east to west. 'The United States', he writes, 'lies like a huge page in the history of society. Line by line as we read this continental page from West to East we find the record of social evolution.'[26] Again:

Stand at Cumberland Gap and watch the procession of civilization, marching single file—the buffalo following the trail to the salt springs, the Indian, the fur-trader and hunter, the cattle-raiser, the pioneer farmer—and the frontier has passed by. Stand at South Pass in the Rockies a century later and see the same procession with wider intervals between.[27]

The succession, he admitted, was by no means uniform in its effects. Yet he argued that, in the eastern states especially, its general outlines were clear enough: beginning with the Indian and the hunter, it is followed by the trader, then by the rancher, by extensive arable farming, intensive cultivation, and finally 'the manufacturing organization with city and factory system'.[28]

On Russia's southern frontier a sequence can also be found, although one that is greatly distorted by the military situation. The cossacks have already been mentioned as constituting the 'outer edge of the wave'. Although they were at first concerned principally with self-sufficiency, their hunting and foraging activities soon stimulated trade with the Russian settlements and towns to the north. Between the cossack lands and the frontier towns were specially designated areas (*ukhozhai*) which were rented by the tsar to individuals and groups for hunting, fishing, and beekeeping. In Voronezh district in 1615, for example, no fewer than seventeen such *ukhozhai* lay to the south of the towns and populated areas, and these were being rented for sums ranging between five and fifty roubles.[29] Again this stimulated the activities of traders, which the tsars sought to control, especially in times of famine, plague, or military insecurity. Finally, to the north, came the military towns with their supporting agricultural lands.

During the course of the seventeenth century, as the military frontier moved south, the cossack territories and *ukhozhai* were gradually overrun. Thus several of the *ukhozhai* listed in the 1615 cadastre for Voronezh district disappeared with the construction of the Belgorod Line and associated towns and villages in the middle of the century. Others, such as the Tolucheevka and Boguchar territories in the south of the future Voronezh province, lasted until the end of the century, when the area began to be settled by the Ostrogozhsk Regiment and other groups.[30] At roughly the same time Peter the Great began to settle the valleys of the Bityug and Ikorets with court peasants, and soon afterwards, in the wake of the Bulavin rebellion, expelled the upper Don cossacks from the valley of the Khoper. As official military and associated settlement moved south, the hunting and foraging frontier, and also the system of patrols and guards keeping watch over Tatar movements, were pushed forward in their turn.

Turner's frontier in the American West was pre-eminently a commercial frontier on which the pioneers soon came to find themselves involved in a burgeoning market economy. The same cannot be said of Russia's southern frontier, where market relations were less well developed and where military considerations were dominant in the seventeenth century. It is therefore hardly surprising that Turner's frontier sequence should be considerably distorted in the Russian case. Two particular differences from the American frontier, as described by Turner, stand out. First, there was no marked livestock or ranching frontier between the hunting-foraging frontier and the agricultural frontier. The reason was a simple one: extensive livestock-raising would have been far too vulnerable to attack by the Tatars. The early Russian settlers seem for the most part to have had only small numbers of livestock, such as one or two horses, which were required for service and draught, and a few cattle, sheep, and pigs for domestic needs.[31] Nevertheless, even in the seventeenth century there was a growing trade in horses, leather, and hides between the cossacks, Tatars, and Kalmyks on the one hand and the Russians on the other.[32] In the following century, as the south became safer, a genuine livestock frontier did grow up as the still extensive steppe grasslands were subdivided by large estate-holders. This livestock frontier is discussed in Chapters 3 and 8.

The other important difference from Turner's frontier is the fact that the towns, far from being the end product of a sequence of frontier occupances, were in fact 'spearheads of the frontier'.[33] Even for parts of the American West, Turner's sequence has been questioned by such scholars as Meinig and Wade.[34] In the case of southern Russia in the seventeenth century the pioneering role of the towns is beyond doubt. The fortified towns commonly consisted of a walled enclosure containing the headquarters of the military governor together with administrative offices, military stores, a cathedral church, and places for housing the populace in case of attack. This was generally surrounded by the settlements of the military servitors, sometimes also fortified. An example is shown in Fig. 1.4 The towns formed the linchpin of Muscovite colonization policy in the south. As a result of long experience of the forest-steppe and steppe lands, gathered in the course of sending out patrols, escorting ambassadors, and many other activities, Moscow had a detailed knowledge of the topography of the unsettled areas. The extent of such knowledge is well illustrated in such documents as the Book of the Great Map, dating from the sixteenth century, and instructions issued to the steppe guards and patrols concerning their duties and dispositions.[35] It was therefore a comparatively simple matter, when policy dictated it in the 1580s and 1590s, or again the 1630s to 1650s, to send out a party of surveyors to look for a suitable site for a town, followed later by a second party actually to construct it. In general, the policy was to site the towns in defensible locations from where military forces could be sent to harry the Tatar raiders advancing along their favoured routes. Some of the priorities for siting a town are mentioned in the construction book for Korotoyak, which was built on the bluffs overlooking the Don, close to its confluence with the Tikhaya Sosna, in 1647.[36] The book states that Korotoyak is located in a 'suitable' and 'strong' place, near water and woodland, and easily defensible against attacks by enemy forces. Moreover, the view from the town extended to more than 20 *versty* to the east across the Don (the 'Nogal side') and to more than fifteen southwards across the Tikhaya Sosna (the 'Crimean side'). The book goes on to list the many buildings and lands belonging to the town, together with the personnel living there. Detailed instructions regarding the siting and construction of a town were commonly issued by Moscow to the leader of the construction party at the time the party set out. Thus, when Bogdan Belskii and Semen Olferev were sent with a large military company to build Tsarevborisov near the confluence of the Oskol and Severskii Donets in 1599, they took with them a document instructing them how the company was to behave *en route*, where and how the town was to be built, and how its future inhabitants were to be recruited.[37] Once a military town was built, its entire life was under the control of its military governor, an official appointed by Moscow and with whom the

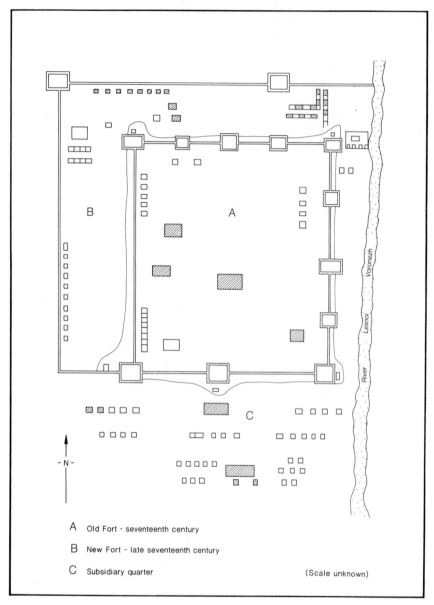

Fig. 1.4.  The town of Kozlov in the seventeenth century

Moscow government was usually concerned to maintain regular and detailed correspondence.

Details concerning the way land was held and used on the southern frontier appear in Chapter 2. All that needs to be emphasized here is the fact that the situation was very different from the frontier of commercial agriculture envisaged by Turner. Land was typically held in exchange for military service and cultivated, under the supervision of the military governor, mainly for subsistence purposes, though any surpluses could be sold. Sometimes the produce was taxed. In certain towns part of the land was held directly by the state and cultivated by military servitors as part of their duty to the sovereign.[38] The produce of this land went to the state to feed visiting military personnel or officials, or for store. A certain amount of land in the south was held by monasteries, churches, court peasants, and even by a handful of great landowners such as the boyar N. I. Romanov, kinsman to the Tsar.[39] But these were exceptional instances in a human landscape where defence was the prime consideration.

A rather different frontier sequence from the one proposed by Turner is recognizable in the variations in the way land was held in the south and in the type of person holding land. As explained more fully in Chapter 2, the virgin land of the south was regarded by the tsars as their property and was distributed to military servitors on a service rather than a hereditary basis. For various reasons the land was distributed in relatively small holdings. Thus, such towns as Tula and Ryazan, which stood close to the frontier line in the late sixteenth century, were at that time populated to an important degree by lesser servitors, and their surrounding districts largely by middle-ranking servitors holding only small service landholdings.[40] Higher-ranking servitors or lords with large estates and numbers of serfs were rarely found on the vulnerable frontier. South of Tula and Ryazan, in the more exposed towns such as Yelets, Voronezh, and Oskol, the importance of the middle-ranking servitors diminished until finally, in the most vulnerable towns such as Valuiki and Tsarevborisov, only lesser servitors were to be found holding land. Beyond these places, of course, cossacks held land on an unofficial basis.

Later in the seventeenth century a sequence of settlement and landholding frontiers moved south across the Central Black Earth Region. While districts such as Tula and Ryazan were colonized by larger estate-holders, the middle-ranking servitors moved south towards the Belgorod Line, where they gradually coalesced with the lesser servitors.[41] Meanwhile the cossacks either migrated even further south or found their position 'regularized' by entry into state service. By the century's end the more prominent owners of estates and serfs began acquiring land even along the Belgorod Line, while there was some outmigration of their social inferiors. In the next century the larger estate-holders also began to take up land in the southernmost parts of the Central Black Earth Region and there was a gradual intermingling of Russian- and Ukrainian-owned estates in this area.

## THE FRONTIER AND THE SOCIAL PROCESS

One of the most significant effects of the American frontier, according to Turner, was the promotion of democracy. 'The frontier', he wrote, 'is productive of individualism. Complex society is precipitated by the wilderness into a kind of primitive organization, based on the family. The tendency is anti-social. It produces antipathy to control, and particularly to any direct control.'[42] Along with the 'selfishness and individualism, intolerant of administrative experience and education, and pressing individual liberty beyond its proper bounds',[43] Turner listed other supposed traits born of the frontier:

that coarseness and strength combined with acuteness and inquisitiveness; that practical, inventive turn of mind, quick to find expedients; that masterful grasp of material things, lacking in the artistic but powerful to effect great ends; that restless, nervous energy; that dominant individualism, working for good and for evil, and withal that buoyancy and exuberance which comes with freedom.[44]

Individualism, as Turner noted, was not necessarily always to be welcomed, however. On the frontier it often gave rise to 'all the manifest evils that follow from the lack of a highly developed civic spirit'.[45] Lawlessness was as much a child of the frontier as was democracy.

Turner's sentiments about individualism and democracy seem a far cry from the regimentation of life on the southern frontier of Russia with its state-controlled towns acting as pivots of frontier society under the control of Moscow, its lands carefully allocated according to norms laid down by the state's expanding bureaucracy, and its social stratification, with each individual allotted his due position in the social order. And yet such a characterization can easily degenerate into a caricature. John Keep has argued that the burdens of service laid upon their subjects by the tsars resulted in part from attempts to rule and defend a poor and undergoverned realm.[46] The tsar's efforts at securing centralized control, never entirely successful, were particularly likely to be frustrated on the open frontier. This failure was perhaps most apparent in the case of the cossacks, who to a considerable extent originated as refugees from the tsar's own dominions. The cossacks were proud of their independence, were always liable to go their own way in relations with the Tatars and Turks, and would for a long time only have dealings with the tsar's foreign chancellery. In cossack society there existed a kind of rough-and-ready democracy, contemptuous of social status and of any form of deferential behaviour. Unlike Turner's hypothesized pioneer family, it was at first a society of individuals, although this soon changed. Cossack women, however, were said to be much freer than their Russian counterparts.[47] Cossack traits were particularly strong among the Ukrainian migrants who began to move into Muscovy from Poland, especially from the 1630s. In Muscovy they were able to negotiate some important concessions from their new tsarist masters, including an autonomous status in their own regiments, the tax-free distillation of liquor, and rights of intake in the virgin land.

Cossack traditions frequently went hand in hand with the kind of lawlessness and brigandage that have characterized so many of the world's frontier regions. Documents of the period contain many references to 'thieves' and outlaws and their many depradations, which are not always easily distinguished from those of the Tatars.[48] The Moscow government was aware that, even by the standards of turbulence common in its realm, the southern frontier was liable to be especially troublesome. In the Time of Troubles at the beginning of the century, for example, the south was often in turmoil, most notably during the Bolotnikov rebellion in 1606–7. Later on the mid-century disturbances that shook many Russian towns and rural districts were well represented in the south, while the major rebellion of Stenka Razin engulfed several towns and districts along the Belgorod Line in 1670–1.[49] Other disturbances included those in the 1680s, which were partially linked with the revolt of the *streltsy* (musketeers) in Moscow, and also the Bulavin rebellion in the early years of the following century.[50] In all the major rebellions a leading role was played by the cossacks, particularly the poor cossacks of the upper Don region, and the rebellions then tended to spread among disaffected elements living in various towns and districts close to the frontier line.

The southern frontier, therefore, fully accorded with Turner's belief that the existence of 'free land' constituted a major challenge to the forces of law and order. The south was also a 'safety valve'. The free migration of peasants from the interior hoping to escape economic and social oppression finally became illegal with the Law Code of 1649. It continued nevertheless and, since it threatened the economies of the central estates and thus the taxation and defence policies of the state, Moscow could not ignore this drain on its resources. Numerous commissions were therefore sent out to search for and return the refugees. Yet, since the migrations did help the settlement and development of vulnerable regions, the policy was not always applied consistently, much to the annoyance of those lords who lost their serfs.[51]

Despite the existence of a close network of frontier towns, Moscow's control was far from complete. For one thing, the sheer distance of the region from Moscow and the poor state of communications ensured that the local military governors were all too often lords unto themselves, greatly given to corruption and arbitrary behaviour.[52] The governors' control over their towns and surrounding districts also left much to be desired. The town dwellers, as noted already, were often unruly and unreliable and were also much given to flight in times of hardship.[53] The authorities were particularly concerned about the latter since it threatened the whole basis of its settlement and defence policy. Wherever possible Moscow strove to settle its newly constructed towns with volunteers, particularly from the cossacks or from the relatives of servitors seeking land for themselves. However, it was occasionally forced to resort to compulsory settlement, and this was almost always followed by mass flight. When the towns of Tsarev-Alekseev[54] and Verkhososensk were built on the Belgorod Line in 1647, for example, it proved necessary to populate them with servitors compulsorily trans- ferred from other, more northerly towns. But this policy met with scant success. In October 1647 the military governor of Tsarev-Alekseev reported that 1,091 people had been settled there by order, yet by late December he had to report that only 795 remained. Similarly, of 395 people originally settled at Verkhososensk, there were soon only 54 left. Interestingly enough, the government was soon sending exiles to populate these towns.[55] Peter the Great's decision to populate the Ikorets and Bityug valleys with court peasants from various central provinces proved equally unsuccessful. Of the 1,021 households (4,919 people) settled there in 1701, only 159 households remained three years later.[56]

In view of the harshness of frontier life the apparent frequency of flight, which seems to have been common even among the middle- ranking servitors,[57] is hardly surprising. Tatar raids, the demands of the military life, the sheer hard work involved in building, cultivating, and adjusting to a new environment, all took their toll. Famine and disease were regular visitors to the frontier communities and the death-rate appears to have been high, especially in new settlements.[58] When government demands reached exceptional levels, as during the building of the Belgorod Line and subsequent recruitment into the Belgorod Regiment, or at the end of the century, when Peter the Great gathered together masses of people for work in his shipyards on the Don, flight became a grave problem indeed.[59] Yet the frontier also offered countless opportunities for individual freedom and even advancement. Migrants might join the cossacks, sign up for service in one of the frontier towns, or gain wealth through trade or plunder. They could receive land in exchange for service; they might also occupy it without service, illegally, until this came to the attention of the authorities.[60]

Muscovite society, with its hierarchy and rigidity, found considerable difficulty in reproducing itself on the frontier. As noted already, the great lord with his estate of many serfs was a rarity in the south. The typical landholder had few serfs or even none, relying solely on his own labour or on that of his family. Even serfdom itself seems to have been less rigid in the south than elsewhere, at least until late in the century. Many southern servitors came from humble backgrounds. The government was anxious to populate and defend its new frontier, and in doing so sought settlers among the cossacks and volunteers of other kinds. In spite of decrees to the contrary, even former serfs were sometimes accepted as servitors and, in the first part of the century at least, entry into the ranks of the supposedly hereditary junior boyars could also be made available to the lowly.[61]

The flexibility of frontier society also affected the towns. Here the most striking feature in the early part of the century was the relative unimportance of the *posad* trading people by comparison with the minor servitors. Much of the trade was in fact in the hands of the latter. In morphology and function the frontier towns had some of the characteristics of central and northern towns, but the most obvious difference lay in the centrality of their military functions. They lacked the social and institutional complexity of many interior places, a fact which is well reflected in the many urban inventories which were undertaken for military purposes.[62]

## The Frontier and the Economic Process

That Russia's southern frontier lacked the commercial character of Turner's 'frontier' has been stressed already. Turner regarded the West as a land of opportunity in which the trader, farmer, and entrepreneur could quickly acquire wealth based upon enterprise and hard work. Such opportunities were frequently lacking in the Russian case, where military exigencies, feudal social norms, and backward technologies were the order of the day. Nevertheless, even in Russia the existence of 'free land' did present commercial opportunities for those willing and able to take advantage of them. Two examples will be mentioned here. First, the *ukhozhai*, as noted above, were unsettled territories rented by the tsar to individuals, groups, or monasteries. These dangerous and exposed regions were then exploited for their rich supplies of mammals (e.g. bear, elk, deer, roebuck, squirrel, marten, tarpan, aurochs, saiga antelope, beaver), birds, fish, and insects. This activity, supplemented by much hunting and fishing in less exposed areas, greatly stimulated the trade and commerce of the frontier territories. Zagorovskii has suggested that the system of renting *ukhozhai* to the highest bidder, irrespective of social origin, is more in keeping with a capitalist than a feudal social order.[63] Be that as it may, the *ukhozhai* are an interesting example of frontier enterprise, although it seems doubtful whether they enriched many individuals in the long term.

A second example of frontier enterprise is to be seen in the trade with the cossacks and other steppe peoples, which was stimulated by the proximity of the complementary economies of agricultural and grassland areas and by the presence of the Don river artery. Trade with the Don cossacks developed as the century advanced and was only occasionally interrupted by famine, disease, or social disturbance.[64] Moreover, as more and more land was settled trading links with Moscow and the central regions also grew, particularly in such districts as Orel, Tula, and Tambov. In 1694, for example, sixty-two traders visited Moscow from towns in the Central Black Earth Region bringing a variety of agricultural produce.[65] Many southern towns, which earlier in the century had served primarily as military garrisons, benefited from the growth of larger and more active commercial quarters. Vazhinskii has analysed southern trade in this period and describes the emergence of a permanent class of merchants, some of whom traded wholesale over considerable distances.[66]

## The Long-Term Significance of the Frontier

Turner's most celebrated and contentious contribution to American historiography lies in the central role he believed the frontier played in determining American development. 'The frontier', he wrote, 'is the line of most rapid and effective Americanization.' 'Little by little', asserted Turner, the pioneer 'transforms the wilderness, but the outcome is not the old Europe. . . . The fact is that here is a new product that is American.'[67]

The significance of the frontier in Russian history is a vast topic lying well beyond the scope of a single chapter, particularly one which considers only one frontier, and that over a mere one-hundred-year period. Yet a few points can be made, by way of conclusion, to emphasize the importance of the southern frontier to Russia's development in the seventeenth century.

In the first place, it is probably the case that the southern frontier in the seventeenth century helped to promote both the centralization and the militarization of Russia. The need for an effective system of defence against an enemy both tenacious and elusive helped to promote new experiments in urban government and control (the system of military governors sited in towns was first introduced on the frontiers in the previous century), in provincial administration (witness the Belgorod Military District in the middle of the century), in landholding (the service landholding system was introduced earlier but reached its apogee in the south in the sixteenth and seventeenth centuries), and in military reform (the 'new formation regiments', introduced in the seventeenth century, were a response to challenges coming from the west, but the south was one of their principal testing grounds). The southern frontier probably also promoted centralization since it was such a drain on resources. Effective systems of taxation were demanded because of the expense entailed in defending the frontier, and the burden of defence had the broadest implications for the economic development and well-being of the Russian people as a whole. The turbulence and lawlessness of the frontier seem to have encouraged the state to assume stricter policing of frontier regions, since such rebellions as that of Razin constituted the most serious threat to its security. Serfdom, too, was undoubtedly strengthened because of the frontier— the constant southward drain of manpower was more than a centralizing state could possibly endure.

That the frontier also acted to some extent as a 'safety valve' has been mentioned already. For many runaway peasants it provided a promise of freedom, while the frontier also enabled the state to exile its dissidents, and later to reward its supporters with new and fertile estates. In this way it may have acted as an instrument whereby potential trouble was diffused. It is also conceivable that the frontier had a subtler and more harmful long-term effect: by diverting energies southwards, it could have delayed technological development and agricultural intensification in the centre, with all the potential consequences for political and social change. These are matters which demand far more scholarship than they have hitherto received.

Within the Central Black Earth Region, as seen in the chapters that follow, the frontier helped to promote some distinctive social and settlement traits which had long-term consequences for the character of the region. Turner may well have regarded this ultimate product of a particular frontier experience as a type of 'section'.[68]

Turner's assertion that the most essential characteristics of American society are traceable to the effects of the frontier has been severely criticized. In particular he has been attacked for failing to give due weight to the European heritage in American development.[69] Similarly, it is difficult to distinguish between the effects of 'free land' on Russian development and those effects which came from a multitude of other sources and influences. None the less, it is apparent that Russia developed along a very different path from the United States, at least as far as the character of its society and polity is concerned. Cossack freedoms and traditions, and the manifestations of individualism discussed above, were eventually extinguished by the advancing Russian state. Turner's genius lay in being able to describe so vividly those features of the frontier which are well-nigh universally applicable: the lawlessness and relative lack of central control, the need for self-reliance, the adaptation to the environment, the possibilities for acquiring wealth, the abundance of resources, the gradual way in which human societies settle and appropriate their surroundings. The particular circumstances attending the frontier experience and its long-term repercussions were different in every case. This chapter has endeavoured to demonstrate the relevance of Turner's thesis even to such a distinctive frontier as that of southern Russia in the seventeenth century.

# 2
## The *Odnodvortsy*

UNTIL their separate status was abolished in the 1860s, the *odnodvortsy* formed a distinctive element in the Russian rural landscape. Although in economy and way of life they were virtually indistinguishable from the mass of the state peasants, they retained certain characteristics which were memorials of their former role as state servitors, guarding the frontiers from the enemies of the tsar. One of these distinctive traits, which they retained almost to the end, was their legal right to own serfs, otherwise a monopoly of the nobility. Only a minority, however, were in a position to exercise this right. In the 1780s, for example, when there were about three-quarters of a million male *odnodvortsy*, there were a mere 22,000 male *odnodvortsy* serfs. The vast majority of the *odnodvortsy* were small farmers, possessing only one dwelling or *dvor* and as poor as most of their neighbours. Their erstwhile role as servitors had once conferred upon them a superior social status, but by the early nineteenth century, for most *odnodvortsy*, this was a mere folk memory.

Compared with the Russian peasantry as a whole, the *odnodvortsy* were only a small group. In the 1740s, at the time of the second revision, there were about 453,000 male *odnodvortsy*. By the 1830s their numbers had grown to some 1.3 million males. At the latter date they thus constituted around 12 per cent of the non-seignorial peasantry and less than 6 per cent of the peasantry as a whole. There was, however, a marked concentration in their distribution, and in the four black-earth provinces of Kursk, Orel, Tambov, and Voronezh they formed between 25 and 40 per cent of the peasant population. Two-thirds of the *odnodvortsy* lived in these four provinces. Quite considerable numbers were also to be found in the provinces of Tula, Ryazan, Penza, Saratov, Kharkov, Orenburg, and Stavropol, as indicated in Fig. 2.1. In these areas the *odnodvortsy* imparted a distinctive social geography which was itself a product of the way these regions had been settled.

Turner's frontier thesis, which was examined at some length in the previous chapter, was promulgated at a period when many Americans became concerned about the possible negative consequences that might ensue for their society once the frontier closed. Turner himself taught that, once a former frontier region becomes fully settled, it will be integrated into the rest of the nation and will change in consequence. Yet he believed that such a region would nevertheless retain some of its

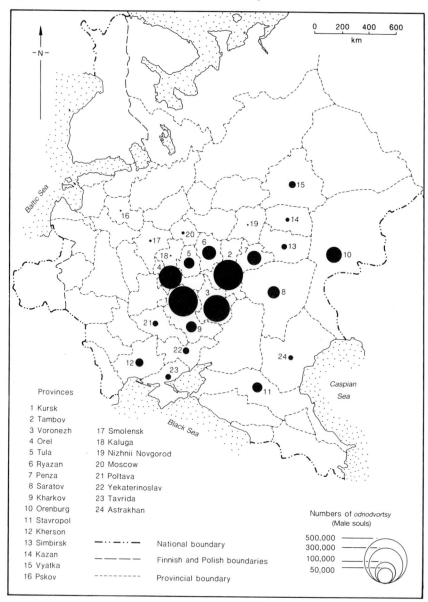

Fig. 2.1. The distribution of *odnodvortsy* in the 1830s

frontier characteristics.[1] In tracing the origins and evolution of the
*odnodvortsy*, the present chapter explores the extent to which their way of
life altered once the frontier passed into history, and the extent to which

the erstwhile frontier retained its significance for their long-term development.


## PRECURSORS OF THE *Odnodvortsy*

The special character of Russia's black-earth provinces derived from the application to them of landholding policies which had arisen with the unification of the Russian state from the late fifteenth century. Moscow's reliance upon military power in circumstances of considerable insecurity and relative poverty, its need to contain the pretensions of the old nobility and boyars (including the former royal princes), and its determination to build up a class of loyal dependents led gradually to the adoption of a system of service or *pomeste* land tenure. Under this system those who held *pomeste* estates were obliged to serve the tsar, usually in a military capacity. As this new form of tenure spread by means of the many redistributions of land which accompanied the consolidation of the Muscovite state, the core of the servitor class came to be the junior boyars (*deti boyarskie*). This mixed group of people, some of whom were former servants of the old royal princes, many the offspring of impoverished noble families, and some even coming from quite humble backgrounds, formed the basis of Moscow's armed strength. To supervise and control the new servitor class an entire bureaucracy came into being. By the end of the sixteenth century a complex hierarchy of ranks was recognized, supported by an equally complicated system of remuneration scales (the *oklad*), which specified the amount of land and other recompense to which a servitor of a specific rank was entitled. In practice the scales tended to indicate the maximum rather than the actual amount of remuneration a servitor might expect. Also accommodated within this system were the holders of hereditary estates (*votchiny*), who, by Ivan the Terrible's decree of 1556, were likewise obliged to serve the sovereign.[2]

It was on the peripheries of the state, which lacked the centre's monastic properties and the hereditary landed estates of the high nobility with their long history of loyalty to Moscow, that the service-land policy could be applied most completely. To the south, where the frontier gradually edged forward under the protection of fortified towns and the *zaseki*, lands were distributed by the government on a service basis. In many southern districts, such as Orel, Dedilov, and Tula, hereditary estates were virtually unknown in the late sixteenth century. Only in such places as Kaluga and Ryazan, where settlement had occurred earlier or life was a little safer by the late sixteenth century, were some hereditary estates to be found.[3]

South of the Oka the open and unsettled nature of the frontier and the danger of Tatar incursions encouraged the government to utilize the distribution of frontier lands as a means of enhancing its frontier defences. Those who received *pomestya* in the south were mainly the lowest-ranking junior boyars, the 'town' (*gorodovye*) junior boyars. What their rank indicated was that they held their land, (and also served) in one and the same district, under the direction of the local military governor (*voevoda*), who was based in the nearby town. Their service included having to man the system of frontier patrols and guard-posts, which was reorganized in 1571. The militarization of life made the southern frontier very different from life closer to the centre of the country, and the character of the service demanded from the southern servitors was also more immediate and burdensome. As Chapter 1 made clear, the southern territories were not at first particularly attractive to wealthier servitors, nor to nobles with large estates and numbers of serfs. In the south the serfs were vulnerable to Tatar raids, and were also liable to desert their masters whenever the opportunity arose.

The fact that the junior boyars holding service estates in the south generally belonged to the lowest-ranking category is probably a testimony to their mixed social origins. In the late sixteenth century it was fairly common for junior boyars without land or with only small estates to seek out new lands in favourable southern locations, and then to petition the government that these lands be allocated to them on a *pomeste* basis. This might be done by individuals but also, increasingly it seems, by groups. Where the government required settlement in furtherance of its policy on frontier defence, it could sometimes resort to compulsion or appeal for volunteers. In the 1570s, for example, junior boyars from Tula and Kashira were resettled in the districts of Venev and Yepifan in order to strengthen the frontier there.[4] In other cases, however, the government had to adopt a pragmatic policy, accepting into the ranks of the junior boyars those who had no hereditary claim to that status. Thus groups of cossacks or lower-order servitors such as musketeers, serving under the leadership of a junior boyar or his equivalent, were sometimes granted *pomestya*, as in Dedilov district. At Yepifan in 1585, 300 cossacks were promoted to junior boyar rank and given small *pomestya* in the nearby district. Even groups of Tatars in the service of the tsar were given estates. This happened, for example, in Kashira, Pronsk, and Zaraisk districts.[5] Both in the late sixteenth century and later it was not uncommon for individual 'free' people, some of whom were runaway serfs, to be directly enrolled into the ranks of the junior boyars.[6]

The insecurity and harshness of life on the frontier was therefore usually attractive only to the poorer social elements, including the poorer junior boyars, whose ranks were augmented by members of other social

groups. This pattern was reinforced by the smallness of most of the
estates allocated in the south, certainly by comparison with those
common in the centre. Not only were the southern remuneration scales
frequently less generous than those towards the centre, but the amount
of land actually allocated to the servitors (the *dacha*) was often even less
than these scales specified.[7] Scholars are not unanimous on the reasons
for this situation. No doubt the relative poverty and the dubious origins
of many southern servitors were factors. The size of the land grants
would reflect the low status of most southern servitors, marking them
out as distinctive by comparison with many servitors in the centre.[8]
Since the southern servitors were for the most part poor and with few
serfs, they would be incapable of working large estates in any case,
especially given the virgin character of much of the territory. The
government was in all probability well aware of the fertility of the black-
earth soils, and of the fact that servitors could support themselves from
quite modest land allotments. Another factor seems to have been the
wish of the government to preserve the frontier lands for settlement by
as many servitors as possible. N. Pavlov-Silvanskii and S. V. Rozhdestvenskii have used the evidence of numerous landless or poor junior
boyars in certain southern towns to argue that there was a lack of land
suitable for granting as service estates in the late sixteenth century.[9]
What servitors really desired was fertile land, preferably protected from
the Tatars, and land that was already being cultivated by peasants. On
the southern frontier there was obviously a shortage of land of this type.
On the other hand there was virgin land which could be used to satisfy
the most needy servitors (particularly if special dispensations such as
temporary reprieves from service were also granted), and at the same
time to further the government's settlement policies.

The junior boyars, whatever their origins may have been in reality,
were regarded as having hereditary rights to their status. Therefore
historians frequently refer to them as servitors 'by patrimony' (*po
otechestvu*). The southern frontier, however, was characterized also by
the very great importance of the lower class of servitors, those 'by
contract' (*po priboru*). These included the musketeers (*streltsy*), artillerymen, harquebusiers, gatekeepers, and others, who were principally
responsible for guarding the walls and fortifications of the towns and
military lines. They also included the various categories of cossack, who
were usually engaged on cavalry duties. Even more than the southern
junior boyars, this group had disparate origins, being recruited from
landless junior boyars, runaway serfs, free cossacks, and others. They
often formed a dominant element in southern towns, undertaking much
of the trade and also engaged in local agriculture. Serf-owning was far
less common among these servitors than among the junior boyars, and

they were consequently even poorer,[10] but they were regarded as of higher status than either the peasants on the one hand or the *posad* townspeople on the other. In the most southerly and vulnerable towns they were often the only servitors. Their importance enhanced the distinctive character of the southern frontier at this period.

### VORONEZH DISTRICT IN THE EARLY SEVENTEENTH CENTURY

As an example of servitor settlement and landholding on the frontier in the early period, the district of Voronezh may be taken (as illustrated in Fig. 2.2). The town of Voronezh was founded in 1585, during the period of Muscovy's thrust to the south which followed the disastrous burning of Moscow by the Crimean Tatars in 1571. The settling of the town and its surroundings, especially along the well-protected Voronezh River and its tributary the Usman, took place in the latter years of the sixteenth century. After a pause during the Time of Troubles, settlement resumed early in the following century. The present discussion is based upon the cadastres (*pistsovye knigi*) compiled in 1615 and 1629.[11]

The population of Voronezh town in 1615 has been estimated at about 6,000. Most of the population (about three-quarters of the households) consisted of servitors, especially of those 'by contract'.[12] The town itself was in two parts: the fortified inner core (*gorod*), which contained the administrative headquarters, and the bigger *ostrog*, also fortified. Within the latter were several quarters (*slobody*), each inhabited by a specific category of servitor. Thus, there were quarters for the musketeers, regimental cossacks, and *belomestnye atamany* and cossacks ('white place' atamans and cossacks, who were freed from certain impositions), and a mixed quarter for the harquebusiers, artillerymen, gatekeepers, state smiths, guards, and a carpenter. Each category of servitor held land outside the town, allocated to its members. Thus, the twelve artillerymen and the thirty-three harquebusiers had 15 *chetverti*[13] of arable each, while the guards, gatekeepers, carpenter, and smiths, twelve people in all, each had 7 *chetverti*. The arable lands of all the servitors were scattered in various locations. In addition, every servitor had shares in the haylands and forests. Many of the cossacks and some of the artillerymen and gatekeepers had cottars (*bobyli*) living with them. These sometimes lived in the town, but those belonging to cossacks more frequently lived in six settlements which had been founded outside the town, where they tilled the land on behalf of their masters.[14]

Beyond the town of Voronezh and its immediate environs lay the villages and settlements of the district (*uezd*). While several of these villages were populated by cossacks, most contained the landholdings

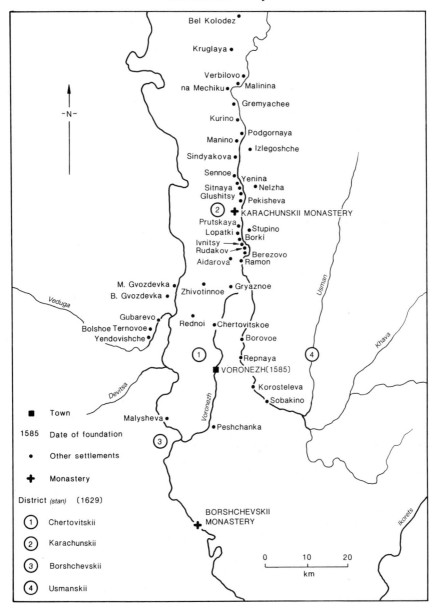

Fig. 2.2. Voronezh district in 1615

and habitations of the junior boyars. Unlike the contract servitors, the junior boyars are listed in the cadastres as individuals since, legally speaking, they held their land by individual grant from the tsar.[15] Even so, the land of most junior boyars was located alongside that of others in the same village. This pattern may have resulted from the granting of a block of land to a whole group of servitors, who were then left to divide it up among themselves, according to the number of *chetverti* specified for each individual.[16] The pattern obviated the need to survey each holding separately and, since it promoted nucleated settlement, it also had advantages for defence. A complicating factor was the tendency for servitors to receive additions to their original holding, as a result of faithful service. Thus, numerous servitors in 1615 and 1629 held land in more than one location.

V. P. Zagorovskii claims that most junior boyars in the early seventeenth-century south were on a remuneration scale of between 100 and 150 *chetverti*.[17] The actual amount allocated, however, was usually much less, as was the case at Voronezh in 1615. Here, out of 275 *pomestya* 223 were less than 70 *chetverti* in size, accounting for 56.5 per cent of all the *pomeste* land in the district. The typical holding in 1615 was in fact 50 *chetverti*, but many servitors had only 30, and some even less.[18]

Mention has already been made of the cottars living among the servitors at Voronezh in 1615. Small numbers were also living among the atamans in the rural areas at this time: forty-three cottars, for example, are recorded as dwelling in the sixty-six ataman households in the village of Sobakino.[19] Possession of peasants and cottars among the junior boyars was on a larger scale, but was still relatively modest. In 1615 only about half of the junior boyars in Voronezh district are recorded in the cadastre as having one or more serfs on their properties.[20] More details come from the 1629 cadastre. For the rural district only at this time the cadastre records 1,570 peasant and cottar households shared among about 300 junior boyars and perhaps 140 atamans. There were also a handful of monastic peasant households.[21] This level of serf-holding was in fact higher than in certain neighbouring districts. At Oskol to the west, for example, there were virtually no serfs at all in 1615 and most servitors are recorded as having 'only one *dvor*'.[22] In the better-protected areas along the upper Don and the tributaries of the Voronezh, by contrast, the numbers of serfs increased and there were even one or two large estates to be found.

Between 1615 and 1629 the number of settlements in Voronezh district had grown by about a dozen to reach approximately sixty. At the first date junior boyars held about 17,000 *chetverti* of land in the rural district, to which may be added some 7,000 *chetverti* held by atamans and cossacks. In 1629 total landholding in these categories was 32,000 *chetverti*. By no means all of this land was actually under cultivation, but the increase does reflect the process of settlement and population growth. New *pomestya* had been allocated in villages appearing between

1615 and 1629, but much land in these new villages, and also in newly formed outlying 'wastes' (*pustoshi*), was held by servitors from the old villages. What had been happening was that long-standing servitors had been adding to their old *pomestya* as a result of faithful service. There appears to have been a shortage of land available for this purpose in the old villages, although some of the *pomestya* recorded in 1615 had fallen vacant by 1629, and had been reallocated to neighbouring landholders. A somewhat surprising feature of the fourteen-year period was the relatively large number of holdings changing hands. Thus, in Chertovit-skii *stan* (a subdivision of the *uezd*), north of Voronezh, only 73 per cent of the landholders recorded in the old villages in 1629 had either been there already in 1615, or had land that had been held by their father or near relative at the earlier date. For Usmanskii *stan* to the east the corresponding statistic was only 49 per cent.[23] No doubt this turnover reflects the harsh and difficult frontier life lived by these pioneers and their propensity to flee from service to the cossacks or elsewhere. It also reflects the willingness of other servitors, suffering from landlessness and poverty, to take their chances and replace those who had left. For the loyal, however, there were rewards such as increases in landholdings and the conversion of *pomestya* to hereditary tenure. The 1629 cadastre records a limited number of hereditary holdings scattered throughout the district.

### SEVENTEENTH-CENTURY DEVELOPMENTS

Even by the late sixteenth and early seventeenth centuries the southern frontier already had a number of quite distinctive social characteristics. Previous sections have commented upon the predominance of small-scale servitor landholding, the considerable importance of contract servitors living in the towns, and the relatively low levels of serf-holding. This distinctiveness was reinforced during the course of the seventeenth century, as more and more southern servitors came to possess only one *dvor*.

The southward and frequently illegal migration of peasants towards the frontier continued during the seventeenth century. As time went on, areas which had previously been recipients of peasant settlers them-selves began to experience the phenomenon of peasant flight. In the late 1630s, for example, the junior boyars of Voronezh district complained bitterly that their peasants were fleeing to enrol as servitors at the new towns of Kozlov and Tambov.[24] Voronezh district itself continued to benefit from an influx of migrants until the 1640s or later. The 1,570 or so peasant and cottar households enumerated there in 1629 had become

2,060 by 1646, but by 1678 there were only 1,088.[25] By this stage many southern servitors had lost all their serfs. In Korocha district there were 134 junior boyar households in 1647 but only twenty-one peasant households. Of the two hundred Orel junior boyars who settled at Karpov in 1648, only three had serfs; of the thirty-eight Kursk junior boyars settling at the same place, only seven were serf-owners.[26] In 1653 the atamans of the village of Usman-Sobakino east of Voronezh protested that they had lost all their peasants and cottars to the new towns then being constructed in the region. Ninety-one of them were serving as dragoons at nearby Orlov.[27] Calculations respecting the entire Belgorod and Sevsk military districts have been made by V. M. Vazhinskii.[28] He estimates that the number of private and court peasant households in these territories grew from 36,000 in 1646 to more than 41,000 by 1678, representing a total population increase from more than 230,000 to over 470,000. However, this increase came about as a result of the settlement of the more northerly districts by lords and serfs, whereas the southern districts actually lost serfs. There was also a tendency for serfs to become concentrated on the larger estates, whose lords could presumably offer greater incentives. In the 1620s, in the twenty districts then existing, there were 6,628 households of servitors and hereditary landholders, of which 43.2 per cent were without serfs. They held just under a quarter of the service land (not including land held by contract servitors). By the end of the 1690s in the same territory 96.5 per cent of the servitors had no serfs and they held 82.9 per cent of the service land.

This loss of serfs among the servitors as a group can be explained by several factors. First, as noted above, wealthier landholders settling at first in the more northerly and safer districts could offer greater incentives to peasant migrants than could the lesser servitors. Throughout much of the century these wealthy landholders were forbidden to acquire land in the actual frontier regions.[29] Secondly, there was the government's wish to settle its new towns and defensive works, especially those along the Belgorod Line from the 1630s onwards. The needs of settlement and service along the southern frontier often undermined the official determination to uphold and strengthen the principle of serfdom. Thus, it was common practice to invite local cossacks and other 'free' people to settle and take up service in the new military towns. Many of the 'free' people thus recruited were not cossacks of long standing, but rather runaway serfs who were successful in concealing their identities. This process continued in spite of official prohibitions and the many petitions from estate-holders unfortunate enough to lose their serfs in this way. An interesting example of the situation is cited by Zagorovskii, who describes events in 1646 when the noble Zh. Kondyrev was ordered to assemble a large force of volunteers at Voronezh for the forthcoming attack on the Crimea. The rush of willing hands far exceeded the number expected and in the event included not only local cossacks but also large numbers of runaway serfs, as well as servitors from far and near.[30] In many of the new towns along the Belgorod Line in the 1630s and 1640s it was occasionally possible for 'free' people to join not only the contract servitors but even the junior boyars.[31]

Where the government was unable to attract sufficient 'free' people to settle its new towns and districts it resorted to the compulsory settlement of servitors from elsewhere.[32] The government was also able to add to the number of its servitors in various other ways. Thus, in the latter part of the 1640s it resorted to the 'militarization' of a number of villages and estates on the upper part of the Voronezh River. The peasants on these estates were registered as servitors.[33]

As the seventeenth century progressed the relationships between the various categories of servitor, and also the military role performed by each group, gradually changed. Thus, while in the 1670s and 1680s entry into the ranks of the junior boyars seems to have become quite difficult for those without a hereditary claim to that status, such promotion may have become easier again by the 1690s.[34] Many junior boyars began to suffer a decline in social and economic standing in this period. Poverty forced numerous junior boyars to take to the plough themselves, in the absence of serfs, and they found it increasingly difficult to undertake their cavalry duties. A decree of 1638 ordered that poor servitors were not to be detained on service in the towns unless a Tatar raid threatened, since 'they are poor *odnodvortsy* and work for themselves'. Junior boyars from Ryazan explained in a petition that some of them were unable to serve away from their homes, and others who served had returned home without permission, out of poverty rather than disobedience.[35] In the latter half of the century the government came to rely less and less upon the old middle-service-class cavalry, of whom the junior boyars formed the core. Increasing attempts were made to enrol the junior boyars into the 'new formation' regiments of soldiers (*soldaty*), dragoons (*draguny*), and cavalrymen (*reitary*) which appeared on a permanent basis at the beginning of the 1650s.[36] Although for reasons of social status the junior boyars greatly preferred to become cavalrymen, the cavalry had a declining role in the new military strategy. It became quite common for the poorer junior boyars to be obliged to join the contract servitors in the ranks of the dragoons and soldiers.[37]

The decreasing military significance of the junior boyars was paralleled by their reduced role in the settlement process. Although they became established around many of the new towns along the Belgorod

Line, only at Kozlov, constructed in 1635, did they constitute a major portion of the servitors.[38] Moreover, the land allotments given to most junior boyar settlers were by now more modest than in an earlier period. Thus, the two hundred junior boyars from Orel who settled at Karpov in 1648 were allotted holdings usually below 50 *chetverti*, and even below 20 in some cases. Similar allocations were made to the thirty-eight junior boyars from Kursk who settled there at the same time.[39] Small landholdings soon came to predominate in such districts as Kozlov and Belgorod, and in other areas as time went on. Of the more than 43,000 *pomestya* analysed by Vazhinskii for the Belgorod regiment for 1697, 60 per cent were below 30 *chetverti* and only about 11 per cent above 70.[40]

The recruiting of junior boyars into the dragoons and soldiers naturally had the effect of eroding their distinctiveness from the contract servitors. The latter half of the seventeenth century also eroded the spatial separation between the two classes. The greater security gained by the building of the Belgorod Line meant that it was no longer necessary to restrict settlement to the environs of the towns or to increase the size of the urban garrisons. As the southern population grew, it became common practice to grant contract servitors lands out in the rural districts, hitherto largely monopolized by the middle service class. Many of the contract servitors so settled were members of the new formation regiments. These regiments now began to receive land for their subsistence, as it became apparent that the government could not afford to sustain the sizeable standing army it had recruited in the 1650s. Soldiers, for example, began to be allocated land on the same *pomeste* basis as the junior boyars. This happened to a group settled on the right bank of the Usman River in 1657.[41] Materially they were thus better off than the musketeers they were intended to replace, although their service was both longer and harder. Zemlyansk district, which was separated from that of Voronezh in the late 1650s, was settled almost exclusively by contract servitors.[42] The description of Tambov and its region in 1678 indicates that cossacks, soldiers, and other servitors were living with their families and cottars in considerable numbers in the various villages.[43] A similar picture of soldiers, cossacks, musketeers, and 'urban' servitors living in the villages of Verkhososensk district is gained from a reading of the enumeration book of 1709.[44]

The poverty of most southern junior boyars, and their intermingling with the contract servitors, naturally tended to undermine their claims to equal status with the serf-holding servitors. The latter, whether belonging to the 'town' junior boyars or to the higher-service ranks of Muscovite society, gradually consolidated to form a cohesive noble class. This was a process which was completed in the eighteenth century. Most southern servitors, on the other hand, moved closer to the peasantry. In landholding arrangements, for example, it became common for southern servitors to hold their arable lands in intermingled strips, in the manner of the peasantry. This arrangement may have arisen first among the urban-based contract servitors, but it also became the norm among the rural servitors, especially as population increased. Only the wealthier serf-owning landholders were able to divorce themselves to some extent from such practices. As time went on and population grew, it became ever more difficult to distinguish between the landholdings and land use of the servitors and those of the peasantry.[45]

Towards the end of the seventeenth century the southern parts of what were to become the Central Black Earth provinces began to experience the settlement of wealthier servitors and their serfs. In northern areas such as Tula and Ryazan this process had occurred much earlier, and had encouraged southward migration on the part of small-scale landholders.[46] Towards the south, however, such factors as the relative stability of the frontier line over a considerable period, the loss of serfs among the junior boyars, the importance of the contract servitors, official policy preserving land for settlement by minor servitors, and its subsequent mass settlement in the middle and latter half of the seventeenth century gave the region a special character which endured even when colonization by wealthier lords affected this area.

Certain Soviet historians have argued for an increasingly 'feudal' exploitation of the southern servitors as the seventeenth century advanced.[47] As evidence they cite the loss of serfs, the reductions in land allocations, the increases in taxation and other burdens, and the civil unrest which characterized the latter part of the century. However, as against that, the pressures of Tatar raiding, which added to the problems of service and of farming, certainly eased later in the century, as defensive systems became more effective. The demands of service probably became lighter for the urban-based servitors, and although wartime service was difficult and demanding for the new formation regiments, those serving in such regiments usually received more generous land allotments than had the musketeers, artillerymen, and other lower-class servitors. Of course, those junior boyars who had lost their serfs were worse off than before. While civil unrest was endemic in Russia, and on the frontier especially, there is no real evidence that it increased as the century progressed. What can be said is that the southern servitors were certainly not among the privileged groups in Russian society, and this became even more apparent in the next century.

## THE *Odnodvortsy* IN THE EIGHTEENTH AND NINETEENTH CENTURIES

By the early eighteenth century the advance of Russia's southern frontiers had reached a point where many of the erstwhile southern servitors were no longer living in militarily vulnerable regions. The old concept of the servitors forming a local frontier force therefore fell gradually into abeyance, and more and more of them took to the agricultural life. In recognition of these facts the southern servitors, including junior boyars, had been gradually subjected to financial exactions since the middle of the seventeenth century.[48] Finally, under Peter the Great, a new status was defined for these groups. Whereas the wealthier servitors now began to form a noble class in accordance with the new Table of Ranks, and were obliged to perform service accordingly, the impoverished southern servitors were subjected to taxation and the recruitment levy, like the peasants. According to Peter's decree of 22 January 1719, the male *odnodvortsy*, as they now officially became known, were to be registered with a view to paying the poll tax.[49] The decree embraced all the contract servitors and also many junior boyars too poor to undertake noble service. Some serf-holding junior boyars, anxious to escape service, also entered the ranks of the *odnodvortsy*. In 1723 the *odnodvortsy* began to pay a special tax (*obrok*) to the state, by analogy with that paid by private peasants to their lords. A decree of 1724 gave them the title 'state peasants', although this decision was reversed by Empress Anna in 1731.[50]

The word *odnodvortsy* had frequently been used in seventeenth-century documents to describe servitors who had no serfs;[51] the 1719 decree gave the term full legal status. Henceforward the *odnodvortsy* were recognized as a social category in their own right, to be distinguished from the nobles. By a succession of decrees in the 1720s, they were subjected to restrictions on their movements and to other controls, like the state peasants. But until late in the century their erstwhile special status was also granted some recognition. For example, they were obliged to serve not in the regular army like the peasants, but in a special frontier force known as the Landmilitia. In the 1730s this force consisted of twenty regiments, which were ordered to settle on the new Ukrainian Line, a project that was never fully carried out. Until 1783, when the Landmilitia was abolished, the *obrok* payments which the *odnodvortsy* made (in this case earmarked specifically for the maintenance of the Landmilitia) were generally less than those made by other state peasants. After that year the *obrok* and the recruit levy were the same for both. Like state peasants the *odnodvortsy* were liable to be drafted for special tasks such as road-building and the repair of fortifications, but unlike virtually all the other state peasants they had the privilege of serf-ownership. However, as noted already, few did in fact own serfs. After 1754 the *odnodvortsy* were allowed to sell their serfs only to other *odnodvortsy* who lived in the same district.

Eighteenth-century government policy towards the *odnodvortsy* was essentially designed to preserve them as a group who would be in a position to pay the required taxes and render necessary services to the state. In essence this was a continuation of seventeenth-century policy. A decree of 1724 attempted to free those *odnodvortsy* who, through poverty, had fallen into dependence on lords. In 1727 it was ordered that the *odnodvortsy* could not sell their lands lest this undermine their tax-paying powers. This order was repeated on subsequent occasions, most notably in the Survey Instructions of 1754 and 1766. The latter decree attempted to divide the *odnodvortsy* lands into several categories, according to their origins, and to restrict rights of alienation.[52] Yet, despite such legislation, the *odnodvortsy* continued to regard their lands as their own, with rights of sale to whomsoever they pleased. By this means, and also by a variety of other processes both fair and foul, their lands fell into the hands of lords, whose possessions thus frequently became intermixed with those of the *odnodvortsy*. The General Survey failed to survey the *odnodvortsy* lands separately from the lands of the lords—and thus left the door open to the further diminution of *odnodvortsy* landholding. The *odnodvortsy* instructions to their delegates attending Catherine II's Legislative Commission, written in 1767, are full of complaints about the acquisition of *odnodvortsy* lands by members of the nobility.[53] The instructions make clear that lords acquired lands not only through purchase, but also through the forcible seizure of outlying plots. In this they were aided by the lack of documentary proof of ownership among many of their victims. The *odnodvortsy* became notorious for their disputes over land. As Catherine II observed, 'They are a people much given to argument.'[54]

The *odnodvortsy* instructions to the Legislative Commission contain numerous pleas that *odnodvortsy* rights to the various privileges of the nobility, including those of promotion in the army, be recognized. Such pleas were no doubt a reaction to the increasingly exclusive privileges which the nobility had recently been obtaining for themselves. Since earlier in the century only those *odnodvortsy* with documentary proof of noble origin, and those who attained over-officer rank in the army, were granted noble status. However, according to an analysis of the instructions coming from the black-earth provinces undertaken by M. T. Belyavskii, only those *odnodvortsy* with junior boyar origins made such pleas to the Commission.[55] In this and in certain other respects the *odnodvortsy* were very much divided in their attitudes, and failed to form a cohesive group.

The processes of inheritance, land subdivision, sale, and so on gradually reduced numerous *odnodvortsy* to poverty and even landlessness. Evidence of this is already to be found in the 1767 instructions, but became even more apparent as time went on. The General Survey Instruction of 1766 attempted to deal with this by proposing that each household be allowed a holding of 32 *desyatiny*, taken if necessary from state land, with an additional 28 *desyatiny* to allow for future growth of the family.[56] However, in many cases state land did not exist sufficiently close to the *odnodvortsy* settlements to allow this to happen. Some of the *odnodvortsy* escaped poverty by enrolling in the towns, an action made considerably easier by Catherine's administrative reforms.[57] Finally, there was always the possibility of resettling in new lands. Settlement along the Ukrainian Line had been government policy since the 1730s, and then and afterwards many *odnodvortsy* had migrated unofficially to other frontiers.[58] Lands were provided for *odnodvortsy* settlement on the Orenburg Line and, later in the century, in such regions as Yekaterinoslav and Stavropol. Some migrants undoubtedly changed their official status on moving, and so it is difficult to be certain about the exact scale of such migrations, but it is a remarkable fact that the majority of the *odnodvortsy* remained in the black-earth provinces. No doubt part of the reason for this situation was government policy, allocating new lands in secure areas to the nobility in preference to groups like the *odnodvortsy*. Strong cossack traditions of acquiring land by intake (*zaimka*) may also have favoured the Ukrainian and Don cossacks in the competition for land. Finally, later in the century when the frontier had moved far to the south and east, sheer poverty undoubtedly discouraged many *odnodvortsy* from migrating such distances.

Towards the end of the eighteenth century the *odnodvortsy* began to find a partial answer to land shortage in communal land-ownership and repartition of land on a more equal basis. The wealthy *odnodvortsy* long opposed repartition, seeing it as an essentially peasant procedure which was directly contrary to the traditional *pomeste* form of landholding. A proposal to institute repartition among the *odnodvortsy* of Kursk province in the 1780s was opposed, in part for this reason. Eventually, however, the movement to institute communal landholding, which affected most of the Russian peasantry from the eighteenth century onwards, began also to influence the *odnodvortsy*. The government and its local officials encouraged, but probably did not instigate, this tendency. Until 1850, when individual rights to *pomeste* land were recognized, transfer could occur by simple majority vote, but transfers continued even after that date. Statistics collected at the beginning of the 1850s indicate that over

60 per cent of the *odnodvortsy* were holding their land communally at that stage, the majority of whom had transferred from *pomeste* landholding.[59] According to N. A. Blagoveshchenskii, who investigated the former *odnodvortsy* communities in the 1890s using *zemstvo* materials, smaller *odnodvortsy* communities, especially those whose forebears had originally settled on the basis of a land allocation to only a handful of servitors, seem to have found transfer difficult because of legal obstacles and the strong traditions of inheritance existing among the close-knit *odnodvortsy* families.[60] Many such communities thus retained the traditional hereditable form of landholding (referred to as *chetvertnoe zemlevladenie*) until the time of the 1917 revolution.[61]

## THE VORONEZH *Odnodvortsy* IN THE LATE EIGHTEENTH CENTURY

In the period between the early seventeenth century and the middle of the eighteenth, *odnodvortsy* had settled in most districts of Voronezh province. The settlement of areas to the north of the Belgorod Line such as Zemlyansk has already been noted. Settlement of areas to the south began only towards the end of the seventeenth century, and here the *odnodvortsy* were in competition with the Ukrainians, and later with Great Russian lords. Even so, some sizeable *odnodvortsy* villages did develop in these parts, and such villages eventually began to sprout out-settlements. The distribution of such settlements at the time of the General Survey is illustrated in Fig. 2.3.

As with the *odnodvortsy* elsewhere, those in Voronezh province were largely without serfs. Protorchina's analysis of the results of the first revision in the 1720s indicates that by this stage private peasants were only to be found in those districts of the province already settled in 1615–29.[62] Even in these districts, however, the private peasants were greatly outnumbered by *odnodvortsy*, and there were many settlements which had had serfs in 1615–29 but which now completely lacked them. Elsewhere in the province there were no private serfs at all among the Russian servitors in the 1720s.

Later in the eighteenth century the acquisition of estates by members of the nobility affected much of the rest of the province. These estates were subsequently settled by the lords' own serfs, who often found it convenient to live in villages alongside or near the *odnodvortsy*. The economic notes to the General Survey therefore record many mixed settlements, or at least mixed survey units. The 1785 topographical description of the province states that 'the nobility for the most part

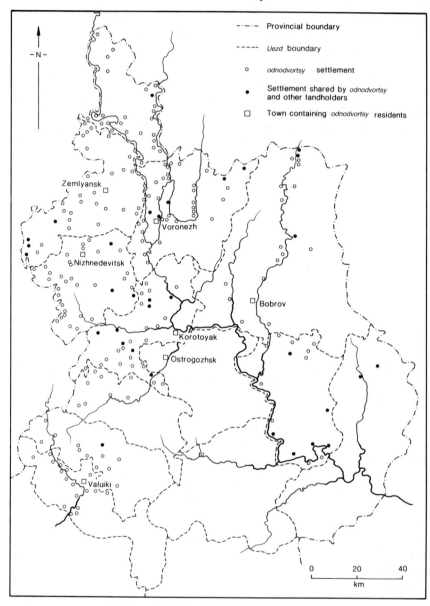

- N -

| | |
|---|---|
| — · — · — | Provincial boundary |
| — — — — | *Uezd* boundary |
| ○ | *odnodvortsy* settlement |
| ● | Settlement shared by *odnodvortsy* and other landholders |
| □ | Town containing *odnodvortsy* residents |

Zemlyansk

Voronezh

Nizhnedevitsk

Bobrov

Korotoyak

Ostrogozhsk

Valuiki

0    20    40
km

FIG. 2.3. The distribution of *odnodvortsy* settlements in Voronezh province at the time of the General Survey (late eighteenth century)

exercise an intermixed [*chrezpolosnyi*] landownership with the *odnod-vortsy*.[63] Disputes about land-ownership were consequently common, and evidence of such disputes is easily found in the economic notes.[64]

An important question concerns the extent to which members of the seventeenth-century junior boyar class had joined the nobility as against the *odnodvortsy*. For a thorough investigation of this topic, a comparative survey of the cadastres, enumeration books, deeds, charters, and decrees of the seventeenth century, together with the revision and General Survey materials of the eighteenth, would be necessary. In the present instance, a comparison of the surnames of landholders living in villages in Voronezh district in 1615–29 with those of the nobility holding lands in the same villages at the time of the General Survey is all that has been possible. The degree of continuity between the two is in fact minimal. There were, of course, some exceptions. Thus, in 1615 the village of Sindyakova was dominated by the Rukin family, five members of which owned two-thirds of the land in the village. Four members of the Rukin family held land as nobles in the village in the late eighteenth century. In Pekisheva A. G. Pilyugin held 100 *chetverti* in 1615. In 1629 the estate was held by his son, I. A. Pilyugin. In the late eighteenth century land was held in the village by two members of the Pilyugin family, as well as by A. P. Shishkin, undoubtedly a descendant of P. M. Shishkin, Voronezh siege-supervisor and important landowner between 1615 and 1629. In Glushitsy O. Ye. Yartsov held 70 *chetverti* in 1615. Fourteen years later the estate was held by his four children. At the time of the General Survey members of the Yartsov family held land in Glushitsy, Bolshoi Mechok, Nelzha, Demshino, and elsewhere.[65]

Other examples could be given, but the point is that these are exceptions. Generally, the surnames had changed over the 150-year period. Several possible explanations suggest themselves. The original junior boyars could have moved elsewhere, as indeed many did between 1615 and 1629. Some surnames found in villages in 1615–29 are found in completely different villages in the following century. Surnames also changed over time—the 1615–29 cadastres record alternative surnames in a number of cases. The sale, exchange, or abandonment of estates, the extinguishing of lines of inheritance, or inheritance through the female line could also have led to changes in surname. Alternatively, the junior boyars may have entered the ranks of the *odnodvortsy*. Most unfortunately, the General Survey materials do not record the latter's surnames and so it is impossible to check to what extent this occurred. However, there is considerable direct evidence from the work of scholars such as Yelfimova and Shcherbina, and indirect evidence from a comparison of villages in 1615–29 and the late eighteenth century, as indicated in Table 2.1.[66] Many villages with only junior boyar landholders in the earlier period were populated by considerable numbers of *odnodvortsy* in the latter. From Kursk province comes evidence that members of the same family holding land side by side in

TABLE 2.1. *Landholders in selected settlements of Voronezh district in the seventeenth and eighteenth centuries*

| Village | 1615[a] | | 1629[b] | | | 1779/80[c] | | | Gen. Survey[d] | |
|---|---|---|---|---|---|---|---|---|---|---|
| | at | jb | at | jb | pc | od | op | pp | nl | |
| Bolshoi Mechok | | 2 | | 1 | | 160 | | | 10 | od |
| Glushitshy | | 12 | | 12 | | 47 | 4 | 71 | 17 | od |
| Gremyachee | | 13 | | 15 | | 85 | 3 | | | |
| Izlegoshche | 19 | 13 | 16 | 13 | | 672 | | 16[e] | 3 | od |
| Kruglaya | | 6 | | 6 | | 44 | | 133 | 5 | od |
| Kurino | | 7 | | 4 | | 32 | | 41 | 6 | od |
| Malinina | | 2 | | 2 | | 305 | 5 | 79 | 10 | od |
| Manino | | 10 | | 12 | | 64 | 4 | 54 | 7 | od |
| Nelzha | | 2 | | 2 | | | | | 4 | od |
| Ostapova | | 4 | | 5 | | 31 | | 38 | 8 | od |
| Pekisheva | | 3 | | 3 | | 23 | 8 | 58 | 7 | od |
| Peskovatoe | | | | 11 | 18 | 325 | | 115[e] | 5 | od |
| Podgornaya | | 8 | | 7 | 3 | 123 | 19 | | 3 | od |
| Sennoe | | 16 | | 13 | | 119 | 30 | 5 | 9 | od |
| Sindyakova | | 8 | | 8 | 3 | 63 | | 29 | 7 | od |
| Verbilovo | | 6 | | 6 | | 111 | 35 | 68 | 6 | od |

*Key*: at = ataman;  jb = junior boyar;  nl = noble  landowner;  od = *odnodvortsy*; op = *odnodvortsy* peasant; pc = *pomestnyi* cossack; pp = private peasant.

[a] TsGADA, f. 1209, d. 614.
[b] Ibid., d. 615.
[c] Ibid., f. 16, yed. khr. 654. (This column designates male population, not landholders.)
[d] Ibid., f. 1355, yed. khr. 51, 97, 249, 263.
[e] Statistics from the economic notes to the General Survey.

the same village were either nobles or *odnodvortsy* at the time of the General Survey.[67]

It seems a fairly safe assumption that the core of the *odnodvortsy* in many villages of Voronezh district in the late eighteenth century was based on junior boyars of the pre-Petrine era. The *odnodvortsy* instructions to Catherine II's Legislative Commission give a similar picture for other areas, such as Livny and Yelets.[68] Of course, some of the original junior boyars had moved away, others had become nobles, and there had been a continuing influx of new settlers. Without much more detailed archival work it is not possible to be sure about the degree of continuity over this long period.

As for the former contract servitors, there can be no doubt at all that they had become *odnodvortsy*. Settlements where atamans had lived in the seventeenth century, such as Sobakino, Stupino, and Izlegoshche, contained large numbers of *odnodvortsy* at the time of the General Survey. Where whole regions had been settled by contract servitors of the new formation regiments, the General Survey shows the complete dominance of the *odnodvortsy*. The soldiers, cossacks, and guards recorded in the villages of Tambov district in 1678 were almost universally *odnodvortsy* a century later.[69]

An earlier section has commented on the growing poverty of many *odnodvortsy* in the late eighteenth century. V. I. Semevskii gives figures for average landholding among the *odnodvortsy* of several provinces. However, because of the character of the General Survey, he can only give data for the minority who held their land separately from other categories of landholder. His figures vary from 5.8 *desyatiny* per male soul in Penza province, and 7.6 in Kursk, to 19.3 in Voronezh.[70] The latter province was of course still in process of settlement. Even so, it is clear from the General Survey that holdings varied greatly in size throughout the province. On the one hand, in Usman district (part of Tambov province by this stage) 4,394 *odnodvortsy* held an average of 8.8 *desyatiny* per man.[71] In the same area, out of 607 *odnodvortsy* holdings investigated by V. I. Nedosekin, 202 consisted of only the land under the homestead.[72] On the other hand, some especially large holdings were found in the southern part of Voronezh province. In Boguchar district, for example, the average male *odnodvorets* had 39.7 *desyatiny* in Nikolskoe and 41.5 in Rudnya.[73]

Because of problems in data comparability, it is difficult to say whether the eighteenth-century *odnodvortsy* who actually held land were obviously worse off than their predecessors of the previous century. However, this seems most likely in view of the abundance of land in the earlier period. Making some reasonable assumptions, for example about the number of males in each seventeenth-century household, it can be estimated that each male cultivator in Voronezh district had 27.6 *desyatiny* in 1629 and, in identical villages, 14.5 in the late eighteenth century. Comparing data for the *odnodvortsy* villages in Tambov district in 1678 and under the General Survey, a similar picture of reductions in holding-size emerges.[74]

## CONCLUSION

By the latter part of the eighteenth century the *odnodvortsy* as a group had acquired an unenviable reputation for sloth, poverty, and ignorance. The 1785 topographical description, for example, comments that they 'are not greatly given to toil, and for this reason most of them are in

straitened circumstances and also uncouth'.[75] When *odnodvortsy* delegates to the Legislative Commission complained of the depradations of
the nobility on their land, the noble delegates retorted that the
*odnodvortsy* had been reduced to poverty not because of the actions of the
nobility, but because of 'their negligence of agriculture and their
laziness'.[76] Writing in the 1850s, G. Germanov also noted their poverty,
but ascribed it to the harshness of frontier life, their lack of leisure in the
past to learn such peaceful pursuits as trade, and a certain 'severity and
brutality' of character, passed down from generation to generation.[77]
The alleged idleness of the *odnodvortsy* was sometimes ascribed to their
quasi-noble origins. Suggestions of this idea are found in Vasilii
Levshin's description of Tula province, written at the beginning of the
nineteenth century.[78] Baron von Haxthausen, who travelled through
Russia in the 1840s, observed that formerly the *odnodvortsy* had 'never
mingled nor married with the surrounding peoples, and they seldom do
so even now. How innate is the feeling of aristocracy in all nations!'[79]

It is apparent, therefore, that the spirit of individual enterprise, which
Turner claimed as a characteristic of the American pioneer and which, he
felt, had been passed down in some measure to his descendants, was not
particularly noticeable among the *odnodvortsy*, at least not among those
living in the late eighteenth and nineteenth centuries. In many respects,
indeed, the *odnodvortsy* had moved closer to the peasantry, displaying
several of the characteristics of the latter group as described by A. V.
Chayanov and subsequent scholars.[80] Clearly the closure of the frontier
had diminished the differences between the former servitors and their
neighbours. Yet it is important not to over-simplify this process, or to
see their evolution as a case of simple social decline. It has been argued
above that many, possibly a majority, of the servitors had lowly origins in
any case, and that life for most of them was harsh indeed, even while the
frontier existed. It was not a simple case of descent from quasi-noble to
peasant. Neither was it a case of the *odnodvortsy* losing every vestige of
their frontier heritage. According to Semevskii, it was a certain stubborn
individualism which helps explain their long resistance to communal
landholding,[81] and the unplanned and even chaotic appearance of their
villages, a product of every individual acting as his own master, was
noted by such eighteenth-century observers as Andrei Bolotov.[82] The
*odnodvortsy* were a product of a particular type of frontier impinging
upon a particular environment; once the frontier had passed by, so did
their claims to a special status. Yet the end-result of this process was not
a uniform social landscape: rather the *odnodvortsy* made their own
unique contribution to the variety which characterized rural Russia in
the pre-revolutionary period.

# 3
# The Province of Voronezh under Catherine the Great

DURING the latter half of the eighteenth century, the materials available for the historico-geographical study of Russia become much more abundant than for any earlier period. This comparative wealth of sources is related to the economic advance of the time, the cultural and scientific progress which Russia was experiencing, and the increasingly marked tendency for the government to interest itself in the economic life of the realm. Under the influence of the encyclopaedists and the physiocrats, the latter seeing agriculture as the principal source of national wealth, the government busied itself collecting a wide range of statistics and general information on local economies. Its activities were supplemented by those of other official bodies, such as the Academy of Sciences and the Imperial Free Economic Society, which was founded in 1765. Provincial governors and local officials were required to answer questionnaires concerning trade, agriculture, prices, markets, and similar matters, and to write reports on local affairs. This, moreover, was the great age of topography, when a systematic effort was made to record and describe the physical and cultural features of the landscape and to map the national territory.[1] In the 1760s the government, worried about the rural unrest resulting from land disputes, embarked upon the General Land Survey, which entailed the mapping, recording, and describing of groups of estates, village by village, district by district. Travellers roamed the countryside and recorded their impressions; scholars wrote local histories and topographical surveys. In short, it was the period when geographical description came of age. It is therefore an appropriate one in which to examine the economic and social geography of a single province.

As an attempt to reconstruct some of the fundamental characteristics of the physical, social, and economic geography of Voronezh province under Catherine II, with particular reference to the last quarter of the eighteenth century, this chapter stands in the tradition of cross-sectional analysis which has a long and respected history in Western historical geography.[2] The aim is to illustrate the geographical use of the wide range of primary sources available for the period. A related aim is to show how the various processes which were influencing the province's development at the period were moulding a landscape of considerable complexity, one which can only be partially and imperfectly described with the available source materials. An earlier section of this book discusses the concept of the peasant ecotype,[3] which seeks to explain how different peasant life-styles and economies arise in response to different environmental circumstances. Here the concept is extended to illustrate the way in which the natural environment, particular settlement processes, and interrelationships with other areas help form the geography of a whole region and help govern the way in which that region develops over the long term.

The territory which forms the focus for this study is the province of Voronezh, as defined by the administrative reform of 1775. This is illustrated in Fig. 3.1. The reform took effect in Voronezh province in 1779, when a new province (*namestnichestvo*) was defined with its capital at Voronezh, consisting of fifteen districts (*uezdy*). However, the two southernmost districts of Belovodsk and Kupyansk are closely associated historically with the region of Slobodskaya Ukraina and will therefore be given only brief consideration here.

A major source for the study of the human and economic geography of Voronezh province in the late eighteenth century is the economic notes to the General Survey. The Survey itself was undertaken in this province in 1778–81, and the data for the economic notes were gathered by the surveyors at the same time.[4] They subsequently appeared in manuscript and tabular form. The notes record areas of land use in each survey unit (*dacha*) under the following headings: land under settlement, arable, hayland, woodland, and unusable land. Additional details such as population totals, social status of landholders, quality of the land and of harvests, state of the woodland, mills, factories, fairs, markets and trade, handicrafts, public buildings, and other matters are given. From the viewpoint of the geographer, their major drawback is the fact that they relate only to the *dacha* and not to individual estates, which were in fact often grouped together for survey purposes. Hence, it is frequently impossible to say how much land belonged to each landholder, or to distinguish, say, between demesne land and peasant land, or between noble land and state peasant land. The reason for the coarse-grained nature of the Survey was the immensity of the task (as it was, the Survey was still incomplete in the next century) and the political problems that would have resulted from trying to demarcate too strictly between the lands of the different owners. For full value, the economic notes need to be read in their several versions together with the manuscript atlases of the General Survey. Unfortunately, this was not possible for the present writer. The economic notes have been described and evaluated as a research tool by L. V. Milov.[5]

A second group of sources used for this study are the topographical

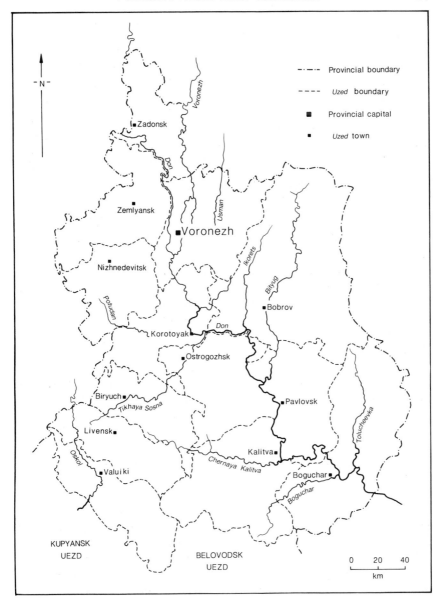

FIG. 3.1. Voronezh province: administrative subdivisions after 1779

descriptions and questionnaires which, as already indicated, were distributed from time to time among various local officials. Use has been made of several manuscript descriptions of the 1770s and 1780s, including the important 'Topographical Description of Voronezh Province' of 1785.[6] The archives in the Soviet Union contain other material of this kind which could not be consulted, although the returns to three questionnaire surveys of the 1760s, sent out by the Academy of Sciences (under the influence of M. V. Lomonosov), the Noble Cadet Corps, and the Senate, were analysed by B. D. Grekov in a publication of 1929.[7] Similarly, returns to the Imperial Free Economic Society's questionnaire of 1768 for Ostrogozhsk district were published in volume 8 of that Society's *Works*, while answers to the Academy of Sciences' questionnaire of 1781 were published by L. B. Veinberg towards the end of the last century.[8] These and other survey materials form the basis of such published descriptions as Bacmeister's *Topographical Information* (1771–4), Shchekatov's *Geographical Dictionary* (1801–8), and Zyablov-skii's *Description* (1810), which have also been consulted.[9]

A third valuable source is Yevgenii Bolkhovitinov's *Historical, Geographical, and Economic Description of Voronezh Province*, published in 1800.[10] Bolkhovitinov was a noted churchman and scholar, tutor at the Voronezh Theological Seminary, and later Metropolitan of Kiev. His work is in the tradition of amateur topography which was becoming established at the time, and was the first history and description of the province to be published. Although written after the death of Catherine II, much of Bolkhovitinov's material in fact relates to the period before 1796.

Use has also been made of a number of other sources, most notably of S. G. Gmelin's travel account, dating from the 1760s, and of a variety of secondary sources and cartographical materials.[11]

## VORONEZH PROVINCE: THE NATURAL ENVIRONMENT IN THE LATE EIGHTEENTH CENTURY

Voronezh province forms part of the huge plain of European Russia characterized by relatively low relief, virtually no point being more than 250 metres above sea-level. The most prominent geographical feature is the broken terrain of the Central Russian Uplands, situated in the western part of the territory on the right bank of the Don River. The Uplands are heavily dissected by river valleys, whose sides are often steep. The Kalach Uplands, lying east of the Don in the south-eastern portion of the province, are a similar physiographic feature. Here again the terrain is broken, although cultivation is possible even on the highest parts.

The north-eastern part of the province offers a pointed contrast to the areas described above. This region lies within the low but gently

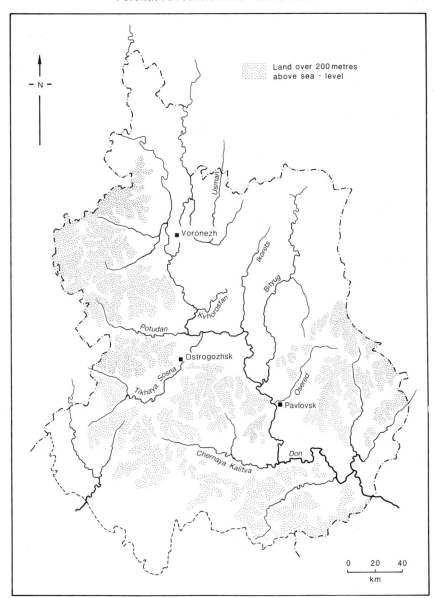

Land over 200 metres
above sea - level

Fig. 3.2. Voronezh province: relief features

undulating Oka–Don Plain, crossed by numerous valleys. The plain, which in some sections is badly drained, is characterized by Tertiary and Quaternary fluvial and marine deposits. The main physiographic features are illustrated in Fig. 3.2.

Drainage is dominated by the system of the Don and its major tributaries: the Oskol (Severskii Donets), the Tikhaya Sosna, the Bityug, the Voronezh, and the Khoper. The river valleys provided the initial sites for settlement and in the past, before the removal of much of the natural vegetation, rivers were of great significance as waterways. In the eighteenth century the Don was a major waterway, providing communication with the Black Sea, while such rivers as the Voronezh and the Boguchar were also navigable. However, the returns to the 1781 Academy of Sciences questionnaire indicate that both natural silting and the construction of water-mills had reduced navigation on many rivers compared with a previous period.[12]

The climate of the province is moderately continental and subhumid. Precipitation decreases to the south and east, and continentality increases, but the climatic differences are not great. The returns to the 1781 questionnaire indicate that rivers generally froze in November or early December, and that the thaw came in March or early April. Spring is characteristically short in this area—April, according to the returns, being the time when trees come into leaf. By contrast, autumns are long and humid, with leaf-fall from September onwards.

The rapid spring thaw can be a major problem for human activity, especially since in the past it was often accompanied by floods. Several of the 1781 returns mention damage to mills and fields, while that for Ostrogozhsk records the particular damage done to mills, houses, and river banks when the spring flood along the Don coincided with that along the Tikhaya Sosna.[13] Annual fluctuations in temperature and precipitation were another problem, especially for agriculture. The dry, scorching *sukhovei* wind, for example, often induced harvest failure, although annual precipitation on average is between 450 and 550 mm. Winters can likewise be variable. Average January temperatures at Voronezh are about −10°C but they can on occasion fall much lower. On the other hand, winters can be extremely mild. S. G. Gmelin, who visited the area in 1768–9, noted the mildness of the autumn and the poor conditions for sledging even at Christmas.[14]

The boundary between the forest-steppe and steppe vegetation zones trends south-west to north-east across the province. Pedologically, as shown in Fig. 3.3, this is pre-eminently the land of the chernozem, which covers most of the central and southern parts, including both the low-lying Oka–Don plain and the watersheds of the Central Russian and Kalach Uplands. Local variations in topography, vegetation, or geology, however, induce some changes. Thus, higher precipitation in the north-west has produced many leached chernozems, while grey forest soils are found on the high right banks of certain rivers, often forming oak woodlands. The terraces of river valleys are frequently

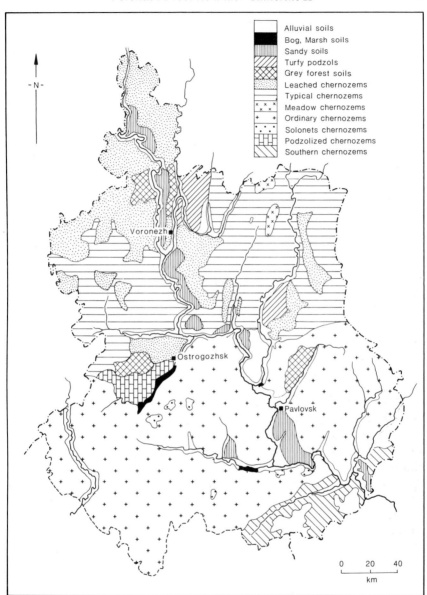

Legend:
- Alluvial soils
- Bog, Marsh soils
- Sandy soils
- Turfy podzols
- Grey forest soils
- Leached chernozems
- Typical chernozems
- Meadow chernozems
- Ordinary chernozems
- Solonets chernozems
- Podzolized chernozems
- Southern chernozems

Fig. 3.3. Voronezh province: soils

associated with sandy soils, while there are alluvial soils in the valley bottoms. In the south-east around Boguchar are areas of poorer 'southern chernozems', produced by the higher temperatures of the summer. In the economic notes to the General Survey the soils of Boguchar district are regularly referred to as 'black-earth', 'chalky', and 'solonets' soils. More rarely, they are 'sandy' or even 'boggy'. 'Sandy' soils are much more frequently recorded in the northern district of Zadonsk, and in Zemlyansk 'clayey' soils are also common.

A mixed vegetation was to be found within the forest-steppe belt. Broad-leaved forests, especially oak, were found on certain upland areas and especially on the upper valley slopes. On the sandy river terraces, pine woodlands often predominated. In Voronezh district an extensive area of oak woodland was located close to the town of Voronezh on the high right bank of the river. Further east were the Usman and Khrenov pine forests, on the sandy right bank of the Usman and left bank of the Bityug rivers respectively. Wooded areas were also to be found in favourable locations much further to the south, where the climate is drier. In Pavlovsk district, for example, was the Shipov forest of broad-leaved species on the right bank of the Osered River. According to S. V. Kirikov, this forest was much utilized during the war with Turkey in 1770, when, in continuation of a tradition established in previous wars, warships were built at Pavlovsk shipyard.[15] Even so, by the end of the century Shipov forest still measured 30 *versty* by 15. In the economic notes to the General Survey Shipov forest was recorded as occupying almost 29,000 *desyatiny*, consisting of oak, linden, and aspen species.[16] Forests such as these had been of great significance in the settlement of the steppe lands of the south, providing much-needed timber and fuel.

Many of the watersheds of the Central Russian and Kalach Uplands, as well as the broad, flat divides of the Oka–Don plain, were occupied by steppe grasslands. The steppe grasslands varied in composition from the broad-leaved, meadow grasses to be found mainly in the northern areas, through the feather-grass (*Stipa* species) steppe, to the semi-arid species, found in some parts of the south. Shrubs were also common. In the eighteenth century, just as the forests were being cleared for construction, boat-building, and other purposes, including fuel for distilling, so the steppe grasslands were subject to continual attrition. Ploughing up and cultivation of the steppe, for example, were common, as was the burning of the grasses to improve them for grazing.[17]

While settlement had reduced the areas of naturally occurring steppeland and forest by the second half of the eighteenth century, such activities as hunting and fishing were still important. The 1781 returns for Zemlyansk district note the importance of wild geese, ducks, and other birds, as well as wolves, foxes, rabbits, hares, ermine, and so on.[18] The skins of bears, wolves, hares, and other animals were sold in the fairs of Livensk district.[19] At the end of the 1760s Gmelin saw wild horses near Bobrov, and claimed that they had been common near

Voronezh itself only twenty years previously.[20] The economic notes to the General Survey assert the importance of fishing on many rivers and water bodies of the province. The 1781 returns for Zemlyansk district, which included stretches of the Don, record the catching of pike, carp, tench, perch, and other species in the area.[21] At Korotoyak on the Don, sturgeon, sterlet, catfish, bream, sander, and other species are recorded.[22] Fish were used for personal consumption or sold on the market at Ostrogozhsk, Pavlovsk, and other locations.[23]

### POPULATION, SETTLEMENT-FORM, AND LANDHOLDING

Since Voronezh province was still undergoing settlement in the latter half of the eighteenth century, it had population densities below those in neighbouring provinces. For example, according to Kirikov's calculations, population densities in the forest-steppe provinces of Orel, Tula, and Ryazan, and also the northern parts of Tambov province, still exceeded those in Voronezh province at the end of the century.[24] Even the newly settled Kursk province had higher densities. Only the southerly districts of Tambov province were at all comparable with Voronezh. Within Voronezh province, however, population densities varied quite significantly. The basic divide was between those districts to the north and west, which had been largely settled in the previous century (with persons per square *versta* in brackets): Zadonsk (22.7), Zemlyansk (19.6), Nizhnedevitsk (19.1), Biryuch (25.9), Voronezh (16.7), and Korotoyak (18); and those districts to the south and east still being settled at the end of the eighteenth: Ostrogozhsk (11.1), Pavlovsk (18.1), Bobrov (11.1), Kalivta (unavailable), and Boguchar (19.1).[25] Thus, the more densely populated districts were for the most part those that had been settled in the late sixteenth and early seventeenth centuries, in association with the building of the Belgorod Line in the middle of the seventeenth century, or (in the case of the first three districts named above) by intensive settlement behind the Line after it was completed. South-western districts such as Biryuch, Livensk, and parts of Valuiki and Ostrogozhsk had experienced the Ukrainian influx of the latter half of the seventeenth century. As shown below, these areas were in some ways intermediate in character between the more northerly and older areas, and the newer districts to the south.

An analysis of the social structure of the province in the 1780s has been undertaken by V. M. Protorchina.[25] Table 3.1, showing rural males only, is based upon her work. Unfortunately, the data appear to be incomplete in a number of important respects, and minor social groups are omitted. Even so, the table does give some idea of the distribution of

TABLE 3.1. *Voronezh province: Social structure at the ti*

| District | Social category | | | | | | | | |
|---|---|---|---|---|---|---|---|---|---|
| | 1 | 2 | 3 | 4 | 5 | 6 | 7 | 8 | 9 |
| Voronezh | | | 346 | | | | 10,948 | 8 | |
| Bobrov | 10,240 | | 206 | | | | 7,933 | 25 | |
| Pavlovsk | 843 | 3,695 | | | | | 9,013 | 65 | |
| Boguchar | | | 876 | 719 | | | 790 | | 11,7 |
| Kalitva | | | | | | | | | |
| Belovodsk | | | | | | | | | |
| Kupyansk | | | | 890 | 740 | | | | 13,6 |
| Valuiki | | | 827 | | 933 | | 3,671 | | 1,5 |
| Livensk | | | | | | | 4,894 | 346 | 8 |
| Biryuch | | | | | | | 3,501 | | |
| Ostrogozhsk | | | | | | | | | 9,3 |
| Korotoyak | | | 9,029 | | 855 | | 10,974 | | 1,4 |
| Nizhnedevitsk | | | | | | | 14,761 | 101 | |
| Zemlyansk | | | | | | | 11,781 | 182 | 1,5 |
| Zadonsk | | | 1,207 | 7 | | 229 | 7,175 | 137 | |

*Key*:
1. Court peasants
2. Court Little Russians
3. Economic peasants
4. Economic Little Russians
5. State *cherkasy*
6. Assigned to the state
7. *Odnodvortsy*
8. *Odnodvortsy* peasants
9. Military residents
10. Retired soldiers
11. Non-estate persor
12. Colonists
13. Postal peasants
14. Voluntarily releas
15. Private peasants

*Source*: V. M. Protorchina (ed.), Voronezhskii krai v XVIII v. (Voronezh, 1980), 76–7.

major groups. Numerically the most significant group was the *odnod-vortsy*, who constituted over 30 per cent of the province's population in the 1780s. Their distribution, as might be expected, was associated with the older settled districts to the north and west, although considerable numbers were also to be found in Bobrov and Pavlovsk. Their presence in Valuiki, Livensk, and Biryuch, to the south-west, can be explained by their migration from their original homes along the Belgorod Line. Another important group were the military residents, Ukrainian descendants of the *cherkasy* of the Ostrogozhsk Regiment, constituting about 17 per cent of the population in the 1780s.[27] They were especially important in Ostrogozhsk district, from where they had migrated in the early part of the eighteenth century into other areas, such as Boguchar and Kupyansk. Their special cossack-type autonomy had been rescinded by Catherine II in 1765.[28] Ostrogozhsk had thus reverted to the status of a normal civil town, and the former cossacks were increasingly regarded as a category of state peasant.

The *odnodvortsy* and military residents, therefore, formed almost 50 per cent of the population of the province. An important niche in the *f the fourth revision, 1780s (rural revision males only)*

| 10 | 11 | 12 | 13 | 14 | 15 | 16 | 17 | 18 | 19 | 20 |
|---|---|---|---|---|---|---|---|---|---|---|
| 245 |   |   | 104 | 6,275 |   |   |   |   |   |   |
| 29 | 23 |   | 29 | 10 | 416 | 7,805 |   |   |   |   |
| 19 |   |   | 55 |   | 358 | 12,517 |   |   |   |   |
| 15 |   |   |   |   | 55 | 2,230 |   | 111 |   |   |
|   |   |   |   |   |   | 15,561 |   |   |   |   |
|   |   |   |   |   |   |   | 18,302 |   |   |   |
|   |   |   |   |   | 60 |   | 9,905 | 34 | 408 | 13 |
|   |   |   | 564 |   | 1,539 |   | 13,313 |   |   | 13 |
| 7 |   |   | 75 |   | 1,734 |   | 15,450 |   |   |   |
| 27 |   |   |   |   | 502 |   | 13,830 |   |   |   |
|   |   | 112 |   |   | 65 | 3,417 |   |   |   |   |
| 15 |   |   |   |   | 627 | 814 |   | 22 |   |   |
| 50 |   |   |   |   | 2,889 | 1,998 |   |   |   |   |
| 157 |   |   | 215 |   | 6,167 |   |   |   |   |   |
|   |   | 12 |   |   | 9,516 |   |   |   |   |   |

16. Dependent Little Russians
17. Dependent *cherkasy*
18. Assigned gypsies
19. Monastic *cherkasy*
20. Church *cherkasy*

province's life was also occupied by the enserfed population, accounting for 35 per cent of the registered inhabitants. The private peasants, who were mainly Great Russians, were liberally scattered throughout the province, but were especially to be found in the older districts to the north and west. By contrast the dependent *cherkasy* and dependent Little Russians predominated to the south and east. There were also many in the south-western districts. These people were part of the seventeenth- and eighteenth-century Ukrainian influx who had subsequently fallen into dependence upon noble landholders. The distribution of the enserfed population, contrasting with that of the ex-servitors, reflects government settlement policy from the late seventeenth century onwards. Lords began to take up estates as far south as the Belgorod Line by the end of that century. However, since much of the best land in these areas was already taken by the former servitors, estate size was limited. During the reign of Peter the Great the regions south of the Line were gradually opened for settlement. Although both the *odnodvortsy* and the military residents were able to acquire land in these regions, government policy favoured the granting of estates to members of the nobility. In Voronezh province it was particularly the Ukrainian nobility who benefited from this government largess, and who subsequently settled their lands, often by agreement, with Ukrainian peasants.

Because the General Survey did not separate the land of the *odnodvortsy* and military residents from that of the nobility, it is impossible to say how much land was held by the different groups. However, from incomplete data Semevskii was able to give some figures for land provision.[29] Thus, among a selection of the *odnodvortsy* of the province he gives an average holding of 19.3 *desyatiny* per male soul (8.3 *desyatiny* of arable). For peasants on large *obrok* estates he estimates 15.7 *desyatiny* (5.6 of arable); for *barshchina* peasants on large estates 21 *desyatiny* (8.3 of arable). By the standards of other provinces these were reasonably high figures.

As in the other provinces of the Central Black Earth Region in the late eighteenth century, most lords in Voronezh province held only small estates. There were, however, some major landholders. According to Semevskii, the most important was Count P. B. Sheremetev with 10,191 male serfs. Then came Count P. A. Buturlin with 6,956, Prince Yu. N. and Princess Ye. N. Trubetskoi with 6,088, and the Counts Vorontsov with 6,677.[30] If Semevskii's figures are accurate, these landowners accounted for over 20 per cent of the serfs in the province in the 1780s. Their estates were mainly located in the south-west and in the more recently settled districts: Sheremetev had land in Biryuch and Korotoyak districts, the Trubetskois in Valuiki, Buturlin in Bobrov, and Vorontsov in Pavlovsk. Of course, those with the most serfs did not always have the most land. For example, in Valuiki district the two Princes Kurakin had over 114,000 *desyatiny* between them, while A. A. and I. A. Bezborodko held 99,420 in Bobrov.[31] Large but weakly settled estates were common in areas to the south and east still undergoing colonization.

In the longer-settled parts of the province noble landholders were both more numerous and also poorer. The economic notes, for example, list 292 noble landholders in Zemlyansk district, 236 in Zadonsk, and 288 in Nizhnedevitsk, compared with only 57 in Bobrov, 43 in Boguchar, and 33 in Pavlovsk. In the northern regions the usual settlement pattern was one of medium-sized villages of a few hundred people whose land was sometimes shared between *odnodvortsy* and several noble landholders. The latter would each have only a handful of peasants, and these were often scattered between two or more settlements.

A typical example of this pattern of landholding and settlement was Zadonsk district. The General Survey divided this district into sixty-five survey units corresponding to one or more settlements. Forty-four units were shared between noble landowners and *odnodvortsy*. Forty-two of these had at least two lords, usually many more. As a typical example, repeated throughout the district, the two villages of Utkino and Verkhnee Kazache may be taken. These villages, constituting a single survey unit, were situated on the Don just below the town of Zadonsk.[32] They had a total male population of 289, with 4,061 *desyatiny* of land shared among eleven noble landholders and an unspecified number of *odnodvortsy*. Several of the landholders held land only here, others in one or two additional places shared with other landlords. On the rare occasions where lords held a survey unit without the *odnodvortsy*, they were likely to hold it individually (in nine cases out of fifteen in the district). Only three survey units in Zadonsk were held purely by *odnodvortsy*, but such units were more common in other districts. All this, however, is not to deny the presence of a few large landowners, even in the older parts of the province. The most prominent example is the Venevitinov family who held land in Voronezh, Zemlyansk, and Zadonsk.

In other districts of the province the pattern was rather different. Districts such as Boguchar, Pavlovsk, and Bobrov, for example, had fewer but larger estates and settlements with simpler social structures. Typically, these were held by state peasants or noble landowners only (in the latter case either individually or at most in twos or threes), or sometimes they were mixed settlements with state peasants and only one or two noble landlords. In Boguchar district, out of twenty-three survey units corresponding to settlements, twelve were held by *odnodvortsy*, military residents, or other state peasants, five by noble landowners only, and six were mixed. Of the latter, only one, corresponding to the settlement of Staraya Melovaya, is at all reminiscent of the situation in Zadonsk. Staraya Melovaya was held by twelve noble landlords and an unspecified number of military residents.[33] The south-western districts of Valuiki, Ostrogozhsk, Livensk, and Biryuch were intermediate in type between north and south, with some large estates, many small ones, and a considerable number of mixed survey units.

These regional differences in landholding reflect the nature and timing of the original colonization and subsequent events. In areas of older settlement population growth, inheritance, and other factors had served to produce some quite complex patterns of landholding, with the intermixing of land held by different categories of landholder and with numerous small estates. The General Survey failed to unravel the intricate patterns of landholding. In the south, by contrast, population densities were still low, few estates had yet been split by inheritance, and there was less intermixing of land belonging to different landholders.

Although the average settlement in the northern districts was of medium size, there were also many small hamlets. The oldest villages were located close to the rivers, especially along the Don and its tributaries. In these valleys houses and settlement continued to expand, so that by the nineteenth century there was in some places an almost continuous band of human habitation. New settlements were still being founded in the late eighteenth century in less accessible territories away from the rivers. Thus, the General Surveyors would frequently record one or two central villages and several hamlets or daughter settlements within one survey unit.[34] The economic notes also refer to this process of internal colonization in a more direct way. Thus, in Voronezh district reference is made to the 'newly settled' hamlet of economic peasants on the Voronezh–Tambov road, and to the 'newly settled' hamlet of Dmitrievskoe belonging to Petr V. Dmitriy.[35] Likewise the notes for Zemlyansk district record 'the hamlet of Novoselenaya settled on the waste [*pustosh*] of Ostrov', 'the newly settled hamlet of Uspenskoe on state land now sold', and 'the hamlet of Nikolskaya on recently purchased state land'.[36] Many similar examples could be given.

In the southern districts the situation was much less advanced. In Boguchar district, for example, the population was concentrated almost entirely along the Don and its tributaries, the Boguchar and the Tolucheevka. Since aridity was greater than in the north, settlement was more concentrated, and most villages had populations of over a thousand. Some were very large indeed, almost like towns. An example was Kalach, with 4,000 inhabitants, all military residents, and 57,000 *desyatiny* of land.[37] In settlement structure this district was divided into a number of distinct regions, with noble-owned villages predominating to the west, mixed noble and state peasant villages in the centre, around Boguchar itself, and mainly military resident villages in the north-eastern part.

The effects of the old cossack tradition of individual or group intake (*zaimka*) from the virgin land was still evident in those parts of the province with cossack associations, and in Ostrogozhsk district in particular. Here the landscape was dotted with the small hamlets known as *khutora*, which were often owned by a single individual and generally populated by a handful of serfs (although some could be large). Ostrogozhsk itself, for example, had sprouted up to thirty *khutora* belonging to several of its military residents and situated up to 30 *versty* away.[38] *Khutora* could be associated with arable land or with hayland only. The sources frequently make reference to cattle *khutora*. Such settlements were also to be found in other districts with a Ukrainian population, but they were less common there. It may be that the relative individualism that *khutor*-founding implies only flourished where the cossack regimental traditions were strongest. Elsewhere the cossack nobility were more able to conduct themselves on the Russian model and acquire estates with large numbers of serfs.

## LAND USE AND AGRICULTURE

Fig. 3.4, based upon the economic notes to the General Survey, suggests a general conformity between land use and population density. Thus, the more densely settled districts to north and west have large percentages of arable land. An exception is Voronezh district, whose eastern and southern portions were still being settled. By contrast, Bobrov, Boguchar, and Kalitva have less arable relative to other types of land, especially hayland.

Some of these trends are confirmed by a study of land provision. Thus, Bolkhovitinov provides statistics for the 1790s, by which time the province had lost some of its southern districts. Nevertheless, from these data it is clear that such districts as Zadonsk (with an average of 7.6 *desyatiny* of land per rural male, or 4.5 of arable) or Biryuch (with 8.4 and 4.4 respectively) were considerably worse off than Pavlovsk (with 11.6 and 4.4) or Bobrov (19.6 and 4.7).[39] For the 1780s, when the province still retained its southern portions, Semevskii calculates that there were 15 *desyatiny* of land for every male in the province.[40]

Unfortunately these basic statistics tell us little about the details of land use at this period. It is evident that throughout the province land provision was still relatively generous compared with more densely populated regions further north.[41] It is also likely that this fact reflects the extensive systems of arable farming being practised in many parts of Voronezh province at this time. These were the long fallow (*perelog*) or field grass (*zalezh*) systems, in which land would be continuously cultivated for several years and then left for perhaps a decade before being used again. From an analysis of different versions of the economic notes and other materials of the General Survey, Milov has shown that extensive farming systems were practised in such different districts as Zemlyansk, Korotoyak, Ostrogozhsk, and Kalitva.[42] In other words, even some of the more densely populated areas were affected.

What is unknown is the actual extent of extensive farming systems compared with more intensive forms such as the three-field system. It seems sensible to suppose that the intensive systems would have been more common in the northern districts. Semevskii implies that extensive systems would have been included in the statistics for arable.[43] From the above figures, however, it is clear that the figure for arable per head is not notably greater in the south than it is in the north. Here Milov's

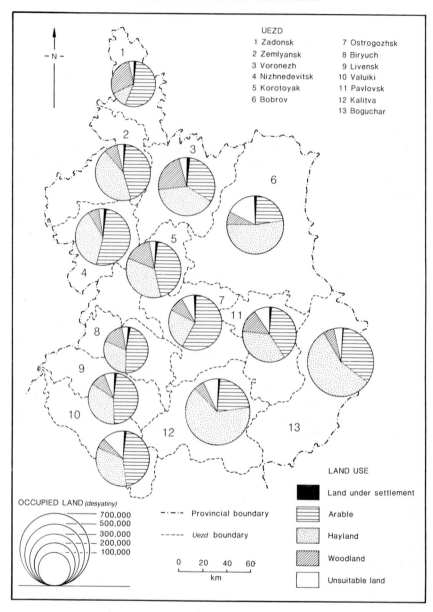

UEZD
1 Zadonsk
2 Zemlyansk
3 Voronezh
4 Nizhnedevitsk
5 Korotoyak
6 Bobrov

7 Ostrogozhsk
8 Biryuch
9 Livensk
10 Valuiki
11 Pavlovsk
12 Kalitva
13 Boguchar

OCCUPIED LAND *(desyatiny)*

700,000
500,000
300,000
200,000
100,000

– · – · – Provincial boundary

– – – – *Uezd* boundary

0    20    40    60·
km

LAND USE

Land under settlement

Arable

Hayland

Woodland

Unsuitable land

FIG. 3.4. *Voronezh province: land use according to the General Survey (late eighteenth century)*

findings are helpful.[44] From a study of the variant materials of the General Survey he shows that extensive systems could have been recorded by the surveyors either as arable or as hayland. In this connection the abundance of hayland in the south might imply that some of it at least was used for *perelog* or *zalezh*. It also gave rise to an important livestock sector.

The sources make frequent reference to rotations. Returns to the 1781 Academy of Sciences questionnaire, for instance, note the use of spring and winter grains, and also fallowing. In a more detailed account of agricultural practice in Voronezh district, the 1785 topographical description states that winter and spring crops are succeeded by a year of fallowing, during which livestock are allowed to graze on the land and thus manure it.[45] However, these are extremely generalized statements which tell us little about the details of farming practice. From Milov's work, for example, it is apparent that even under a three-field agricultural system all the areas listed as arable might not have been utilized all of the time.[46] *Perelog* could exist side by side with more intensive cropping systems in a fairly complex interrelationship, even under what is normally described as a three-field system.

A variety of crops were grown, with a predominance of rye and oats. In Voronezh district, for example, both winter and spring rye and also oats were the most common crops.[47] Other crops were less usual in this area: only about one-third as much buckwheat was sown, one-tenth as much spring wheat, and even less winter wheat. Small amounts of millet, barley, poppies, peas, hemp, and flax are also recorded. In 1781 in Ostrogozhsk district 72 per cent of the grain sown was rye, 16 per cent oats, and the rest made up of buckwheat, wheat, barley, hemp, and small amounts of millet, peas, and flax.[48] Wheat was rather more important in the south. Bolkhovitinov's statistics for 1797 suggest that the rye and oats harvested in the province in that year each exceeded wheat by five to six times.[49] However, this was at a time when southern districts such as Boguchar were not part of the province. Rubinshtein analysed the governors' reports for the period 1785–95 and showed that only about one-third as much wheat was sown as rye and oats, with lesser roles being played by barley and buckwheat. Flax, hemp, peas, and millet were relatively insignificant.[50]

As regards yields, Rubinshtein's data suggest only modest achievements and considerable annual fluctuations. Thus, in 1785 rye yielded seven times the seed, in 1786 three times, in 1787 one and a half times, in 1788 four times, in 1789 three times, in 1790 five times, in 1791 five times, and so on. Similar fluctuations are evident for the other main crops. As averages the 1785 topographical description suggests six to seven times the seed for winter grain in good years, rarely up to nine or ten times, four times in medium years, and twice the seed in poor years.[51] For Voronezh district it is suggested that the best years record ten times the seed for rye and oats, fifteen times for buckwheat, eight times for wheat, peas, and barley, six times for flax and hemp, and twenty times for millet.[52] The 1781 returns give more modest yield ratios. In general, it seems that yields for Voronezh province were usually somewhat higher than in more northerly regions, as might be expected from the fertility of the black-earth soils, but they were unspectacular by European standards, and the annual fluctuations must have given rise to many difficulties.

Methods of cultivation were largely traditional. In areas inhabited by Russians variants of the *sokha* were generally used. This was a forked plough with or without wheels, usually drawn by horses. The *sokha* in common use appears to have been the type able to cut and turn a sod. Thus, in Pavlovsk district the *sokha* with an iron coulter was regularly used with three horses.[53] However, this was an area of steppe grasses and fewer horses may have been usual further north. The Ukrainian tradition was to use the heavy *plug* plough drawn by up to six or eight oxen, especially advantageous on virgin land. The *plug* was able to cut a deep sod but, judging from Shafonskii's discussion of its use in Chernigov province, its main disadvantage was the fact that many peasants had insufficient oxen and hands to use it on their own, and so had to band together with neighbours.[54] The amount of time required to use this heavy plough meant that the land was often insufficiently prepared for sowing, and the light wooden ploughs and harrows used by the Ukrainians were incapable of breaking up the sods left behind by the *plug*. Thus, this author commended the *sokha*, by which the ground could be ploughed twice, as was apparently common in Voronezh district.[55] Shafonskii claims that in Voronezh and neighbouring provinces, the *sokha* usually led to higher returns than the heavy *plug*, despite the fact that the fertility of the steppe soils favoured the Ukrainians.

Manuring was not apparently very common. The 1781 returns for Livensk district state that winter rye was the basic crop but 'the fields are not manured—the land is fertile'.[56] The same point is made in the returns for Ostrogozhsk district.[57] In 1811 the governor of Voronezh stated in a report that 'the soil here is largely black earth and demands no manure and rarely do the landowners use dung. In a good harvest the labour of the husbandman is greatly rewarded.'[58] It may be that many small peasant holdings lacked sufficient livestock for effective manuring of the arable.

Grain was harvested by means of the traditional sickle, although the scythe was evidently used in some places.[59] Once the harvest was placed in ricks, the grain could be removed, taken to the threshing floor, and later dried in kilns, threshed, and milled, occasionally by hand but more usually in water- or windmills. One source declares that it was rare for grain to be stored for longer than a year.[60]

As already noted, the growing of vegetables was not an important activity. Apart from peas, the 1781 returns for Biryuch mention cucumber, water-melon, melon, turnip, beet, and cabbage. Lentils were grown in some places.[61] At the German colony of Rybensdorf near Ostrogozhsk, which was founded in 1766, grain farming was the dominant activity with the colonists using ploughs 'of the type common in Wurttemberg and Brandenburg'. However, the colonists were also noted for their apples, red and white cabbage, and potatoes, and their Virginia-type tobacco was traded far and wide.[62] But the Germans were exceptional and do not appear to have had a great impact on local cultivation practices.[63]

Peasants would typically have small numbers of domestic livestock on their holdings. Thus, as noted already, the horse was the main draught animal for the Russians, the ox for the Ukrainians. The *odnodvortsy* were obliged to supply horses for use on the postal relays, but the military residents and dependent Little Russians, having no horses, paid a special tax instead. The 1781 returns describe horses, cattle, pigs, and sheep being kept in small numbers in the more densely settled districts, generally for domestic purposes. Where the grasslands were abundant, however, livestock farming was much more important. The 1781 returns make it clear that this was true for much of the east, south, and south-east. Livestock were grazed on the open steppe when the grasses were suitable. Use was also made of meadowland along rivers, strips of grassland between and on the edges of arable fields, long fallow lands, and other suitable areas.[64] On the steppe lands of the east and south some large estates were devoted almost exclusively to livestock farming. Almost half of the Buturlin estate at Buturlinovka, for example, was recorded in the economic notes for Bobrov district as being devoted to hayland.[65] Also in Bobrov district was the huge Khrenovskii stud granted to Count A. G. Orlov in the 1770s. The stud occupied an area of about 100,000 *desyatiny*.[66] The 1781 returns describe herds of between fifty and one hundred cattle, and flocks of between one hundred and five hundred sheep being kept by court and economic peasants and *odnodvortsy* on the steppelands of the district. Livestock were sold locally, and cattle were sold to the Don and Ukrainian cossacks. Between June and August herds of between one thousand and seven thousand cattle were driven northwards to Moscow, St Petersburg, and Riga.[67] The 1781 returns paint a similar picture for the other southern and eastern districts. However, they also point out the local nature of much of the cattle trade. The larger enterprises seem to have been exceptional.

The breeding of horses was a significant local activity, particularly among the Ukrainians. Studs existed in the districts of Voronezh,

Bobrov, Kalitva, Boguchar, and especially Ostrogozhsk. The 1781 returns for the latter district describe the specialist breeding of German, Polish, Turkish, Danish, and some English horses by members of the nobility, although Russian breeds constituted the chief type.[68] Horses were an important item of trade, especially in view of the long tradition of horse-rearing among the cossacks, Tatars, and Kalmyks. For example Arab, Crimean, Turkish, and Prussian horses are described as being sold at the Valuiki fairs, while Kalmyk and Russian breeds were sold at Zemlyansk.[69]

The overall importance of market relations in agriculture is difficult to ascertain. There can be no doubt that subsistence agriculture remained widespread, although many kinds of agricultural production did enter into trade. The generous land provision and soil fertility must obviously have had the effect of enabling people to produce beyond their immediate needs. Data assembled by Bolkhovitinov suggest that up to a quarter of the rye flour, half the oats, and 60 per cent of the meat produced in 1797 were exported outside the province.[70] These figures seem a little high. Nevertheless, there was an important market for grain southward in the new territories bordering the Black Sea, and also, though to a lesser extent, among the Don cossacks. This trade was conducted mainly by barks sent downriver each year. Cossacks also visited the markets and fairs of the province to purchase grain, liquor, and other products. Voronezh was, of course, not well placed to participate in the trade with the central regions, but grain and livestock did travel northwards by the more difficult overland route. Grain, livestock, meat, and liquor entered into local trade. Those who for some reason earned insufficient from their own agricultural holdings hired themselves out. Common occupations included those of carrier, worker on the Don navigation, hay-maker on the steppes, and labourer in mills and factories. But such seem to have been minority pursuits.

Milov has analysed the incidence of *obrok* and *barshchina* on estates as a means of assessing the degree of noble participation in the market-place.[71] From the economic notes he finds that *barshchina* was well developed on medium and small estates in Zemlyansk district. From this he concludes that less wealthy lords were active participants in the market-place. In general, on the estates he was able to study, demesne occupied about 40 per cent of the arable area. However, in some other districts *barshchina* was relatively unimportant, especially where settlement was less dense. For example, on several Boguchar estates demesne occupied only 5.4 per cent of the total arable area. There is evidence that hired labour was often used in such areas. For Voronezh province as a whole, only 36 per cent of the enserfed peasantry were on *barshchina* at this time, well below average for the black-earth provinces. The gap closed in the first half of the nineteenth century. One factor behind this change might have been the increasing availability of labour, making direct production for the market more viable for many lords than it had been in the previous period.

## TRADE AND TOWNS

In reviewing the towns of Voronezh province in the 1760s, B. D. Grekov wrote:

A closer acquaintance with these towns leads me to the conviction that almost all of them, in the middle of the eighteenth century, carried the name of town by tradition only. They were all former forts, developed in their time for purposes of defence, especially from the Tatars, but were now needed by no one. Their fortifications were falling into ruin, yet no one complained about it; their former value was forgotten; tradition was overlooked. The new generation had other interests.[72]

Economic life, he points out, now flowed around the towns as a result of the closure of the frontier. Only Voronezh and Ostrogozhsk could be called towns in the full sense.

It was particularly those towns along the old Belgorod Line constructed in the middle of the seventeenth century which had suffered eclipse in this way. Thus, at Verkhososensk in the 1760s the fortifications were described as 'in decay'. The houses and churches all lay outside the old fort, and all were of wood. There were no markets, fairs, or factories. The inhabitants 'mainly cultivate grain and sustain a good livelihood from this pursuit'.[73] Twenty years later, in the economic notes to the General Survey, the 'former town' is described as possessing 'a small earth and wooden fort, dilapidated' and as being occupied by *odnodvortsy* and private peasants, all agriculturists.[74] Likewise, Userd, which in the 1760s had 'extremely dilapidated fortifications', parts of which had been 'burnt down, totally without remains', had neither markets nor fairs. There was a saltpetre mill, established by the government in 1731 and sold to a noble family in 1751, but most of the inhabitants 'cultivate grain which is of the finest and there is a sufficiency of bread'.[75]

Catherine II's administrative reform, which was introduced to the province in 1779, significantly changed the urban network by downgrading several of the least lively towns. The towns which lost status at this time—Orlov, Demshinsk, Kostensk, Uryv, Olshansk, Userd, and Verkhososensk—were places where no resident merchants were recorded in 1727.[76] The newly designated district centres, by contrast, were all market towns or the sites of fairs. One or two of the other new centres also had little else to commend them. Thus, Zadonsk had three annual fairs, attracting merchants from Voronezh and Yelets, but, according to the economic notes, the inhabitants 'have no handicrafts but cultivate grain on their land'.[77] Likewise Nizhnedevitsk, with a single annual fair, is described in the 1785 topographical description as having only poor wooden buildings and as lacking any handicrafts or industrial activity.[78]

These facts are testimony to the difficulty of finding suitable administrative centres in a largely agricultural milieu, particularly in view of the government's policy of avoiding the designation of privately owned settlements. In fact some of the liveliest trading centres were either privately owned or at least had noble landholdings in them. Examples included the villages of Vorontsovka and Aleksandrovka in Pavlovsk district, belonging to the Counts Vorontsov, both of which had several annual fairs and a flourishing trade in fish, grain, and wooden goods.[79] Similarly Verkhotishanka was one of the major market centres in Bobrov district, with two annual fairs and a lively trade in grain and peasant handicrafts. It was a mixed village of court and economic peasants and dependent Little Russians.[80]

Much of the trade and economic activity in the province, therefore, went on in the countryside. The province naturally benefited from the growing population and expanding markets which characterized Russia in these years. Thus, whereas the economic notes to the General Survey record seventy-five fairs in the province during the course of the year, Bolkhovitinov at the end of the century records 113.[81] Most of these took place outside the towns. Likewise, the countryside was the site of many manufacturing concerns, the majority of which were small. Among such concerns might be mentioned several cloth-mills, located in Voronezh and Zemlyansk districts, and about 300 distilleries with a particular bias towards the southern, Ukrainian areas,[82] where the military residents enjoyed tax-free rights of distillation. Such rights were an important stimulus to trade.

It is ironic that, despite the evident importance of the countryside to the economic life of the province, over 8 per cent of its population lived officially in the towns.[83] This is a much higher percentage than that officially recorded for most other provinces at the time. The explanation lies, of course, in the agricultural character of many of the towns. Important exceptions to this pattern of small towns, however, were Voronezh, the provincial capital, and Ostrogozhsk. Voronezh was a town of almost 15,000 people at the fourth revision in the 1780s. After a big fire in 1773 a new plan had been designed by the celebrated architect I. Ye. Starov and approved by the Empress. The plan divided the town into three parts with broad, regular streets, and provision was made for the construction of new administrative quarters for the province. In 1779 a trading row was built on the main square with places for 375 shops. Additionally, a merchants' hall with 48 stalls was constructed in 1786. By the end of the century, as reported by Bolkhovitinov, many new private homes 'in the latest architecture' had been built by merchants and nobles.[84] Also significant was the town's manufacturing base. Most important were the ten cloth-mills recorded as in operation in the 1780s. The town's cloth industry dated back to the time of Peter the Great, since when the manufacture had been directed especially towards the army. Some cloth, however, was traded as far afield as Moscow, Orenburg, and even Siberia. Better-quality wool was purchased in Little Russia, but most was acquired locally, from the Don cossacks, or from Tambov province. By the century's end the cloth-manufacturing enterprises utilized over 250 looms and employed over 2,000 hired workers, many working during the winter season only, and over 500 assigned and possessional serfs.[85] In addition, Voronezh had the usual broad span of manufacturing activity which was typical of the towns of the period. Sources for the 1790s list such activities as dye-making, sugar-making, malting, brewing, tanning, tallow-melting, and glue- and soap-making.[86] Many of these small-scale activities were oriented towards local consumption.

South of Voronezh was Ostrogozhsk, a town of over 10,000 people in the 1780s and a major centre for distilling. 'It is incredible how great a quantity of liquor is distilled here each year,' wrote Gmelin in the 1760s,[87] and much the same was true twenty years later. The 1781 returns list sixty-eight distilleries in and around the town, plus three of the 'English variety'.[88] Grain for this purpose was purchased far and wide, including in Orel and Kursk provinces. Ostrogozhsk was also noted for its saltpetre production and some fine handicrafts. However, the overall orientation was towards distilling, and the agricultural trade and its three annual fairs were points of exchange for high-value and manufactured goods coming from the north, and for agricultural products from the south and the local area.

An earlier section has already commented on the province's trading connections during this period, especially for agricultural products. An official inventory of the province's products and exports for 1797 was published by Bolkhovitinov.[89] This shows the province's most significant exports to have been grain (rye and oats), livestock products (including meat), and liquor. Cloth was also important. The previous section also noted the southward orientation of many trading linkages and the growing connections to the north.

CONCLUSION

The rich variety of sources available for the latter half of the eighteenth century constitute a potentially invaluable and certainly insufficiently exploited means of reconstructing the human and economic geography of Russia at that time. For no earlier period of Russian history could such comprehensive detail be obtained. Admittedly, the materials are all too often imprecise or inaccurate, but they are symptomatic of the burgeoning interest in topography and local economies that was characteristic of the day, and they permit a surprisingly large amount of statistical and qualitative data to be assembled.

The picture that emerges from the assembling of such data for Voronezh province is of an area still undergoing intensive agricultural settlement and whose resource base was thus still generous. This situation was to change markedly in the next half century, as a growing population impinged ever more noticeably upon the land resource, and as a traditional economy continued to hold sway. In the mean time Voronezh province became increasingly integrated into the Russian economy, enjoying developing trading relations both to the north and to the south, and finding her own patterns of specialization as a consequence of the character of her natural environment and of a society and culture moulded by two centuries of human settlement. The complex story of a single province is one that could be repeated for other provinces, no doubt underlining the diversity of social and economic patterns to be found in pre-industrial Russia, patterns which continued to exercise their influence long after the pre-industrial period had ended.

# 4
# Agricultural 'Culture Islands' in the Eastern Steppe
## The Mennonites in Samara Province

TRAVELLERS who visited the southern and eastern steppe of European Russia in the eighteenth and nineteenth centuries commented upon the variety of national or religious groups of different descent settled in the area. Apart from the Russians, who had come to the south during the protracted conquest of the steppe, peoples were to be found there of German, Swedish, Armenian, Bulgarian, Serbian, Walachian, Moldavian, Polish, Jewish, and Greek origin, together with descendants of the traditional steppe dwellers, the Tatars, Bashkirs, Chuvash, Kirgiz, Kalmyks, and Mordvinians. The ethnic diversity of the settlers in the steppe was matched by the diversity of their cultural mores and religions. The southern steppe became a refuge for nonconformist groups from Europe, such as the Moravian Brethren, Hutterites, and Mennonites, or from within Russia, such as the Dukhobors and the Old Believers. These coexisted in the steppe with members of the Orthodox religion and the 'pagan' former nomads. The Duc de Richelieu in a report to the tsar of a tour of the southern steppe at the beginning of the nineteenth century made the following observation:

Never, sire, in any part of the world, have there been nations so different in manners, languages, customs and dress living within so restricted a space. The Nogais occupy the left bank of the Molotschna, families from Great Russia the right bank, then higher up, are the Mnemonists, facing the Germans, half Lutheran, half Catholic; higher up again at Tolmak, the Little Russians, members of the Greek religion, then a Russian sect, the Dukoboitsi.[1]

As the Duc de Richelieu's report indicates, the various ethnic and religious groups inhabiting the steppe retained their cultural distinctiveness and, as each group generally occupied a separate settlement, they remained geographically separate as well. Such was the cultural and spatial distinctiveness of these groups that they can be said to have formed 'culture islands' in the steppe similar to the ones observed by W. M. Kollmorgen in the American south.[2] Thus, on an investigative tour of Russia, the Baron von Haxthausen observed of the German settlers in the Volga basin:

We felt at once transported to the valleys of the Vistula, in Westphalia, so thoroughly German was everything around us: not merely the people, their language, dress, and dwellings, but every plate and vessel, nay even the domestic animals, the dog, cow, and goat, were German. These colonists have even succeeded in giving a German aspect to nature itself throughout the whole district; a landscape-painter might very well call the scenery German.[3]

Half a century later P. Semenov-Tyan-Shanskii was able to make the same sort of observation:

In their material and cultural life the Germans of the Volga colonies constitute, as it were, a state within a state—they have an absolutely special and self-sustaining life and they are distinguished from the surrounding Russian population by their beliefs, their language, their physical type, their buildings, their clothes, and their farming practices.[4]

Inevitably, comparisons were made between different cultural groups. The German villages, and more particularly the Protestant ones, were written about by most observers in glowing terms. P. S. Pallas in his account of his travels in 1793–4, for example, observed that Sarepta, a village settled by Moravian Brethren, was very prosperous, the inhabitants making a good living out of vine and mulberry cultivation and weaving, but that Catholic German villages were less prosperous, and their populations less industrious.[5] M. Holderness, writing in 1827, praised the Greeks and 'Monnonists', describing the latter as a 'most industrious and religious class of people deservedly held in high estimation',[6] but criticized the Swabian Germans for being poor farmers and drunkards.[7] By the middle of the nineteenth century mythologies had been constructed about the various cultural groups. Successful farming was associated with qualities such as diligence, sobriety, love of order, integrity, culture, and morality, and poor farming with a low cultural level, a partiality for alcohol, and the too frequent celebration of feast-days. These latter characteristics were the ones most frequently attached to the Orthodox Russians, while the former characteristics were most frequently to be found among particular groups of Germans.[8] The greatest praise was reserved for the Mennonites, a religious sect that arrived from the North German plain from the eighteenth century, as illustrated by the following extracts from the works of contemporary observers. A. Klaus, author of the most comprehensive work on foreign settlers in Russia, described the land farmed by the Mennonites thus:

In the steppes, where previously there was neither water nor any sign of woodland, quite miraculously one after another robust villages have flowered each with an abundance of well-drawn water, groves of fruit, mulberry, and timber-bearing trees, rich, well-cultivated cornfields, whole flocks of sheep, and excellent breeds of horses and cattle.[9]

The inside of their houses was, for another author, a more telling indication of the fruits of Mennonite labours:

But one must enter their houses to appreciate the habits of order and industry to which they owed not only an ample supply for the necessaries of life, but almost always a degree of comfort rarely to be found in the dwellings of Russian nobles . . . You may be certain of finding in every house a handsome porcelain stove, a glazed cupboard containing crockery, and often plate, furniture carefully scrubbed and polished, curtains to the windows, and flowers in every direction.[10]

Inevitably, the Mennonites were held up as a model for other farmers. Baron von Haxthausen, for example, argued that 'The Mennonites may serve as a model, to the Government and people of what may be effected by industry, morality, and order.'[11] The 'demonstration' value of the Mennonites had for some an importance beyond the material world, however, as was pointed out by E. Henderson, travelling in southern Russia with the British and Foreign Bible Society:

Their industry, and the prosperity and neatness of their villages . . . have frequently called forth the panegyric of the traveller; but, although we could not but admire these features in their colonies, we felt disposed to contemplate their establishment in a much higher point of view . . . [as] agents of Divine Providence who put them there to spread the scriptures.[12]

The positive assessment of the contribution that the Mennonites could make to Russia continued until the 1917 revolution, and they were investigated in detail at various times throughout the nineteenth century and early twentieth. Although they were something of a special case among the cultural groups settling in southern Russia, they were not so different from others as to make them atypical. They are a good subject to use to develop a profile of a 'culture island' in the Russian steppe.

## THE HISTORY OF MENNONITE SETTLEMENT IN THE SOUTHERN AND EASTERN STEPPE

The migration of Mennonites from the Low Countries to the steppe formed part of a general flow of people to the southern steppe after its gradual incorporation into the Russian state, a process that was completed by the annexation of the Crimea in 1783. The height of foreign migration came in the eighteenth century and was associated, in particular, with Catherine II. Coming under the influence of the strong current of 'populationist' thought that swept the European states in the aftermath of the Seven Years War, Catherine II sought ways of expanding the size of Russia's population and of channelling settlers to the newly acquired territories to the south and east.[13] But she wanted to do this without challenging the existing social order. One solution was to invite foreign settlers to come and take up land in Russia, offering them favourable terms to do so. The first manifesto containing the invitation to potential settlers was issued in 1762, and active encouragement of foreign colonization continued until 1819, when Russia's borders were closed to all but a few select groups of foreign settlers.

The invitation to foreigners to settle in Russia was greeted favourably in a number of European countries. The first wave of immigrants was dominated by people from south and west Germany, with 26,000 arriving in Russia between 1764 and 1767. The majority of these (23,000) were directed to the region of the lower Volga river between Saratov and Tsaritsyn, and a minority to Voronezh and Chernigov provinces on the former Belgorod Line, and to St Petersburg province and Liflandia in the north-west. From the 1780s New Russia or Novorossiya, the vast area in the southern steppe obtained at the expense of the Tatars and cossacks, became the focus for settlement. The foreigners who arrived in New Russia at this time came from northern Prussia, the Baltic, Minorca, Albania, Italy, Greece, and the Transcaucasus. The initial period of settlement was difficult as the in-migrants were entering virgin territory which could only be brought into productive use by an exceptional effort.[14] Groups of migrants were settled in discrete colonies, or *kolonii*. As Figs. 4.1, 4.2*a*, and 4.2*b* show, such foreign colonies, consisting of both the original settlements and their later offshoots, had come to occupy large tracts of the steppe by the 1860s.

The Mennonites were the spiritual descendants of the Anabaptist movement of sixteenth-century Europe and followers of Menno Simon (d. 1561), who was a radical theologian of the Reformation.[15] They believed that they were part of the kingdom of God manifest upon earth and set about building a community, *Gemeinde*, that consisted of a voluntary and exclusive following of true believers who agreed to live their life separately from other groups, and according to a firm set of rules. A central tenet of their religious belief was a prohibition on the shedding of other people's blood. Hard work and simplicity were encouraged, and excessive consumption and contact with outsiders discouraged. Marriage to an outsider resulted in the application of 'the ban'—expulsion from the community. J. Urry, a historian of the Mennonites, has shown that there were strong currents of egalitarian-ism, communalism, and anti-intellectualism in the Mennonite com-munities that arrived in Russia, but that by the 1870s these had given way to an acceptance among many of Western bourgeois values,

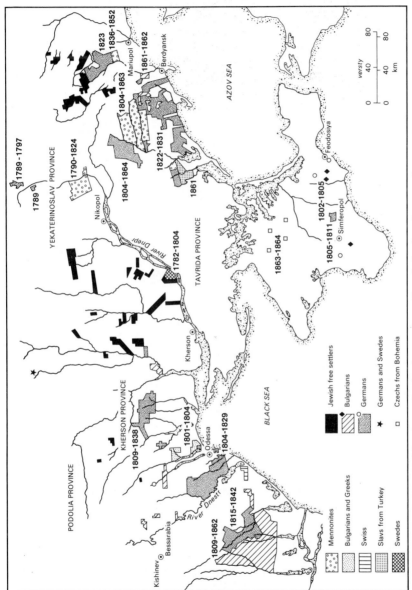

Fig. 4.1. Distribution of foreign colonies in southern Russia in the nineteenth century

*Agricultural Culture Islands*

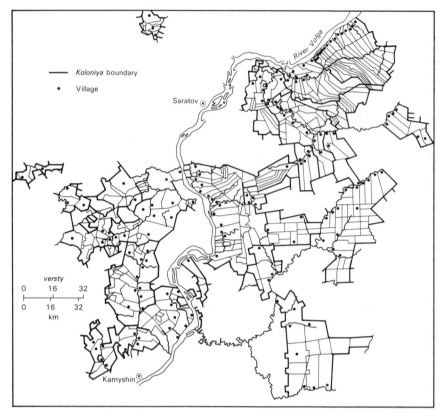

Fig. 4.2*a*. Distribution of foreign colonies in the Lower Volga in 1860

including those of individual achievement, private wealth, and liberalism in education.[16] For the first half-century in Russia, however, government policy helped the Mennonites to preserve their separateness: 'the colony system, the articulation of the villages and the building of the houses and partitioning of the land, all provided both a mental and a physical system which crowded the people together and made them dependent upon one another.'[17]

The Mennonites who settled in Russia originated in the area around Danzig where they were successful dairy-farmers. Their move to Russia took place in response to an Imperial edict of 1785 inviting foreigners to settle in New Russia, which came at a time when they were being put under increasing pressure by the Prussian government, particularly on account of their pacifism. The terms under which the Mennonites settled in Russia were carefully negotiated, and on the whole they were very favourable to them.[18] Land was to be allocated to the Mennonite

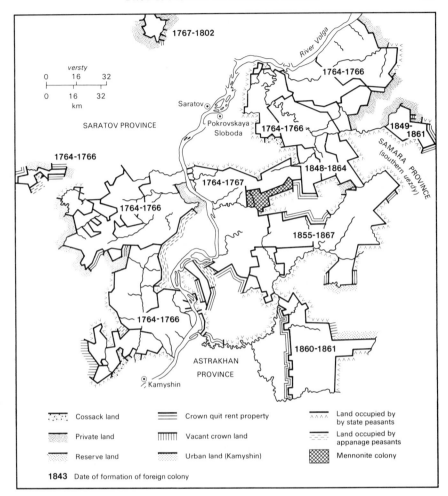

FIG. 4.2*b*. Ownership of land abutting on foreign colonies in the Lower Volga
in 1860

colonies in perpetuity in a place of their choosing in the steppe. Every
member household of a colony was to be allocated 65 *desyatiny* of land in
inalienable ownership, with rights to hand this on to a single heir, and
surplus land was to be put in a reserve to be allocated in the future.
While free to sell to one another, the settlers were not allowed to dispose
of land to anyone outside the colony. The migrant families could bring
with them tools, equipment, and capital up to certain limits. A certain
number of artisan households could be included in each migrant party.
On arrival they were to establish a system of self-government and select

a representative to liaise with the Russian government agencies, but they were not to be subject to state taxation or military service. They were free to practise their religion without interference from the Russian authorities, and to organize the education of their own children. Although these various conditions meant that the Mennonites had a high degree of autonomy in running their affairs, the systems of landholding and self-government they had to adopt were new to them, and were essentially the creation of the Russian government.[19] The government established a clear hierarchy of authority with which its agents could deal. As Urry has observed, the *Schult*, the equivalent of the *mir*'s elected headman, was a 'mini-autocrat' with powers over the distribution of land, the hiring of shepherds, and the agricultural timetable.[20]

The Mennonites arrived in Russia in four migrant parties in 1789, 1804, 1853, and 1859, establishing a new colony on each occasion. The first two colonies, Khortitsa and Molochna, were in New Russia, and consisted of seventy-nine villages, and the last two, Am Trakt and Alek-sandrtal, were in Samara province on the Volga left bank, and consisted of eighteen villages. Each of these colonies in turn spawned new daughter colonies at other sites as far afield as Khiva in Central Asia, the Caucasus, and Siberia. Table 4.1 charts the foundation of Mennonite colonies and their daughter settlements in the period between 1789 and 1859. Like all foreign settlers in the steppe, the Mennonites experienced hardships in the first years after their arrival. The difficulties of the first settlers to arrive were compounded by the fact that they could not settle in the place that had been selected for them by their scouts.[21] Instead of finding themselves in an environment bearing some resemblance to the wooded lowland of the German plain, the settlers were forced to start farming in the exposed and arid site of Khortitsa. But it was the lot of all migrants to come to terms with the steppe environment, and the Mennonites turned out to be somewhat better at this than most. Once the first difficult years were behind them the two pioneer colonies in New Russia prospered, and the Mennonites developed a reputation as innovative farmers. On the strength of this reputation, in the middle of the nineteenth century the tsarist government invited further Mennonites to settle in Russia, despite the fact that by this time its policy of encouraging foreign migrants had ended. The two Samaran colonies were established as a result of this later invitation and, like their New Russian counterparts, they developed a successful agricultural economy, having overcome the initial hardships associated with bringing the virgin steppe into production.

Historians of the Russian Mennonites have shown how their faith and way of life came under various stresses in the new setting of Imperial

TABLE 4.1. *Mennonite colonies founded in Russia, 1789–1859*

| Name of colony (province) | Year of foundation | Number of constituent villages | Acreage | Population | No. of daughter settlements | Provinces and regions in which located |
|---|---|---|---|---|---|---|
| KHORTITSA (Yekaterinoslav) | 1789 | 19 | 89,100 (1789) 405,000 (1917) | 2,888 (1819) 13,965 (1914) | 43 6 7 2 7 14 ⎯ 79 | Yekaterinoslav Kherson Tavrida Voronezh Saratov Orenburg |
| MOLOTCHNA (Tavrida) | 1804 | 60 | 324,000 (1835) | 6,000 (1835) 17,347 (1926) | 17 25 19 2 4 19 8 1 15 6 ⎯ 116 | Yekaterinoslav Tavrida Kherson Kharkov Stavropol Ufa Orenburg Samara Terek Central Asia |
| AM TRAKT (Samara) | 1853 | 10 | 44,134 (1897) | 1,176 (1897) | 1 ⎯ 1 | Central Asia |
| ALEKSANDRTAL (Samara) | 1859 | 8 | 26,500 (1870) 53,500 (1917) | 1,144 (1913) | 4 1 ⎯ 5 | Samara Saratov |
| *Others*[a] | | | | | 2 4 2 101 20 ⎯ 129 | Stavropol Kharkov Kuban W. Siberia E. Siberia |

[a] Daughter settlements made up of Mennonites from more than one parent colony.

*Source*: adapted from *Mennonite Encyclopaedia*, iv. 386–7.

Russia. The very success of the first settlers mounted a challenge to traditional values which resulted in a period of conflict within the communities between traditionalists and progressives. As Urry has observed, the Mennonites were divided between those who had come to Russia with the aim simply of refounding the old order and those who saw Russia as offering new opportunities.[22] Agrarian developments were in part the catalyst of change in the Mennonite communities. The Russian government was keen to see new branches of agriculture developed in the southern steppe, and it was able to persuade some of the new Mennonite colonists to act as the innovation leaders in its experiments. The name of Johann Cornies in particular was associated with agricultural innovation. Cornies became president of the Agricultural Union, a government-founded organization whose purpose was to introduce new farming methods to all the colonies in south Russia. The new directions introduced in agriculture were, from the 1790s, sericulture and merino sheep-farming and, from the mid-nineteenth century, commercial wheat production. Cornies's own farm specialized in merino sheep and he, along with a number of other Mennonite farmers, became very wealthy indeed. Urry writes of a class of 'millionaire' farmers, who married among themselves and had estates of tens of thousands of *desyatiny*, having emerged from among the ranks of the Mennonites by the beginning of the nineteenth century.[23] The colonies also became involved in trade, and there grew up a group of successful Mennonite entrepreneurs and industrialists. Peter Lepp, for example, who began work in the colonies as a watchmaker, became the owner of a large and profitable agricultural machinery works employing a labour force of 150 people and supplying the whole of south Russia. Meanwhile, other Mennonites quit the colonies and moved to Berdyansk on the Azov Sea, where they became merchants and middlemen.

The emergence of successful Mennonite capitalists could not but cause tensions in the colonies, since their existence seemed to contradict the very principles of austerity, equality, and isolation upon which the 'Mennonite commonwealth' had been founded. The tensions erupted in the early nineteenth century over the question of land. Considerable landlessness had developed in the colonies as a result of the combined effect of population growth and inheritance rules which forbade land division between heirs. The provision of reserve land in the original allocation to colonies should have meant that newly formed households were catered for, but much of this land had been appropriated by the new class of agrarian capitalists.[24] After the intervention of the Russian government in the 1860s, the reserve land was distributed among the landless on the basis of 12 *desyatiny* per person, and, when it became clear to them that the government was not going to allocate additional reserve lands, the colonies developed a policy of buying up new land in order to found daughter settlements. The principle of joint responsibility for the members of their order was thus re-established among the Mennonites. Inequalities in the colonies remained, however, as did the basic tension between supporters of the old closed order and those who favoured a new, more open order. In 1871 the colonies were once again plunged into a period of crisis, when the tsarist government abolished the special status all foreign colonists had earlier enjoyed and made them subject to conscription into the tsarist army. This was of particular concern to the pacifist Mennonites. Although the colonies were able to negotiate with the government to fulfil their service obligation by working in civilian concerns, for some the change was taken as the signal to leave. In the period after 1871 between one-third and one-quarter of Mennonites in Russia's south left for the New World, while numbers of those on the Volga, led by Claas Epp and inspired by millenarian ideas, migrated to Central Asia to await the Second Coming there. The Mennonites who departed were either the poor or the ideologically strict, and they left behind them colonies that were balanced in favour of the progressive and liberal elements.[25]

Although by the second half of the nineteenth century the Mennonite colonies in Russia were more open and willing to change than they had been at the time of their foundation, they none the less remained geographically separate and culturally distinct from the groups around them. On the eve of the First World War there were more than 400 Mennonite settlements in Russia which remained closed to outsiders, and in towns Mennonites congregated into distinct quarters.[26] As Urry has argued, the Mennonites' success in retaining their separate identity was in part a product of their ability to adapt to Russia on their own terms. The policy of setting up daughter settlements, for example, depended upon the creation of sufficient wealth in the mother colonies to purchase land outside, and it was as a result of the creation of these geographical offshoots that the Mennonites were able to grow numerically and reproduce their culture. The 400 settlements that existed in Russia in 1914 represented a quadrupling of the number of pioneer villages. The sacrifice of some of their ideals seems to have enabled the Mennonites to withstand the pressure of the government's assimilationist policies after 1871 and growing anti-German sentiments at the end of the century. On the eve of the 1917 revolution their economy was generally buoyant and their culture, for the time being, secure.

ALEKSANDRTAL: A PROFILE OF A MENNONITE COLONY IN
SAMARA PROVINCE

*The Establishment of the Colony*

The Mennonites who arrived in Samara province to found two new colonies in 1853 and 1859 were invited to Russia under special circumstances, for, as has already been observed, the policy of recruiting foreign settlers had been shelved by the first decade of the nineteenth century.[27] An exception was made for the Mennonites because of their reputation as good farmers, and the purpose in inviting them to Russia was that they should act as models for the surrounding population and raise the level of agricultural production in the region beyond the Volga River. The terms of the settlement negotiated by the new wave of Mennonite settlers were less favourable than for the earlier groups, but they were still sufficiently attractive to be taken up by five hundred Mennonite families and, since each of these had to lodge a deposit of three hundred thaler with the Russian Embassy in Berlin before departure, it was only the more prosperous who could in fact take up the invitation to settle.[28] The Aleksandrtal colony was investigated twice in the 1880s, and it was included in the household census in 1910 and in another survey made by the *zemstvo* in 1909.[29] A detailed picture of how the colony changed during its sixty years of existence before the 1917 revolution can be constructed. The colony was situated on the River Kondurcha, a small tributary of the Volga, approximately eighty-five miles from the town of Samara. Nearby was a small market town called Koshki, where 'one could hear seven or eight languages spoken on the weekly market-days',[30] and a Russian village, Borma. The land occupied an elevated area of open steppe sloping gently to the south and south-west. According to Yegorev, one of the nineteenth-century investigators, it was in a place that was 'completely free of forest, with pure boundless steppe around, save now and then on the edge of the river or a stream a small clump of trees showing up dark, denoting a village or a hamlet. Wherever you look—a mirror-like surface, and everywhere there is silence.'[31] The site was good for the development of farming. The soils were deep black earths and the river valley was lined with water-meadows which were flooded by the River Kondurcha every spring. Compared with neighbouring villages snow-melt came early to the colony, by some two weeks, which conferred an advantage on the Mennonite farmers, allowing them to start ploughing early. Aleksandrtal had some additional geographical advantages: it had a market near by at Koshki, and at a distance of thirty-five miles the railway at Melekess. The surrounding steppe was well populated with Mordvinians, Bash-kirs, Chuvash, and other steppe peoples whom the Mennonites could hire to work on their farms. Despite the various advantages of the site chosen, conditions in Aleksandrtal were not as favourable as in the colonies in New Russia. Both it and its more southerly neighbour, Am Trakt, had to cope with arid conditions and periodic droughts.

Five hundred households arrived in the two colonies in Samara province between 1853 and 1865, of which 178 settled in Aleksandrtal. By the beginning of the twentieth century the colonies had populations of over a thousand each.[32] Fig. 4.3 shows the layout of Aleksandrtal at the beginning of the twentieth century. Originally it was made up of ten discrete villages, but this number had been reduced to eight as a result of the merging of three villages into a single community called Krasnovka. Individual dwellings were arranged along the main transport artery, and every household held its land in two unequal parcels stretching either side of the road, the larger one of which constituted the

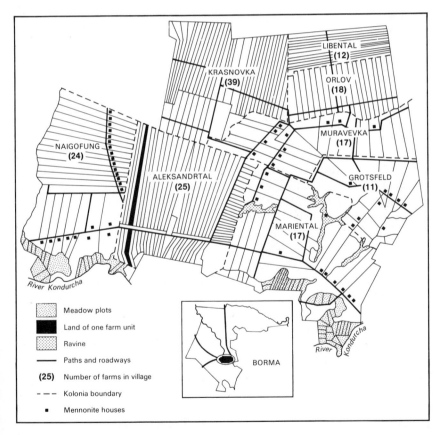

FIG. 4.3. Field layout in Aleksandrtal rural district, Samara province

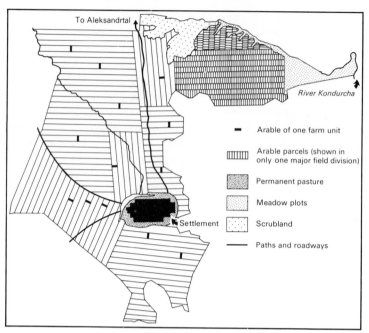

To Aleksandrtal

River Kondurcha

Arable of one farm unit

Arable parcels (shown in only one major field division)

Permanent pasture

Meadow plots

Settlement

Scrubland

Paths and roadways

FIG. 4.4. Field layout in the Russian peasant village of Borma, Samara province, at the turn of the century

arable, and the smaller the permanent pasture. Fields in the arable were of a regular shape and size and each household's was held in a single block. This layout was different from that of the surrounding Russian communes, as a comparison of field patterns in the colony with those of the neighbouring village of Borma in Fig. 4.4 shows.[33] The distinctiveness of the Mennonite villages was even more obvious in the physical type of their dwellings, which reproduced the designs that had been used in Prussia. The Mennonite house was a long, low structure that stood parallel to the road, and was fronted by a fruit-orchard and flower-garden enclosed by a small fence. Each family's dwelling, livestock-shed, and outbuildings were under a single roof, which allowed livestock to be closely supervised and easily tended.[34] Apart from dwellings, there were various public buildings in the colony, mostly to be found in the principal settlement, Aleksandrtal. These included the church, a shop, the flour-mill, and the rural district administration and school buildings. There were also artisans' workshops, making farm implements and consumer goods such as boots and other leather articles.[35]

Some elements of Mennonite landholding imitated traditional Russian patterns. A number of villages, for example, had an area of common pasture that was used jointly by all Mennonite households, and wastes and lakes were held in common. There were in addition certain tasks that householders had to perform together, such as building bridges over ravines and felling timber to heat public buildings. Hereditary ownership of land was, of course, an important way in which Mennonite communities differed from Russian ones, although in the immediate neighbourhood of Aleksandrtal colony there were two rural districts, Kandabulak and Petropavlovka, in which Russian peasants held their land in 'family ownership'.[36] Other Germans in colonies in neighbouring Konstantinovka rural district also held their land in hereditary tenure. To confuse the picture, the Aleksandrtal colony was in fact allocated its land by the Russian government in the joint ownership of all the settlers in communal tenure, unlike its sister colonies in New Russia and Am Trakt, which were allocated their land in hereditary tenure. This seems not to have affected the patterns of land-usership, however. Klaus believed the landholding system of the Mennonite colonists to be a hybrid form, which he called personal-communal tenure (*lichno-podvornoe zemlevladenie*) and which corresponded neither to the communal tenure of the peasants nor to the private landholding of the nobility.[37] The commune was an institution with which the Mennonites were unfamiliar in north Germany. However, as Klaus observed, elements of the commune struck a chord with the Mennonites' religious world-view and sense of order.[38] Other authors saw Mennonite landholding differently. In a 1909 survey of progressive farms in Samara province the Mennonite holdings were classified as *khutora*, presumably because their land was concentrated in a single lot and was in the inalienable possession of the owner.[39]

On the death of the head of household, land in Mennonite colonies was passed undivided to a single heir, who had to compensate co-heirs with a sum of money equivalent to their theoretical share in the inheritance.[40] The purpose of compensating co-heirs was to ensure that they had the means to purchase land for themselves, but the system did not always work well. Co-heirs, for example, could inherit too little to be able to buy a reasonable-sized holding, and the principal heirs could find that the size of compensation payments they had to make forced them to sell up. Few contemporary observers failed to comment upon the rapidity with which the process of socio-economic differentiation took place in Mennonite colonies. In the Aleksandrtal colony changes appeared in the socio-economic composition of the settlers within twenty-five years. Thus, whereas at the time of settlement farms were either 32 or 65 *desyatiny* in size, in 1882 only 12.3 per cent and 22.8 per cent were these sizes respectively. The changes are shown in Table 4.2. More telling was the fact that within twenty-five years of settlement some sixty-four households had had to sell up their farms, so that only 114 of the original 178 settler families remained. By 1909 the number had fallen to eighty-nine. Correspondingly, the average size of Mennonite farms increased, and by 1909 it was 98.8 *desyatiny*. As the author of one investigation commented, 'the Mennonite farms lose their stability, the smaller farmers noticeably decline in numbers and are swallowed up by the group of large landholders who concentrate more and more land in their hands'.[41] A similar process occurred among the Mennonites in the Am Trakt colony to the south, as Table 4.3 shows.

It seems that both Samaran colonies avoided the worst of the conflicts over land that had divided the colonies in New Russia. Aleksandrtal founded two daughter colonies in Samara province—Bezenchuk in 1898 and Bugulma in 1910—and it purchased additional land abutting on Krasnovka from the state. These land purchases enabled the colony to give land to its landless. Mennonite farms of thousands of *desyatiny*, comparable to those in New Russia, had not appeared in Samara province by the turn of the century, but the concentration of land in the hands of ever fewer owners did mean that Mennonite farms were among the largest in the province. In 1910 the average amount of sown land per farm in the Aleksandrtal colony was 53.8 *desyatiny* compared with between 5 and 12.8 *desyatiny* in neighbouring Russian villages.[42]

TABLE 4.2. *Size of Mennonite land-holdings in Aleksandrtal rural district, 1882 and 1909*

| Size of farms (*des.*) | % of households | |
| --- | --- | --- |
| | 1882 | 1909 |
| 1.0–10.0 | 7.0 | 18.8 |
| 10.1–32.0 | 12.3 | |
| 32.1–67.0 | 25.4 | 20.3 |
| 67.1–100.0 | 22.8 | 20.3 |
| 100.1–200.0 | 29.8 | 29.7 |
| >200.0 | 2.6 | 10.9 |
| | 100.0 | 100.0 |

*Sources*: 1882: *Sbornik Statisticheskikh svedenii po Samarskoi gubernii: Otdel Khozygaistvennoi Statistiki, Samarskii uezd*, vyp 1, p. 92.
1909: Samarskoe gubernskoe zemstvo, *Podvornoe i khutorskoe khozyaistvo v Samarskoi gubernii: Opyt agronomicheskogo obsledovaniya*, 1 (Samara, 1909).

TABLE 4.3. *Size of Mennonite landholdings, Am Trakt colony, 1854–1907*

| Size of landholding as proportion of original size of allotment | No. of households | | | |
|---|---|---|---|---|
| | 1854 | 1888 | 1897 | 1907 |
| 0.5 | — | — | — | 22 |
| 1.0 | 211 | 102 | 58 | 91 |
| 1.5 | — | — | — | 3 |
| 2.0 | — | 48 | 53 | 27 |
| 2.5 | — | — | — | 4 |
| 3.0 | — | 11 | 18 | 9 |
| >4.0 | — | 4 | 23 | 7 |
| TOTAL | 211 | 165 | 152 | 163 |
| No. of allotments per farm | | | | |
| One or less | 211 | 102 | 58 | 113 |
| More than one | — | 63 | 94 | 50 |

*Source*: Samarskoe gubernskoe zemstvo, *Podvornoe i khutorskoe khozyaistvo v Samarskoi gubernii: Opyt agronomicheskogo obsledovaniya*, 1 (Samara, 1909).

## Farming in Samara Province

At the time of the Mennonites' arrival on the River Volga, Samara province still occupied a peripheral position in the Russian economy. Half a century later the province had become a major commercial grain region producing wheat for the international market. Both peasant farmers and larger private landowners were involved in supplying the market; peasant farming was more commercialized in the Volga provinces than elsewhere in Russia. The transformation of the province's role into that of a major grain producer had been accompanied by important changes in farming in the province. Former extensive long-fallow systems were replaced by more land-intensive systems of cultivation on peasant land. Whereas in the 1880s approximately 14 per cent of agricultural land in the province had been left under long fallow, by the turn of the century it had disappeared from all but the most southerly district. Short fallow and the near monoculture of spring wheat came to characterize peasant farming in communes in the province, as Table 4.4 shows.

Rotations in which fallow was reduced to one-quarter of the cultivated area and wheat monoculture were not uncommon in the province. These developments had a detrimental effect on yields. While the average yield for spring wheat in the period 1901–15 was 45–50 *pudy*

TABLE 4.4. *Distribution of crops on peasant land in Samara province,*
*1901–1915* (%)

| Years | Winter rye | Winter wheat | Spring wheat | Barley | Spring oats | Potatoes | Flax | Others |
|-------|------------|--------------|--------------|--------|-------------|----------|------|--------|
| 1901–10 | 24.0 | 0.0 | 55.0 | 2.8 | 8.0 | 0.8 | 0.3 | 9.1 |
| 1911–15 | 21.2 | 0.3 | 60.8 | 3.1 | 6.8 | 0.8 | 0.1 | 6.4 |

*Source: Sbornik statistichesko-ekonomicheskikh svedenii po selkomu khozyaistu Rossii i inostrannykh gosudarstv* (Petrograd, 1917).

per *desyatina* for European Russia as a whole, in Samara the average over the same period was 37–42 *pudy*. Variability of yields became a major problem for peasant farmers as the nineteenth century progressed, and harvest failures, in which yields fell to 5 *pudy* per *desyatina*, became notorious in the province. Between 1900 and 1911 the province experienced no fewer than five harvest failures. Describing the 1912 harvest failure, Sev, the province's chief agronomist, wrote:

Millions of hours' work put into the earth has yielded no results; the harvest failure in many places is absolute, with the farmers harvesting less grain than they have sown . . . the peasants' barns are empty now; they do not have enough fodder for their livestock, or enough grain to feed their families, or seed to sow in the fields . . . suffering and confusion—this is what rules in the soul of the Samaran farmer.[43]

As Sev implied in his lament, livestock husbandry had been reduced by this time to a subsidiary concern on peasant farms. At the turn of the century the Witte Commission, set up to investigate agriculture in European Russia, reported for Samara that 'livestock husbandry in this region is not flourishing . . . [owing to] . . . the small area of pastureland, insufficient winter fodder, the difficulty of controlled breeding, and the danger of epidemics', and also that 'the local livestock are small and can hardly be fed in years of abundant harvest, and in years of harvest failure and low yields they are slaughtered for their meat and hides'.[44]

   The problems of Samaran farmers had increased with their involvement in the market. Given the existing agricultural technologies, monoculture was bound to result in the over-use of the soil and make the peasants increasingly susceptible to adverse environmental conditions. The most significant of these was drought. Samara province had an arid environment with an average rainfall of 333–50 mm. in the north and below 300 mm. in the south, with most falling in the late summer, usually in the form of intense downpours. The rainfall regime made wheat cultivation problematic, since most of the precipitation came at the time when the crop was ripening ready for harvesting, and was at its most vulnerable to storm damage. The difference in rainfall totals between late and early summer was marked, 93.4 mm. falling from late July to early September compared with 34.9 mm. from late April to June, when it was needed for growth. In the three years after 1906 the early summer rains were greatly below the average, which helped to account for the harvest failures in that period. The vulnerability of the wheat crop to the rainfall regime was confirmed by experiments conducted at an agricultural research station at Bezenchuk. It was found that in general the regime was only unfavourable for the major spring grains, such as wheat, oats, and barley, which, as shown in Table 4.4, made up the bulk of the peasants' crop. Winter rye was found to suffer from the rainfall regime when grown on late fallow (i.e. fallow ploughed up late the previous summer), while there was a whole range of crops that benefited from the regime, such as sunflowers, beets, and potatoes.[45] The peasants persevered with wheat because it earned a higher price than other crops, and when summer rains did come early the combination of the heat and black earths in the province produced excellent harvests.

Agronomists working in the provincial *zemstvo* in the late nineteenth century recommended ways of reducing peasant farmers' vulnerability to drought. They proposed various approaches to the problem. Sev represented one strand of thinking, arguing that nothing short of a radical restructuring of farming, involving crop diversification and the introduction of complex rotations, was required. He had little confidence in the ability of technical changes, such as the use of new types of implements or the improved working of the soil, to effect great improvements. By contrast, other agronomists, such as those working in the Samara district *zemstvo*, gave first priority to the introduction on peasant farms of such technical changes, while playing down more radical proposals.[46] The split between proponents of the two approaches was really a matter of emphasis, as each side recognized the validity of the other's arguments. There was a measure of agreement about the type of changes that were needed if the peasants were successfully to combat the problem of drought. The most important of these included the rational working of the soil, the use of seed-drills and seed cleaning, the introduction of intertillage crops and ley grasses, and the introduction of new rotations.

Some explanation of the agronomists' recommendations is necessary in order to put the Mennonites' achievement in context. When agronomists referred to the rational working of the soil they were principally thinking of the management of fallow land. The fertility and moisture content of the soil, they believed, would be much improved if a greater area were left under fallow, and if the fallow due to be sown with the winter crop were ploughed as early as possible in the summer (in late May or early June) and then kept 'black', i.e. free from any growth. The problem that this proposal posed for peasant farms was that the fallow was used for livestock grazing, so that ploughing normally took place later on in the summer, when the effect of turning the soil was to expose it to the drying agents of high temperatures and winds. Less readjustment of the existing farming calendar was required by another of the agronomists' suggestions, the use of the regulated seed-drill (*ryadovaya seyalka*), a machine which appeared on the market towards the end of the nineteenth century. This drill, unlike previous mechanized sowers, sowed seeds in deep furrows and, as experiments with the machines in the southern, most arid district showed, it could increase yields on peasant farms by 15 to 30 *pudy* per *desyatina*. The advantages of the machine were said to be that deep sowing protected seeds from drying out and that, when planted evenly, crops had space to grow and could be harvested easily. Seed cleaning, the purpose of which was to separate grain seeds from weeds, was another technical improvement recommended by the agronomists. For this, as well as for the introduction of seed-drills, the outlay on equipment was considerable. The *zemstvo* organized machinery-hire depots and mobile seed-cleaning stations which toured rural districts.

The recommendations for the introduction of intertillage crops and grasses were based on the results of the experiments carried out at the Bezenchuk research station. 'Intertillage (*propashnye*) crops' was the name applied to root- and oil-bearing crops which had a long period of maturation and could thus cope well with the uneven spring and summer rainfall regime. Sev was a particular champion of maize, which had this property and was also 'multi-purpose', providing the peasant family with food and alcohol, and fodder and bedding for livestock. Where grass cultivation was concerned, there was a wide variety of grasses found to be suitable for Samara's arid conditions, including sorghums, lucerne, and vetches. Intertillage crops and grasses could be accommodated in existing field systems by being grown in the fallow field or on a special allotment, but for Sev the only satisfactory way of diversifying was to incorporate the new crops in a proper rotation with cereals, so that the latter benefited from the improvements roots and leguminous plants made to the soil. Sev recommended the following rotations for the north of the province: fallow—winter cereal—spring cereal—grass, potatoes, and other intertillage crops—spring grain; and for the south: fallow—winter rye—spring cereal—spring cereal—grasses and intertillage crops. With such rotations the area of cereals would fall to three-fifths of the arable from its existing two-thirds to three-quarters, but this drop would in theory be compensated for by the increased levels of output associated with diversification.

The improvement of farming in Samara required peasant farmers to make different types of adjustment to their existing practices. Early ploughing and the cultivation of intertillage crops in the fallow, for example, meant that alternative grazing had to be arranged for livestock. The reduction of the area sown to wheat was an obstacle to the introduction of more intensive rotations, whatever the general advantages were said to be. Despite these difficulties some peasant farmers in the province did introduce new practices on their farms, although the majority continued to use traditional systems into the twentieth century.

## Arable Farming in Aleksandrtal

The adjustments the Mennonites had to make to their traditional way of farming on arrival in Samara were considerable. In their homeland in the lower Vistula they had practised mixed farming with a strong emphasis on dairy production, but the climate and remoteness of the steppe meant that they had to develop new skills and specialisms. At the time of the new colonists' arrival the Mennonites in New Russia had already undergone 'the wheat revolution' shifting the emphasis in their farming away from livestock husbandry to cereal production. The Samaran Mennonites followed their example. During the first years of their settlement in Samara, the Mennonites in both Am Trakt and Aleksandrtal colonies cultivated wheat to the exclusion of other crops, but they were soon forced to abandon such monoculture as soil quality declined. In response to this failure the Mennonites diversified into other cereals, such as rye (which Aleksandrtal began cultivating from 1886), spelt, and millet, and into industrial crops such as flax and hemp. Of these only rye was successful, and after 1873 it was grown in more or less equal measure with wheat on Mennonite farms. Later, potatoes were added, and by the twentieth century melons, sunflowers, and grasses were also grown. The diversification of crop production was more marked in the Aleksandrtal colony than in Am Trakt in the south.

Table 4.5 shows how the cropping pattern on Mennonite farms in Aleksandrtal colony changed between 1883 and 1910, and how it compared with the pattern in neighbouring rural districts. It shows that it was principally in the pattern of cereal cultivation, and especially of the spring grains, that Mennonite farms differed from their neighbours. In other respects their cropping profile was not exceptional. Thus, while they put more of their sown land under winter rye, minor grains, potatoes, and grasses than the average for the district as a whole, the same was true of some of their Russian and German neighbours. The pattern of production in Koshki and Teneevo rural districts (where melons, grasses, and sunflowers occupied well above average portions of the sown land) was more advanced than the pattern in Aleksandrtal.

TABLE 4.5. *Distribution of crops on Mennonite, German, and Russian land in eight rural districts in Samara province, 1910*

| | % distribution of crops | | | | | | | |
| | Rye | Wheat | Oats | Other | | | | |
| | | | | grains | potatoes | peas | melons | flax |
| Mennonite | | | | | | | | |
| Aleksandrtal (1883) | 28.6 | 35.7 | 28.6 | | | 7.0 | | |
| Aleksandrtal (1910) | 36.9 | 39.1 | 22.1 | 0.04 | 1.8 | — | 0.02 | 0.0 |
| German | | | | | | | | |
| Konstantinovka | 42.6 | 42.9 | 11.9 | 0.6 | 2.4 | 0.01 | 0.3 | 0.0 |
| Russian | | | | | | | | |
| Kandabulak[a] | 25.5 | 63.7 | 9.3 | 1.0 | 0.05 | 0.3 | 0.05 | 0.1 |
| Petropavlovka[a] | 26.7 | 63.6 | 9.1 | 0.3 | 0.2 | 0.1 | 0.05 | 0.001 |
| Chistovka[b] | 36.8 | 53.8 | 7.8 | 0.8 | 0.5 | 0.2 | 0.03 | 0.01 |
| Koshki[b] | 38.4 | 50.9 | 8.3 | 0.2 | 0.1 | 0.0 | 2.0 | 0.01 |
| Teneevo[b] | 25.9 | 62.8 | 10.2 | 0.6 | 0.3 | 0.04 | 0.15 | 0.04 |
| Zubovka[b] | 46.6 | 46.5 | 5.7 | 0.4 | 0.6 | 0.008 | 0.0 | 0.06 |
| Av. for Samara district[c] | 27.6 | 61.6 | 7.2 | 2.8 | 0.3 | 0.1 | 0.3 | 0.03 |

[a] Peasants in these rural districts held their land in hereditary ownership.
[b] Peasants in these rural districts held their land in communal tenure.
[c] Average for 36 rural districts in Samara district.

*Source*: Otsenochno-statisticheskii otdel Samarskogo gubernskogo zemstva, *Podvornaya perepis krestyanskikh khozyaistv v Samarskoi gubernii: Samarskii uezd* (Samara, 1913).

Commenting on the first two decades of Mennonite farming in the district, a *zemstvo* investigator in 1883 charged the Mennonites with conservatism:

At first the Mennonites sowed flax and hemp on their land, but now they have abandoned these crops; also, they are gradually abandoning the cultivation of millet because yields are low. The Mennonites, mainly through lack of knowledge, have not tried to find out the reason for the failure of these crops, nor have they tried to think up any new, better methods of growing them.[47]

The 1880s, when this was written, were a period of transition in Aleksandrtal during which the colony adopted the three-field rotation on their arable. Yegorev, who investigated the colony in 1881, describes the situation in that year:

. . . the Mennonites have not yet come to a firm decision about which rotational system is better—the existing four-field system . . . which like other multiple-field rotations is beginning to fail as a result of insufficient manure . . . or the three-field system . . . which is less demanding on the soil and, as experience [in the province] over the past twenty years has shown, gives more reliable yields.[48]

TABLE 4.6. *Use of fallow on Mennonite, German, and Russian land in eight rural districts in Samara province, 1910*

|  | Use of arable (%) | | |
|---|---|---|---|
|  | Crops | Short fallow | Long fallow |
| **Mennonite** | | | |
| Aleksandrtal | 71.8 | 28.1 | 0.1 |
| **German** | | | |
| Konstantinovka | 68.4 | 30.9 | 0.7 |
| **Russian** | | | |
| Kandabulak | 60.4 | 24.5 | 15.1 |
| Petropavlovka | 66.6 | 24.1 | 9.3 |
| Chistovka | 70.7 | 22.9 | 6.4 |
| Koshki | 63.0 | 30.4 | 6.6 |
| Teneevo | 69.3 | 23.0 | 7.7 |
| Zubovka | 72.7 | 27.3 | 0.0 |
| Av. for Samara district[a] | 63.0 | 21.4 | 15.6 |

[a] Average for 36 rural districts in Samara district.

*Source*: Otsenochno-statisticheskii otdel Samarskogo gubernskogo zemstva, *Podvornaya perepis krestyanskikh khozyaistv v Samarskoi gubernii: Samarskii uezd* (Samara, 1913).

The four-field rotation, which the Mennonites were in the process of abandoning in the 1880s, involved three-quarters of the arable being sown annually to cereals and one-quarter left to short fallow. It had supplanted the long-fallow system they had used when they first arrived. The change from long fallow to the three-field system, through a transitional four-field rotation, was typical for the eastern provinces. It seems unlikely that the Mennonites were the innovators in this change. According to the *zemstvo* survey of 1883, the Mennonites had in fact adopted the three-field system from their Russian neighbours.[49]

The transition to the three-field rotation was complete by the end of the 1880s and, despite the introduction of new crops in subsequent years, the system remained dominant into the twentieth century. A 1909 investigation of farms in Samara province noted, for example, that grass was generally cultivated outside the main rotation on Mennonite land, and the Witte Commission a few years earlier had noted that the three-field rotation or long fallow prevailed in the two Samaran colonies. Table 4.6 shows the portion of arable left to fallow in the colony in 1910. The practice of short fallowing was still an integral part of Mennonite farming at this time. On the other hand, the use of long fallow was clearly a thing of the past in the colony, whereas it was still used in a few of the neighbouring rural districts. There is more information about rotations in the Am Trakt colony at the beginning of the century.[50] In Am Trakt a smaller range of crops was grown than in Aleksandrtal, but a five-field rotation had been introduced on the arable which, in the way recommended by the province's agronomists, 'spread the burden' of grains between the early- and late-ripening ones. This rotation consisted of fallow—winter rye—spring wheat—spring wheat—oats, barley, and millet. Grass cultivation had been introduced experimentally on forty-six farms (out of ninety-four), but intertillage crops were not grown. In 1909 some farms were recorded as using a four-field rotation, of the sort criticized by agronomists for its potential damage to the soil, and others had partitioned off a separate area, on which wheat was rotated with long fallow. At the same time the investigators observed that the Am Trakt Mennonites cultivated their land on completely different principles from the surrounding population in Novouzensk district but that none had developed a system of farming appropriate to the environmental conditions of Samara province.

If the assessments of the arable rotations used by the Mennonites were ambiguous, there was far more agreement among contemporaries about the advantages of the Mennonites' system of cultivating the soil. Their ploughing regime, for example, attracted the attention of both Yegorev, writing in 1881, and the *zemstvo* investigators in 1883. As was reported in the second of these sources, the expression 'to plough like a German' was current in Samara district at that time.[51] It referred to the Mennonites' practice of repeated and early ploughings of their land. On Mennonite farms the fallow field was ploughed and harrowed four to five times from mid-May onwards, and the spring field was ploughed during the autumn before its planting. The improvement in soil structure which was said to derive from such frequent and early ploughing was further helped by the care the Mennonites put into cultivation. They had a panoply of farm implements which were well suited to coping with the conditions of the steppe. These included two types of plough—the heavy 'Prussian plough', which was used to break up the soil, and a smaller and lighter-wheeled plough for second and third ploughing, a 'couch-grass frame' which was developed by Mennonite craftsmen when they first arrived in Samara province for clearing the steppe of weeds, the iron-toothed harrow to help keep the fallow 'black', and a type of roller made from wood and stone which was used in the spring to close up cracks that developed in the soil with the onset of dry conditions.[52] The result of the Mennonites' careful cultivation techniques, in the words of the *zemstvo* investigator, was that '. . . the fields are almost completely free of weeds and clods and the air can circulate freely in the soil'.[53] Most later observers agreed with these early assessments of Mennonites' cultivation techniques. In the Witte Commission's report for Samara province the Mennonites still had the reputation for working the soil more thoroughly than their Russian neighbours, and the 1909 *zemstvo* investigation of progressive farming in the province also confirmed that they worked the soil better than their neighbours, and in such a way as to conserve moisture in it. However, this last report had critical words for the practice, still common in 1909, of ploughing the fallow field up to four times during the spring and summer, which was now said to be an inefficient use of labour and of doubtful benefit to the soil.

Although by the twentieth century reservations were being expressed about some aspects of Mennonite land management, their ploughing regime allowed them to develop a timetable of sowing and harvesting that gave them some measure of protection against unfavourable weather conditions. As Yegorev discovered, Mennonite farmers in Samara district harvested their winter crop nineteen days before neighbouring Russian farmers, and their spring crop twenty-one days before the Russians.[54] In the case of both spring and winter crops they were able to do this largely because of early sowings, although the favourable situation of their land, already noted, may have played a part. The advantages early sowing conferred were considerable. For the winter crop it protected seeds from damage by early frosts and from their chief pest, 'the winter worm'. Early sowing similarly gave the spring crop some measure of protection from the weather; it was well established by the time of June and July droughts, and harvested before late summer downpours. Further protection from the elements was afforded by the careful management of sown fields. Thus, weeds were removed from the fields as soon as they appeared, cracks developing in a recently sown field were immediately rolled shut, and winter crops that developed too fast before the onset of winter would be kept back, usually by putting out sheep to graze the field.[55] Similar care was taken over the gathering of the harvest and threshing and winnowing of the grain, to make sure that as little of the crop as possible was wasted, and long before *zemstvo* stations were set up the Mennonites were in the habit of sorting their seeds with the help of a primitive cylinder device.[56]

When Yegorev was writing in the 1880s, the agricultural machinery used by the Mennonites was simple and many tasks were done by hand. Thirty years later the Mennonites still used many of the same tools, but these were now supplemented by some of the latest types of agricultural machinery available in Russia, such as regulated seed-drills, threshing machines, and harvesters. Table 4.7 shows the contrast between the Mennonites and their neighbours in relative ownership of agricultural

TABLE 4.7. *Agricultural machinery on Mennonite, German, and Russian farms in eight rural districts in Samara province, 1910*

| | Per 100 farms | | | | |
| --- | --- | --- | --- | --- | --- |
| | Iron plough | Seed drill | Harvester | Thresher | Harrow |
| Mennonite | | | | | |
| Aleksandrtal | 300.0 | 91.01 | 116.8 | 68.5 | 496.6 |
| German | | | | | |
| Konstantinovka | 125.9 | 12.4 | 29.0 | 20.5 | 215.7 |
| Russian | | | | | |
| Kandabulak | 69.8 | 3.1 | 3.8 | 5.0 | 219.0 |
| Petropavlovka | 94.1 | 3.4 | 20.6 | 5.8 | 215.6 |
| Chistovka | 76.2 | 1.9 | 7.3 | 4.4 | 200.6 |
| Koshki | 52.4 | 0.8 | 1.0 | 2.2 | 145.2 |
| Teneevo | 68.8 | 3.2 | 2.8 | 3.4 | 176.8 |
| Zubovka | 69.4 | 0.1 | 0.6 | 2.4 | 147.2 |

*Source*: Otsenochno-statisticheskii otdel Samarskogo gubernskogo zemstva, *Podvornaya perepis krestyanskikh khozyaistv v Samarskoi gubernii: Samarskii uezd* (Samara, 1913).

machinery. The averages conceal important differences in the ownership of machinery between Mennonite farms, but even taking account of these the use of modern machinery was clearly widespread in Aleksandrtal by the twentieth century. In particular, the regulated seed-drill, which was favoured by agronomists for the protection it offered against drought, was gaining ground in the colony among the largest farmers. In Am Trakt forty-three out of ninety-four farmers owned such a drill in 1909, and some had two.

While thorough working of the soil, and later the use of modern machinery, must have compensated to some degree for reductions in the area of land left to fallow, by the 1880s the Mennonites had been forced to start using organic fertilizers on the black earths to try to replenish their fertility. Although in this, as in other aspects of farming technique, their record was considerably better than their Russian neighbours', at no time between the 1880s and the 1910s were they able to treat all their land. The Mennonites fertilized their land either with livestock manure or 'green manure', the latter referring to the ploughing into the soil of flowering buckwheat or chaff. According to Yegorev there was generally sufficient livestock manure on Mennonite farms in Samara district to fertilize fields put under winter wheat, and winter rye was fertilized with green manure.[57] The Samara district census three years later also

TABLE 4.8*a*. *Crop yields on Mennonite and non-Mennonite farms in Novouzensk district, Samara province, 1908*

|  | Average yield in *pudy* per *desyatina* | | | |
|---|---|---|---|---|
|  | Rye | Wheat | Oats | Barley |
| Novouzensk district | 7.4 | 9.5 | 6.2 | 7.8 |
| Best-yielding rural districts | 29.1 | 21.0 | 14.6 | 21.4 |
| Am Trakt | 38.8 | 27.1 | 23.1 | 24.1 |

Source: Samarskoe gubernskoe zemstvo, *Podvornoe i khutorskoe khozyaistvo v Samarskoi gubernii: Opyt agronomicheskogo obsledovaniya*, i (Samara, 1909).

TABLE 4.8*b*. *Comparison of Mennonite and Russian peasant harvests in Novouzensk district, Samara province, 1880–1908*

|  | % of poor and abundant years | | | |
|---|---|---|---|---|
|  | Wheat | | Rye | |
|  | Poor | Abundant | Poor | Abundant |
| Russian peasant farms | 50.0 | 50.0 | 53.6 | 46.4 |
| Mennonite farms | 14.3 | 85.7 | 3.5 | 96.5 |

Source: Samarskoe gubernskoe zemstvo, *Podvornoe i khutorskoe khozyaistvo v Samarskoi gubernii: Opyt agronomicheskogo obsledovaniya*, i (Samara, 1909).

mentioned the treatment of land under potatoes.[58] Unfortunately, none of the sources reveals whether the situation in the Aleksandrtal colony changed in the succeeding years, but in Am Trakt manure was being used by approximately one-third of all Mennonite farmers in 1909.

It is difficult to tell which of the methods (or combination of methods) that the Mennonites used to cultivate their land was the most useful way of conserving the soil. Contemporary accounts did not always agree on the worth of various measures, although it is indisputable that the Mennonites were doing something right. The best data comparing Mennonite and non-Mennonite harvests are for the Am Trakt colony, as shown in Tables 4.8*a* and 4.8*b*. Mennonite harvests were notable for their reliability. Between their arrival in Samara province and Yegorev's 1881 investigation they had had no harvest failures, although there had been five in the province as a whole. 'According to their [the Mennonite] way of thinking,' wrote Yegorev, 'there cannot be harvest failures, except those caused by unexpected events like hailstorms or fire; they do not

TABLE 4.9. *Hired labour on Mennonite, German, and Russian land in eight rural districts in Samara province, 1910*

|  | No. of permanent hired workers per | |
|---|---|---|
|  | 100 farms | 100 *des.* of land |
| Mennonite | | |
| Aleksandrtal | 138.0 | 1.84 |
| German | | |
| Konstantinovka | 9.0 | 0.44 |
| Russian | | |
| Kandabulak | 7.0 | 0.52 |
| Petropavlovka | 6.0 | 0.33 |
| Chistovka | 0.9 | 0.07 |
| Koshki | 1.3 | 0.17 |
| Teneevo | 1.9 | 0.15 |
| Zubovka | 0.7 | 0.06 |

*Source*: Otsenochno-statisticheskii otdel Samarskogo gubernskogo zem-stva, *Podvornaya perepis krestyanskikh khozyaistv v Samarskoi gubernii: Samarskii uezd* (Samara, 1913).

acknowledge harvest failure due to drought and other such reasons . . . this is borne out by twenty years of practice.'[59] The reliability of their harvests meant that Mennonite farmers were in a much better position to innovate than their neighbours since they could afford to take some risks. The risk factor was further reduced by the setting up in the colony of an insurance society for farmers to insure against poor harvests.

Many of the traditional land-management practices employed on Mennonite farms were labour-intensive. The repeated working of the fallow, weeding, and all the other soil cultivation methods would have been impossible had Mennonites not been able to hire labour. From the moment of their arrival in Samara province the Mennonites began using workers from neighbouring rural districts to supplement their own labour on the land, and by the 1880s they had ceased to deploy family labour, relying on hired labour for the bulk of the work in the fields. As the 1883 *zemstvo* investigator observed, the Mennonites 'do not labour on farms; they oversee the work.'[60] The Mennonites preferred to hire Tatars, Mordvinians, and Chuvash from the surrounding area, whom they considered to be better workers than the Russians. In 1883 Mennonite farms in Aleksandrtal hired an average of four workers, the majority on one-year contracts. At peak times in the agricultural calendar, such as at the harvest, temporary labour would be drafted in. The Mennonites' early harvest gave them a distinct advantage over local

TABLE 4.10. *Ownership of horses on Mennonite, German, and Russian farms in eight rural districts in Samara province, 1910*

| | No of horses per farm | | | | | Av. no. of horses per farm |
|---|---|---|---|---|---|---|
| | 0 | 1 | 2 | 3 | ≥4 | |
| Mennonite | | | | | | |
| Aleksandrtal | 14.6 | 11.2 | — | 3.4 | 70.8 | 10.0 |
| German | | | | | | |
| Konstantinovka | 12.0 | 12.2 | 20.1 | 22.2 | 33.4 | 3.0 |
| Russian | | | | | | |
| Kandabulak | 17.8 | 18.3 | 26.3 | 20.7 | 16.8 | 2.2 |
| Petropavlovka | 7.5 | 15.9 | 31.4 | 24.6 | 20.6 | 2.6 |
| Chistovka | 11.1 | 24.5 | 33.7 | 18.7 | 11.9 | 2.1 |
| Koshki | 25.3 | 26.3 | 29.1 | 14.1 | 5.1 | 1.5 |
| Teneevo | 16.8 | 31.4 | 26.8 | 16.2 | 8.7 | 1.8 |
| Zubovka | 16.7 | 32.9 | 31.6 | 11.8 | 6.9 | 1.6 |
| Av. for Samara district[a] | 17.8 | 23.5 | 25.5 | 15.8 | 17.4 | 2.1 |

[a] Average for 36 rural districts in Samara district

*Source*: Otsenochno-statisticheskii otdel Samarskogo gubernskogo zemstva, *Podvornaya perepis krestyanskikh khozyaistv v Samarskoi gubernii: Samarskii uezd* (Samara, 1913).

landowners, enabling them to buy labour when prices were low. Any doubts about the capitalist nature of Mennonite farms must be dispelled by the statistics given in Table 4.9 relating to the hiring of labour in 1909–10.

In addition to requiring large labour inputs, the intense working of the land that characterized Mennonite farming was dependent upon there being sufficient horsepower. The Prussian plough was pulled by two to four horses, depending upon the depth of soil, and the wheeled plough by three to five. Horsepower was also required for the other implements and machines in use on Mennonite farms by the twentieth century. A minimum of four horses per farm and one horse to every five to ten *desyatiny* was required if the Mennonites were to keep to their farming calendar. These requirements were met on most Mennonite farms, and it put them in a quite different category from their neighbours, as Table 4.10 shows. The majority of Russian and non-Russian farms in the neighbourhood of Aleksandrtal had two or fewer horses, or they relied upon bullocks or camels to pull their ploughs, which, while demanding less fodder than horses, were also less powerful. The Am Trakt

Mennonites, meanwhile, developed their own breed of horse, which became well known in the eastern steppe.

The provision that Mennonite farmers made for livestock was vitally important to their ploughing, sowing, and harvesting schedule. Had they been forced to use fallow fields and stubble for grazing—in the time-honoured Russian fashion—they would not have been able to take advantage of the early local spring to begin work in the fields. As has already been shown, the head start they had in the spring enabled Mennonite farmers to time the critical stages of crop development in relation to weather conditions, and it gave them advantages over their neighbours in the market-place. Whether or not livestock had to be put on the fallow thus seems to have made a major difference to farmers' ability to make the most of the local environment. Perhaps, more appropriately, it enabled them to avoid the worst that it had to offer, which the Mennonites were able to do.

### Livestock Husbandry

For the Samaran Mennonites livestock husbandry was a secondary enterprise to cereal production, and work-horses were the most important animals kept on their farms. Dairy production constituted a significant subsidiary economy (the Mennonites were renowned for their fine cheeses, butter, and cream), and pigs and poultry were also sold. Sheep were kept exclusively for domestic needs. The maintenance of ten horses, eight to ten dairy cows, a dozen pigs, and half a dozen sheep, plus poultry and young animals, did not constitute a big problem for the Mennonites, given their large landholdings. For them the use of fallow or of harvested fields to pasture livestock was a matter of choice and not a necessity, as it was for farmers with smaller amounts of land. Most of the Mennonites chose not to put their livestock on the arable. The exception to this rule, already noted, was when sheep and horses were put out on the winter field to keep the crop from developing too quickly in the autumn.

Although in theory they had no difficulty in providing for their livestock, Mennonite farmers were not keen to put valuable cereal-growing land to hay or permanent pasture. By the twentieth century arable constituted over 80 per cent of the usable land in Aleksandrtal. The rest was made up of small areas of natural meadow in the river-valley, pasture, woodland, and the gardens, vegetable plots, and orchards next to the houses. The provision made for the summer grazing of livestock varied. In the Aleksandrtal colony common pastures had been laid out at the time of the original settlement, and they were still being used when the 1883 census was taken. The preference of the

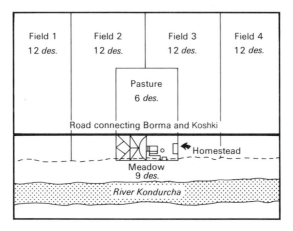

Fɪɢ. 4.5.  Ground plan of G. Epp's farm, Aleksandrtal rural district, Samara province

Mennonites, however, was for livestock to be looked after individually, and for grazing to be controlled. Common pastures were thus supplemented by small areas of permanent pasture made out on an individual's land, as shown in the plan in Fig. 4.5 of G. Epp's farm, or part of the arable next to the house would be separated off from the rest of the land and put under a long fallow rotation. This second strategy was followed in six villages in the Am Trakt colony and consisted of two years of wheat followed by two years of grass. The grass cultivated in separate plots on some farms could also be used to supplement permanent pastures, and every farm had an area, albeit not large, of meadow for grazing. As commentators in the 1880s and 1910s noted, the Mennonites took a lot of care with their pastures, some, for example, sowing them with grass or using fertilizers, and using fences to control feeding. These measures helped to improve their productivity. Despite this, the Mennonites were limited in the number of livestock they could keep by a relative shortage of summer grazing land. This explains the small number of sheep on their farms and the practice of stall-feeding pigs all the year round. The winter stall-feeding of all the animals posed some problems for the Mennonites. Hay was generally used up in the summer as feed for the work-horses and lactating cows, and so straw and chaff constituted the principal components of the animals' winter diet, supplemented with potatoes and any left-over hay. Preferential treatment in winter feeding was given to horses, which helped them to maintain their condition, but by 1909 the Mennonites' dairy cattle were poorly fed in the winter, losing weight and suffering reductions in milk yield as a consequence. On the whole, however, the Mennonites were known, as they were in their arable husbandry, for the care they took with their livestock. Mennonite horses were well groomed and livestock sheds were kept warm.

THE MODEL FARMERS AND THEIR NEIGHBOURS

Within two decades of the Mennonites' arrival in Samara province contemporary commentators began to assess the success of the government's aim of using the colonists to set an example to the surrounding population. On the whole the results were agreed to be disappointing; the impact that Johann Cornies had had in New Russia was apparently not being repeated in Samara. Yegorev, for example, lamented the fact that '. . . that high and laudable aim with which the Mennonites came to Russia and were given land has not yet been understood by the peasants, and remains like a voice in the wilderness'.[61] For Yegorev, the responsibility for this state of affairs lay with the Russians, peasants and landowners alike: 'The Mennonites came to Russia only twenty years ago and have surged ahead; our farmers seeing such a marked difference are distressed . . . and rather than following their example remain as they were and hardly make any progress.'[62] The Russians, Yegorev urged, needed to 'shake off a little laziness' and follow the Mennonites' example. The *zemstvo* investigator surveying Aleksandrtal colony two years later had a different explanation, blaming the Mennonites' isolated existence for the failure of their methods to spread to surrounding populations: 'The isolated, introverted life that the Mennonites lead among the Russians, who could learn something from them, makes it difficult and even impossible for the Mennonite way of farming to have any impact on the Russians.'[63] At that time contacts between Russians and Mennonites were limited mainly to market-days. The outsiders who lived or worked in the colony were non-Russians, and few Mennonites ventured beyond their colony for anything other than trade. Schooling took place in Aleksandrtal, newspapers were German, and the 'ban' was strictly enforced.

The commentators of the 1880s did not doubt that the Mennonites constituted a model that it was desirable, and possible, for the majority of Russian farmers to follow. By the turn of the century this view began to be modified, although the basic belief in the utility of model farms in solving Russia's agrarian problem was not challenged. As the foregoing section has shown, there were a number of respects in which the system of farming used by the Mennonites did not constitute a good example to follow, although particular features of their methods did gain recognition from twentieth-century agronomists. These included the use of the seed-drill, improved breeding of livestock, deep autumn ploughing, the use of early green fallow, well-cared-for working livestock, the feeding of lactating cows, and, where practised, multiple-field rotations. The main thing which prevented the Mennonites from fulfilling the function of innovation leaders, however, was their superior factor endowment compared with the majority of the surrounding farmers. More land, capital, and labour would have been needed on ordinary peasant farms if the Mennonites' methods were to be imitated. This applied even to such a seemingly simple, but potentially very beneficial, change as moving forward the date of ploughing in the spring. Mennonite and Russian farms were the antithesis of each other, the former capitalistic and the latter small commodity producers. It was unlikely that one could affect the other in any but superficial ways.

Contacts inevitably did take place between Mennonites and their neighbours, and the frequency of these increased the longer the Mennonites were in Russia. The removal in the mid-nineteenth century of the special status of colonists and the introduction of laws making education and the keeping of official records in Russian compulsory in all colonies increased the basis for intercourse between the various national groups inhabiting Russia's peripheries. Yet, as travellers' observations made clear, dress, habits, architecture, and farming practices continued to differentiate groups from one another. As far as the Samaran Mennonites were concerned, by the 1880s they had, in the words of the *zemstvo* investigator, adopted 'three-field husbandry and the division of fields by *desyatiny*, the Russian bath, and the samovar'.[64] During the next half-century they adopted more from their neighbours as their rhythm of life became more closely tied in with that of Russia as a whole. According to Urry, by the end of the nineteenth century the Mennonites had adopted Russian holidays and festivals, some Russian preferences and tastes in food, and Russian modes of address when dealing with people of rank, and they had gained knowledge of the Russian language.[65] To repeat a point made earlier, the Mennonites were changed by their experiences in Russia. The identity that they had acquired by the twentieth century was no longer based upon their faith alone, but also in part upon the character of their economic life in the colonies. The agricultural culture island represented by Aleksandrtal was not merely a transplantation into the Russian steppe of an alien cultural group, but was the product of the interplay of the forces of tradition, faith, geography, and the state.

# 5
# Farming Regions of Russia in the Late Nineteenth Century

MOST of the literature on the agrarian history of Russia in the nineteenth century recognizes the existence at that time of the four basic farming types: the long fallow system, the three-field system, the multiple-field system, and industrial and commercial crop systems. Of the four the three-field system was, in one form or another, dominant in most regions of Russia on peasant and noble land alike. Long fallow systems existed in the peripheral provinces of European Russia in the south, east, and north, and intensive systems of farming, involving the cultivation of fodder crops, grasses, and industrial crops, had made their appearance in the western and south-western provinces and around the major cities. For contemporary observers, the reasons for the differences in farming systems between regions became a matter for speculation. Growing problems in the countryside, including harvest failures and famines at the end of the century, fuelled this interest. But knowledge of Russia's farming systems was relatively thin, and only in the last few decades of the nineteenth century did they begin to be understood. Much of the growth in knowledge at this time was due to the work of A. S. Yermolov, whose book on Russian farming systems, first published in 1879, ran into several editions.[1] It was in Yermolov's work that the characteristic features of Russia's basic farming types were first described in detail, and his analysis of their evolution formed the basis of subsequent work on Russian farming systems. In particular, Yermolov influenced A. N. Chelintsev, who worked on the agricultural geography of Russia in the first two decades of the twentieth century, producing a theory that related the evolution of peasant farming systems to the sequence of settling Russia's lands. In a later section of this chapter Chelintsev's agricultural regionalization and the debate surrounding it will be discussed. But we begin with an account of the basic farming types in Russia as they had evolved by the nineteenth century on private and peasant land.

## The Basic Farming Systems of Pre-Revolutionary Russia

### Long Fallow Systems

Long fallow systems were those in which part of the land available for cultivation was ploughed and sown to crops, while the remainder was left under natural vegetation. In European Russia such systems were recognized in two different natural zones in the late nineteenth century: in the coniferous forest belt in the north, where it took the form of a forest long fallow (*perelozhnaya lesopolnaya sistema*) or slash-and-burn (*podseka*), and in the open steppe east of the River Volga, where it constituted a grassland long fallow (*zalezhnaya sistema*). In the north, unworked land was naturally forested, but when abandoned after several years of cultivation it regenerated as thicket. In the steppe, by contrast, vegetation regenerated as steppe grasses. In both cases the fallows were used for livestock grazing. The sequence of events in long fallow systems was for a parcel of land to be ploughed up, cultivated for several years until yields began to decline, and then abandoned in favour of another newly ploughed parcel. In time, arable crops would be resown on the original parcel of land; the length of this interval could vary quite markedly. Yermolov made a distinction between 'unregulated' and 'regulated' long fallows. In unregulated long fallows there were no set intervals for alternating a given parcel of land between crops and fallow. Decisions about the area of land to be left to regenerate to its natural state each year depended upon the level of the previous year's harvest: good harvest years allowed more land to be transferred back into fallow than did poor harvest years.[2] A regulated long fallow, by contrast, was one in which land was alternated between crops and fallow according to a predetermined schedule.[3] Regulated long fallows generally left less land under fallow and for shorter periods of time than the unregulated system. Thus, whereas in classic unregulated long fallow systems it was common for a five- to six-year cropping period to be alternated with twenty to thirty years of fallow, under regulated long fallow systems ten, twelve, or eighteen years of cropping would be alternated with twenty-four, eighteen, and fifteen years of fallow.[4] During the course of the nineteenth century the period of fallow fell repeatedly to as few as three or four years, with a consequent reduction in yield. Systems with such short periods of fallow could only be sustained in the short term, being superseded in the long term by ecologically sounder systems.

Long fallowing allowed extensive livestock husbandry to be combined with the cultivation of a range of subsistence crops. In the nineteenth century the crops grown varied between regions. Flax and a succession of hardy grains commonly followed the fallow in the northern forest regions. In the steppe the succession was usually:

*Year* 1: millet, flax, or pumpkin and water-melon
*Year* 2: spring wheat, millet, or flax
*Year* 3: flax or spring wheat
*Year* 4: peas, barley, or flax
*Year* 5: peas and barley
*Year* 6: oats.

One year of short fallow followed by a year of winter rye was inserted into the succession in some places, and, in another modification, spring wheat and millet were grown for two successive years. In the regulated long fallow system the succession of crops on the arable was much the same, but from the sixth year grasses, such as timothy, might be grown for hay for a number of years, or crops such as maize or sugar-beet cultivated. The combination of crop and livestock products generated by the long fallow systems enabled households to meet their varied needs in food, clothing, animal feed, and motive-power. It also produced surpluses, especially of animal products, for the market. But pressures on the system, such as population growth and high taxes, made it difficult for all these needs to be satisfied without adjustments being made in the system, adjustments which upset the delicate balance the system had with nature. In the nineteenth century the long fallow system was in retreat almost everywhere in European Russia. Nevertheless, it was still found in a number of provinces on the peripheries in its modified forms involving a shortened duration of fallow. Forest long fallow systems were still used in the second half of the nineteenth century in Archangel province, Olonets, Vologda, and the northern districts of Novgorod, Kostroma, Vyatka, and Perm provinces. Steppe long fallow systems were used in Ufa, Orenburg, Samara, and Astrakhan provinces, in the Kuban, the Don basin, and in parts of Tavrida, Kherson, and Yekaterinoslav provinces.[5] A third variant, which Yermolov termed the 'long fallow–short fallow' system (*zalezhno-parovaya sistema*), was found in Siberia. Under this system some fields currently under cultivation would be left to short fallow but, together with the surrounding cultivated land, would be transferred back to long fallow at some stage in the cycle.

*The Three-Field or Short Fallow System*

The three-field system in its classic form consisted of two interrelated elements that distinguished it from the long fallow system: first, the permanent assignment of land to different uses—pasture, meadow, and arable; and secondly, the use of a short annual fallow (*par*) on the arable in a three-year rotation with spring and winter grains. In the early stages of its development in Russia the fallow in the three-field rotation was carefully worked and kept free of all vegetation, in what was termed its 'black' state. In its ancient meaning *par* in fact referred to the thorough cultivation of the soil before sowing.[6] Long before the nineteenth century, black fallow on peasant land had been replaced by 'green' fallow, i.e. fallow left to grow to natural vegetation which could be used for pasturing livestock.[7] In this form the three-field system was found in almost all provinces of Russia at the end of the century but was most dominant in the long-settled central provinces north and south of the dividing-line between the black- and non-black-earth soils.

The three-field system is believed to have been well established in central Russia by the sixteenth century, having evolved out of more extensive systems.[8] The same transition was believed by Russian scholars to be taking place in the nineteenth century in the southern steppe provinces. The system had had to withstand a number of assaults upon it by the authorities, which at various times had decided that it was backward. In Paul I's reign, for example, there was an attempt to replace the three-field system by a seven-field rotation on apanage lands,[9] and in the 1830s the Imperial Free Economic Society, under the presidency of Count Mordvinov, campaigned against the three-field system, blaming it for the harvest failure in 1833.[10] Despite such campaigns, the system continued to be used on peasant and proprietal land into the twentieth century. One geographical explanation for the three-field system's persistence was its suitability for Russia's climatic regime. The late harvest which resulted from Russia's northern latitude meant that, given the levels of technology then available to farmers, there was insufficient time to prepare the soil for sowing a winter crop after the harvest. The insertion of a year between spring and winter crops during which the soil could be adequately prepared for sowing thus made sense. The succession of crops followed in the three-field system reflected this environmental constraint; fallow was succeeded by the winter crop (usually rye, the peasants' staple) and then by a spring crop such as oats, barley, or buckwheat. In the southern Black Sea steppe provinces winter and spring wheat could replace the hardier grains. Under the three-field system a range of large and small livestock was normally kept to meet households' subsistence needs, to provide motive-power to pull ploughs, and to produce organic fertilizer for the fields. The last was a feature mainly in the northern, non-black-earth provinces, where soils were infertile and heavily leached; in black-earth provinces livestock manure was more often used as fuel than on the land.

Although in broad outline the system described above was to be found on much of Russia's farmland, numerous departures from the classic three-field rotation could be encountered. These were minutely recorded and commented upon by Yermolov. They included the cultivation of a different range of crops in the arable from the peasants' staples, the cultivation of forage crops in the fallow, or the creation of a separate field for the cultivation of speciality crops, such as tobacco, melons, or clover. Fallow could also be cut back to less than one-third of the cultivated area. Not all of these adaptations could be absorbed into the three-year rotation without the balance in the system being disturbed, and in extreme cases over-use of the soil could lead to erosion.

*Multiple-Field Systems*

Multiple-field systems, in which cereals or industrial crops were rotated with forage crops and cultivated grasses, were believed by nineteenth-century Russian agronomists to hold the key to overcoming Russia's agricultural backwardness. By the end of the century such systems had made their appearance in most Russian provinces in one form or another, but they can be said to have been dominant only in the north-west and in tsarist Poland. In his 1878 study of agriculture Yermolov described a variety of multiple-field rotations that had been introduced on landowners' estates, while by the end of the century studies of forage crop cultivation showed that multiple-field rotations were spreading on peasant land in the central industrial and northern black-earth regions as well. The most celebrated study of clover cultivation was that by P. A. Vikhlaev in Moscow province.[11] The advantages of forage crop cultivation for farmers currently using the three-field system were twofold: First, the cultivation of forage crops could supplement natural forage sources, allowing livestock herds to be built up, which in turn would increase the amount of manure available for use on the land. Secondly, when cultivated in an appropriate succession with other crops, roots and clover improved the fertility of the soil. The simplest multiple-field rotation, and the one that was most commonly encountered, consisted of a four-year cycle involving the addition of a fourth field sown to clover to the normal three-field rotation:

*Year 1*:  fallow
*Year 2*:  winter grain
*Year 3*:  spring grain
*Year 4*:  clover.

The provinces in which such four-field clover systems were first introduced on to peasant land were Moscow, Kaluga, and Yaroslavl. A modification of the four-field rotation that emerged where clover cultivation was difficult consisted of the cultivation of timothy grass for three years after cereals and fallow in a six-year rotational cycle. While agronomists lauded any moves in the direction of more intensive rotations, it was recognized that simple clover and grassland rotations did not provide a solution to the shortage of livestock feed experienced on many peasant farms. More productive multiple-field systems were found in the north-west provinces of Russia and in tsarist Poland. Yermolov noted their existence on a number of nobles' estates in Smolensk, Moscow, Kaluga, Tver, Yaroslavl, Vladimir, and Ryazan and in the northern districts of Tula. Such systems rotated grasses with roots and cash crops so that approximately half the land was under forage crops and half under commercial crops, with fallow reduced to an insignificant portion or eliminated altogether. Such systems, Yermolov noted, required 'a high level of knowledge and adequate working capital'.[12] Usually, when roots were introduced on peasant land they supplemented but did not replace naturally occurring sources of fodder. Potatoes were an important crop in the northern black-earth provinces, where in addition to providing animal feed they were used in the alcohol industry. Sugar-beet cultivation in the south-western provinces supplied the sugar industry there. New crops could form the basis of longer rotational cycles as, for example, those Yermolov recorded on peasant land in western Orel province:

*Year* 1: fallow
*Year* 2: winter rye
*Year* 3: potatoes, sugar-beet, and lentils
*Year* 4: fallow
*Year* 5: winter rye
*Year* 6: spring oats, barley, and buckwheat.

Other variants of multiple-field rotations which might more properly be considered improved long fallow systems, were encountered in the commercial grain-producing regions of the northern steppe, where cultivated grasses such as timothy and lucerne replaced naturally regenerated steppe grasses in the long fallow. The meadows and pastures thus created formed the basis of commercial livestock husbandry with a particular emphasis on sheep production. On the arable a diverse range of crops was cultivated, including spring wheat, millet, oats, barley, and peas for fodder. Such systems were found in Saratov, Samara, Voronezh, Kharkov, Poltava, Yekaterinoslav, and Kherson provinces and they were especially widespread among German and Mennonite colonists.

## Commercial Systems

For Yermolov, the most advanced farming systems were those in which production for the market entirely replaced subsistence and semi-subsistence production. Such systems were termed commercial or 'unconstrained' (*volnye*) in nineteenth-century Russia, and they were highly specialized and sustained by the use of artificial fertilizers. While there were specialized farming units in Russia, as for example among sugar-beet producers in the south-west or market gardeners in the environs of Moscow and St Petersburg, none had reached the level of chemicalization and mechanization that characterized farming in countries such as Belgium and Britain.

## Degenerate Systems

The thumb-nail sketches given above of the basic types of agricultural activity in the nineteenth century have only hinted at the variety of farming systems in Russia at that time. In reality distinctions between systems were often blurred, especially during periods of their transformation. Not all the changes that took place in farming systems in the nineteenth century were beneficial, and failure to adjust adequately to market and subsistence demands became a hallmark of peasant agriculture in some places. In an attempt to produce more from the land, for example, farmers extended the area of arable cultivation at the expense of meadows and pastures, or cut down on the area of fallow with adverse consequences for soil conservation and fertility. A downward spiral consisting of encroachments on the pastures leading to declining livestock herds, which in turn led to declining cereal harvests and, consequently, further pressure to extend the sown area, was as familiar a response to the need to increase farm production in some regions of Russia as intensification was in others. The increasing pressure on land also had environmental consequences. One development detrimental to soil fertility was the loss of the systematic rotation of crops and fallow on the land: instead of organizing crops and fallow to follow one another in a more or less regular and predetermined succession, farmers began to make *ad hoc* decisions about what to grow, with little regard to the different requirements of crops and land. Whenever this happened *pestropole*, which can be translated as 'mosaic' or 'peppered' fields, was said to have emerged. Characteristically, *pestropole* occurred during the transitional period between long and short fallow systems or, as had become common by the end of the nineteenth century, where peasants had very small amounts of land. As Yermolov noted, in the latter cases the available land area 'was too small for it to be possible to introduce on to it a correct cereal rotation'.[13] He further described these systems as 'totally chaotic'. As a transitional form between long and short fallow, *pestropole* was most widespread in Voronezh, Kursk, Tambov, and Saratov provinces, and as a maladjustment to land shortages it was found in all the densely populated Central Black Earth provinces.


## AGRICULTURAL REGIONS OF RUSSIA ON THE EVE OF THE REVOLUTION

Although the farming systems described above were often found juxtaposed in neighbouring villages, there was a definite geographical pattern to agricultural activity in pre-revolutionary Russia.[14] Contemporary economists and agronomists attempted to explain the patterns by looking at the relative influences of environment and market on crop and livestock combinations. The influence of physical conditions, for example, was stressed in the work of A. I. Skvortsov, who argued that improvements in transportation brought about by the development of canals and railways had reduced the importance of distance and isolation as determinants of production, thus allowing other factors, particularly natural conditions, to make themselves felt on economic activities.[15] In stressing the importance of the environment Skvortsov followed a long line of economists and geographers who divided Russia into regions on the basis of its physical geography.[16] The contrary view, which saw the market as the principal determinant of the geographical pattern of agriculture, was put forward at the beginning of the twentieth century by neo-classical and Marxist economists like G. I. Baskin, P. I. Lyashchenko, and S. N. Prokopovich, and it underlay Lenin's description of emergent regions of agricultural specialization in *The Development of Capitalism in Russia* (1899). However, by far the most comprehensive discussion of the geography of agriculture was associated with the work of the economist Aleksandr Chelintsev. Unlike many of his contemporaries, Chelintsev tried to develop a methodology for identifying a set of 'total' agricultural regions for Russia. In this endeavour he was influenced by the approach of the group of economists and sociologists who in the 1920s constituted the Organization and Production School, and of which Chelintsev was a leading member. The corner-stone of the School's work was the theory of peasant economy developed by its founder, A. V. Chayanov. According to Chayanov, peasant farms constituted a particular type of non-acquisitive enterprise in which decisions about how much to produce were determined by the labour–consumption balance on the farm. This was the balance between the satisfaction of family consumption needs and the 'drudgery of labour' required to satisfy these needs.[17]

Like other members of the Organization and Production School, Chelintsev accepted the concept of the peasant farm as a labour–consumption unit (*trudovoe khozyaistvo*) and he employed it in his own work about the evolution of farming systems. In this he argued for the primacy of demographic factors over market and physical conditions in determining the evolution of farm structure and, consequently, of farming systems.[18] With population pressure as the principal agent of agrarian change, Chelintsev was able to argue that areal differences in peasant farming in European Russia reflected the sequential settling of lands away from the core area of Slavic settlement.[19] For him, agricultural systems were dynamic, changing as population pressures on the land built up. By the same token agricultural regions were dynamic. The title of his first major paper on agricultural geography, 'Agricultural Regions as Stages in the Evolution of Farming' (1910), reflects this belief. In it he wrote, '. . . everywhere in the world the same evolutionary sequence is played out, and agricultural regions express one or other of the stages of the scale'.[20] In attempting to link space and time in his theory of regions, Chelintsev continued an existing tradition in Russian scholarship; a similar approach had been used by Kachorovskii and Kaufman in their works on the development of the peasant land commune, which will be described in Chapter 6. Most contemporary agricultural geographies lacked Chelintsev's dynamic approach, however. While his originality lay in that approach, it also constituted his most vulnerable point, since not everyone could agree with his view of farm evolution. In the final section of this chapter the criticisms levelled against Chelintsev's thesis of regionalization of agriculture will be discussed. It is interesting to note, however, that despite later rejections of his approach in the Soviet Union in favour of the class and economic analysis of Lenin, demographic determinism has continued to constitute a strong current in analyses of the history and geography of farming systems of the world. This is particularly true of the past decades, with the publication of neo-populationist works by Ester Boserup, David Grigg, and Clifford Geertz.[21]

Chelintsev's starting-point in developing a system of agricultural regions for Russia was to define the nature and aims of agricultural production. For Chelintsev, agriculture constituted the arena in which mankind struggled against nature, 'overcoming her inertness, miserliness, adverse influence, and obstacles'.[22] The aim of the struggle was to allow a given population to subsist, from which it followed that the key feature of any agricultural system was the means by which the fertility of the soil was maintained. Chelintsev identified four basic methods of conserving land resources, each of which had different labour require-ments.[23] These were the familiar long and short fallows and organic and inorganic fertilizers. He described their properties as follows:

*Long fallows*: fertility under this system was reproduced by the natural processes of the soil. Long fallowing required the smallest labour inputs of any of the methods of fertility conservation.

*Short fallows*: with short fallows natural processes were supplemented by cultivation of the soil. Short fallows required increased labour inputs but allowed a greater expansion of the cultivated area than long fallows.

*Organic fertilizers*: these were generated internally by the peasant farm as a result of the development of more intensive livestock husbandry, which required further additions to the farm labour force.

*Inorganic fertilizers*: these were produced externally to the farm but in the conditions of nineteenth-century Russia their use required yet further additions to the labour force.[24]

Each of the methods of sustaining soil fertility was shown by Chelintsev to be associated with a particular pattern and intensity of land use, and of crop and livestock production. The nature of the relationships he believed existed can be illustrated by taking the first method, long fallow, as an example. For Chelintsev, the long fallow system existed because low population densities set limits on the amount of land that could be brought under the plough while forcing farmers to use the unploughed land in such a way as to minimize labour inputs. The result was the development of the 'primitive pastoral' systems of agriculture typical of the sparsely populated peripheries of Russia.

Chelintsev's propositions about the evolution of agriculture followed logically from his definition of farming systems: with growth of population the demand to produce more from the land—and the simultaneous increase in availability of labour to work it—would lead to the adoption of ever more intensive methods of soil conservation. Since each method of soil conservation was associated with its own pattern of land use and crop and livestock production, the result of population growth had to be the evolution of any given system to a higher stage. In Chelintsev's own words the sequence of events was described in the following way:

... raising the productivity of agriculture under the influence of population expansion is associated with changes in the structure of agriculture in all its main features. This transformation comes about spontaneously and with an immutable strength as population grows and socio-economic conditions change in response to some elemental force.[25]

As is clear from this statement, population growth was treated as the independent variable in the evolutionary progression. This assumption is a major weakness in demographic explanations of change, as critics of populationist theories point out. In Chelintsev's case, however, the argument must be seen in the light of the history of colonization in Russia, during which successive waves of immigrants filled empty parts of the country and in the process brought new pressures to bear on land in these areas.

The sketch-maps in Figs. 5.1 to 5.12 recreate the stages by which Chelintsev constructed his system of agricultural regions for Russia.[26] They show the distribution of separate branches of production and techniques employed in agriculture which, when combined, produced a set of multi-feature agricultural regions. The branches and techniques which Chelintsev chose to portray cartographically were the same as those he used in his definition of farming systems, namely the methods of maintaining soil fertility, land use, and crop and livestock production.[27] The composite map reproduced in Fig. 5.13 thus shows provinces grouped by the intensity of the system of farming practised in them. From the beginning Chelintsev expected contiguous provinces to fall into the same classes since they generally had similar demographic histories.

Chelintsev's first group of maps (reproduced in Figs. 5.1, 5.2, and 5.3) showed the distribution of long fallow and short fallow and the use of organic fertilizers. In each case provinces in which the particular method of conserving soil was generally practised were identified and plotted cartographically. As the figures show, the pattern was of an east to west and north-west progression of increasing intensity. Thus, provinces with a significant amount of long fallow—in Chelintsev's classification system where they constituted more than 14 per cent of the ploughed area—described a semicircle around a region of short fallows—the group of provinces in which 26 to 33 per cent of arable was under annual fallow. The region in which the most intense application of organic fertilizers took place overlapped in part with the short fallow region, but it extended west into tsarist Poland, north-west to the Baltic littoral, and south into Podolia and Kiev province. Manuring was also practised, although less intensively, in the far northern long fallow region, where poor soils made farming difficult. Comparing the distribution of methods of soil conservation with the labour-force data, shown in Fig. 5.4, Chelintsev noted a direct relationship between the intensity of labour needed for the task of maintaining soil fertility and population density, and an inverse relationship with the degree of industrialization of a province, which he measured by the share of the male population of working age in non-agricultural labour. The least densely populated and least industrialized provinces had the most primitive method of maintaining soil fertility, and the most densely

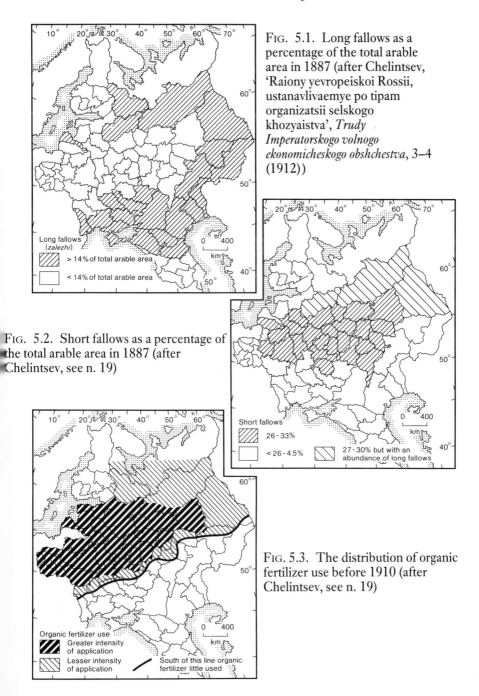

FIG. 5.1. Long fallows as a percentage of the total arable area in 1887 (after Chelintsev, 'Raiony yevropeiskoi Rossii, ustanavlivaemye po tipam organizatsii selskogo khozyaistva', *Trudy Imperatorskogo volnogo ekonomicheskogo obshchestva*, 3–4 (1912))

Long fallows (*zalezhi*)
> 14% of total arable area
< 14% of total arable area

FIG. 5.2. Short fallows as a percentage of the total arable area in 1887 (after Chelintsev, see n. 19)

Short fallows
26–33%
< 26–4.5%
27–30% but with an abundance of long fallows

FIG. 5.3. The distribution of organic fertilizer use before 1910 (after Chelintsev, see n. 19)

Organic fertilizer use
Greater intensity of application
Lesser intensity of application
South of this line organic fertilizer little used

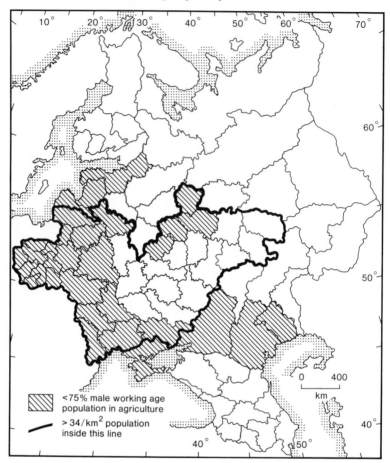

Fɪɢ. 5.4. The density and occupational structure of European Russia's
population in 1897 (after Chelintsev, see n. 19)

populated and most developed provinces had (for nineteenth-century
Russia) the most advanced method. The other facets of agricultural
activity that Chelintsev went on to map fitted in with these basic
relationships.

   Figs. 5.5 to 5.10 show provinces grouped according to the characteristics of their crop and livestock production. Where crops were
concerned, Chelintsev concentrated on plotting the distribution of grass
and root crops, which were more labour-intensive than traditional cereal
crops and were often associated with the development of intensive
multiple-field rotations. The main region of both grass cultivation and
root crop cultivation was in the western borderlands of European Russia

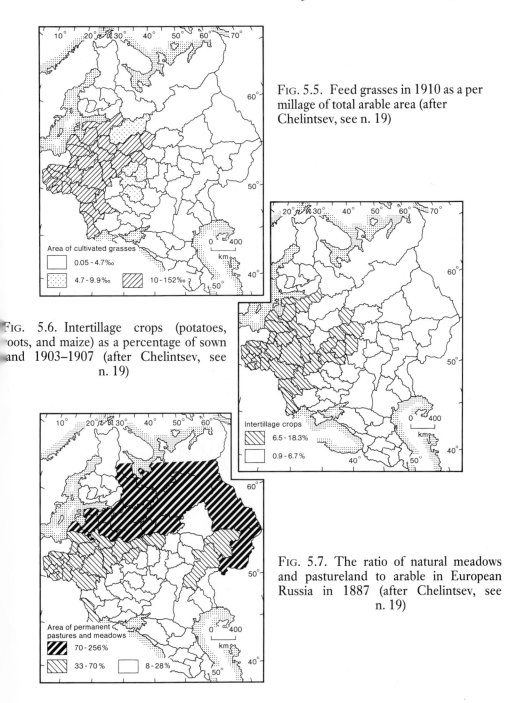

FIG. 5.5. Feed grasses in 1910 as a per millage of total arable area (after Chelintsev, see n. 19)

FIG. 5.6. Intertillage crops (potatoes, roots, and maize) as a percentage of sown land 1903–1907 (after Chelintsev, see n. 19)

FIG. 5.7. The ratio of natural meadows and pastureland to arable in European Russia in 1887 (after Chelintsev, see n. 19)

and extended into the interior of the Central Industrial Region. A comparison of the distribution of grass cultivation and root crop cultivation with the relative abundance of natural sources of forage shows that a shortage of natural pastures and haylands was not a necessary condition for the spread of forage crop cultivation. For Chelintsev, this was proof that population pressure was the driving force behind intensification: 'The necessity to intensify agriculture with the growth of population leads to the introduction and spread of grass cultivation without the expansion of the ploughed area.'[28] In Moscow province, where meadow and pasture accounted for 68 per cent of agricultural land, high values for grass cultivation were recorded. In Vyatka province, where only 24 per cent of all agricultural land was under meadow and pasture, there was little grass cultivation. Chelintsev attributed the difference between the two to the high population density of the former and the low density of the latter.

In his treatment of livestock husbandry, as with crop production, Chelintsev concentrated on identifying signs of intensification of production. Overall numbers of livestock that could be kept were related to the availability of forage, but the type of animals kept and the use to which they were put depended, he maintained, upon the availability of labour. Pigs, for example, commanded a higher price on the market than sheep, but required more labour, while cattle kept for dairying were more labour-intensive than cattle kept for beef. For these reasons patterns of livestock husbandry could be expected to show a strong relationship with patterns of crop production and land use, and with population densities. Chelintsev took the clearest indicator of the intensity of livestock production to be numbers of sheep and pigs, an abundance of the former indicating a low level of intensity and an abundance of the latter a high level. The distribution patterns for the two types of animal are shown in Figs. 5.8 and 5.9. Their numbers are recorded as a percentage of the total head of cattle in each province. With the exception of six provinces, in which there was an abundance of sheep and of pigs, two discrete regions are evident: a western region of pig production and an eastern and southern region of sheep production. These coincided with the regions of more and less intensive systems of arable farming respectively. These types of relationships were more difficult to portray for cattle, since nineteenth-century data did not distinguish dairy and beef herds. Cattle-rich regions thus appear in Fig. 5.10 in both the eastern regions of non-intensive arable farming and western regions of more intensive arable farming. The uses to which cattle were put and the way they were husbanded differed. In the south, east, and north cattle husbandry was in Chelintsev's term 'primitive', relying on either extensive natural sources of forage or long fallow, and

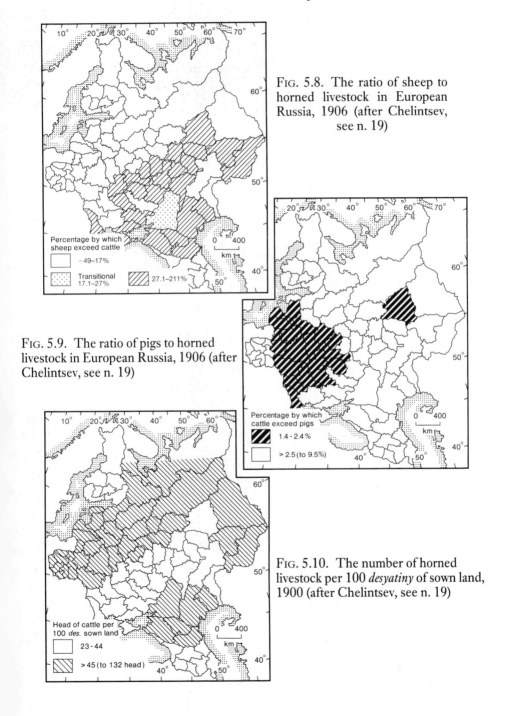

FIG. 5.8. The ratio of sheep to horned livestock in European Russia, 1906 (after Chelintsev, see n. 19)

FIG. 5.9. The ratio of pigs to horned livestock in European Russia, 1906 (after Chelintsev, see n. 19)

FIG. 5.10. The number of horned livestock per 100 *desyatiny* of sown land, 1900 (after Chelintsev, see n. 19)

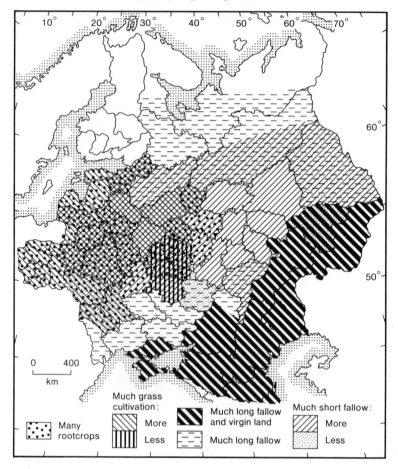

Fɪɢ. 5.11.  Arable farming regions of European Russia (after
Chelintsev, see n. 19)

production was for beef. This contrasted with the western provinces,
where cattle were kept to supply a developing milk and butter industry.

In order to derive a single set of multiple-feature agricultural regions,
Chelintsev amalgamated his various maps, creating two composite maps,
one of arable and one of livestock husbandry (Figs. 5.11 and 5.12). He
then superimposed these upon one another to arrive at the system of
eight regions and sub-regions shown in Fig. 5.13. Each region was
distinguished from others by the degree of intensity of farming. Starting
with the most intensive region and descending in order to the least
intensive, the regions identified were the following:

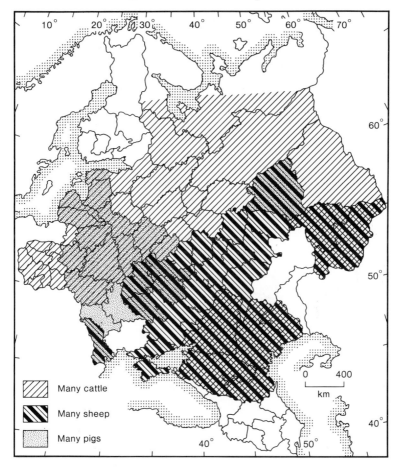

FIG. 5.12. Livestock regions of European Russia (after Chelintsev, see n. 19)

*Region* 1, Tsarist Poland: rotation crop system with many cattle and few sheep and pigs.

*Region* 2*a*, Belorussia and Lithuania: short fallow system with grass and root crop cultivation, many cattle and sheep, and few pigs.

*Region* 2*b*, Moscow and St Petersburg Industrial Regions: short fallow system with grass and root crop cultivation, many cattle, and few sheep and pigs (Kaluga province with many pigs).

*Region* 2*c*, Central region: fallow and grass sowing with fodder crops cultivation and many livestock.

F<sub>IG</sub>. 5.13. Farming regions of Russia as stages in the evolution of farming systems (after Chelintsev, see n. 19) (see text for key)

*Region* 3, Baltic Provinces: grassland rotation system with many cattle and pigs and few sheep.

*Region* 4, South-west: short fallow rotation system with many pigs and few cattle and sheep.

*Region* 5, Kursk, Orel, Chernigov, northern Poltava, Tula, and Ryazan provinces: intensive short fallow three-year system with many pigs and sheep and few cattle (Tula and Ryazan few pigs).

*Region* 6*a*, Black-earth provinces of Tambov, Penza, northern Saratov, Simbirsk, Nizhnii Novgorod, Kazan, and non-black-earth provinces of Vyatka and Kostroma: extensive short fallow three-year system with many sheep and few cattle and pigs.

*Region 6b*, Yekaterinoslav, Kherson, southern Bessarabia, Kharkov, Voronezh, southern Saratov, and southern Poltava: long fallow system with *pestropole*, many sheep, declining numbers of cattle, and few pigs.

*Region 7*, Tavrida, the Don, Kuban, Stavropol, Astrakhan, southern Samara, Orenburg, and Ufa: virgin land long fallow system with many cattle and sheep and few pigs.

*Region 8*, Olonets, Vologda, Perm, eastern Novgorod: northern combined long fallow–short fallow region with many cattle and few sheep and pigs.

Having developed a system of agricultural regions, Chelintsev returned in his 1912 article to his central argument: that these regions must be viewed as stages in the evolution of farming. Thus, he wrote: 'The provinces making up each of the regions . . . are living through their set economic histories. Agriculture in every place, therefore, exists in the economic atmosphere inherited from the recent past and momentarily it approaches its average degree of intensity.'[29] In the discussion that followed his presentation of this system of agricultural regions to a seminar held by the Imperial Free Economic Society, critics pointed out that there were other factors that could exert an influence upon the nature of agriculture in any region, such as the market, natural conditions, government policy, and a variety of random factors.[30] These points were repeated in subsequent years. However, Chelintsev kept to his basic proposition about the role of population in determining the nature of agricultural systems, and to his method of deriving agricultural regions. He denied that he ignored other factors but argued that, if they did exert an influence upon agriculture, they were secondary to, or themselves an indirect consequence of, the operation of demographic factors. In a 1923 book on the agricultural geography of Russia, he devoted several chapters to exploring the influence of the market, the development of railways, the degree of involvement of the population in non-agricultural labour, and the environment on agricultural development, and conceded their importance in some regions. However, none of these factors had, in his view, changed their influence in the period between the turn of the century and the 1917 revolution so much as to affect materially the general pattern of agricultural regions he had already identified.[31] The principal differences between Chelintsev's 1912 and 1923 systems of regions were the division in the second of Russia into two macro-regions corresponding to the black-earth and non-black-earth belts and the movement of some of the boundaries between regions to take account of evolutionary changes in agriculture since the turn of the century.

## LENIN'S ALTERNATIVE:
### The Social and Commercial Character of Agriculture

The relative weight that should be given to market, demographic, and other factors in shaping the nature of agricultural activity remains a matter of debate in the second half of the twentieth century. Chelintsev was merely one early participant in this debate and he has been considered here because of his interest in the regional dimension of agricultural development. Alongside him in arguing the case for taking demographic factors as a primary determinant of agricultural regions were a number of other scholars, some, such as A. F. Fortunatov, N. P. Oganovskii, N. A. Kablukov, and A. A. Kaufman, well-known economists and agronomists of their day. In addition, mention has been made of mid- to late nineteenth-century works that took the major physiographic divisions of the country as the primary building-blocks of the regional economic geography of Russia. The Russians also made use of early formulations of locational theory to explain and identify agricultural regions. Von Thunen's *The Isolated State* sounded a sympathetic chord among Russian economists.[32] Some, such as Chayanov, 'rewrote' von Thunen for an isolated state populated by labour-consumption households.[33] Others used the concept of locational rent to analyse the distribution pattern of Russian agriculture. G. I. Baskin was the clearest representative of the second current. In two works, the first analysing 1916 census data for Samara province and the second data for the whole of Russia, Baskin argued that distance from the market determined the structure of farms and the shape of agricultural regions.[34] In these works he identified regions of decreasing agricultural intensity around Russia's major markets. In an interesting echo of Chelintsev, Baskin argued that each region progressed through an evolutionary sequence of agricultural types. For him, the driving force for change was the outward expansion of the market rather than population pressure: 'Every region which is located at a particular distance from an economic and cultural centre has in its outer neighbour an illustration of its past, and in its inner neighbour an illustration of its near future.'[35]

While the work of Chelintsev and his contemporaries demonstrated a variety of approaches to the theoretical and practical problems of agricultural regionalization, they were all later to be grouped together by Soviet historians as 'subjective', 'bourgeois', and 'anti-Marxist', the more easily to be dismissed.[36] The majority, it was now claimed, had missed the most important point about agriculture: that like all economic activity it exists within the context of a particular mode of production and it is this that determines how and why it changes. Regional schemes, according to this line of thinking, should have aimed to show how far capitalism had progressed in different parts of the country and the character of the emerging geographical division of labour between them. Soviet economic historians maintain that Lenin's work on the Russian economy in late nineteenth-century Russia did precisely this. *The Development of Capitalism in Russia* has been heralded as 'the first application in history of the revolutionary Marxist method to the question of economic regionalization'.[37] While in the absence from Lenin's work of a map or any reference to the problem of regionalization such claims must appear overstated, it is true that the analysis of capitalist development in agriculture which Lenin made in 1899 can be reconstructed to form a set of agricultural regions for the turn of the century.

The author who has taken Lenin's geographical analysis of capitalism furthest is V. K. Yatsunskii, who, in an article in 1953, attempted to portray Lenin's findings in *The Development of Capitalism in Russia* cartographically.[38] He produced three maps, depicting agriculture, industry, and migration, each of which was a compilation of Lenin's own observations, with omissions and blank areas filled in by Yatsunskii. The agricultural map is reproduced in Fig. 5.14. One set of regions in the map (Fig. 5.14a) uses Lenin's work on the socio-economic differentiation of the peasantry. In this research Lenin tried to show that, as a result of the spread of market relations into the countryside, the peasantry was breaking up into two opposing camps: the rural proletariat and the bourgeoisie.[39] It was the regional variations in the degree of such class differentiation that Yatsunskii attempted to capture in his cartographic representation of Lenin's thesis. Lenin had shown such differentiation to be greatest in the western and south-western provinces of Russia and around the two primate cities of Moscow and St Petersburg, and least in the central Russian provinces, where the remnants of the previous feudal social relations between landlords and peasants persisted. The Volga occupied an intermediary position. A second set of regions on the map (Fig. 5.14b) groups provinces by the type of commercial farming developed or developing in them by the late nineteenth century. The basic division is between the northern and western region of commercial livestock husbandry and the southern and eastern region of commercial cereal and crop production. In Lenin's analysis the degree to which agriculture was commercialized was reflected in the social relations of the rural population: the more advanced commercialization was, the deeper the differentiation between wealthy peasant households and landed estates on the one hand, and poor peasant households on the other. Among the landowning classes the transformation from feudal to capitalist enterprises was, according to

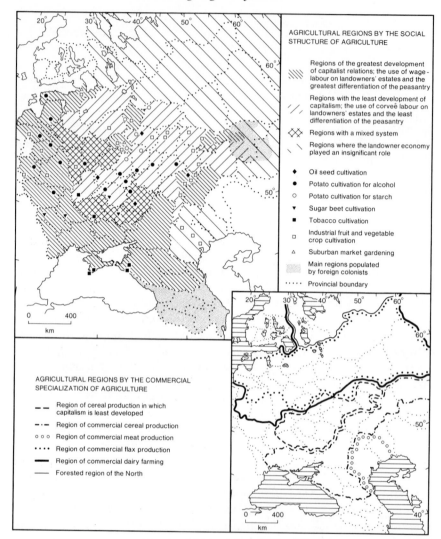

Fig. 5.14. The agricultural regionalization of pre-revolutionary Russia based on the work of V. I. Lenin (after V. K. Yatsunskii, 'Voprosy ekonomicheskogo raionirovaniya v trudakh V. I. Lenina', *Voprosy geografii*, 31 (1953))

Lenin's findings, proceeding apace in the nineteenth century, but it was particularly advanced in the south-western and southern provinces, where landowners used their estates for sugar-beet production to supply local sugar-processing factories, or had turned them into 'cereal factories' producing wheat for export.

Yatsunskii's agricultural geography based upon Lenin's account of the development of capitalism in the nineteenth century is different from Chelintsev's. It draws upon an analysis of the processes taking place in rural society that stressed the primacy of market forces over demographic factors. One consequence of this difference was that while Chelintsev stressed the homogeneity of the system of agriculture practised in any region, Yatsunskii focused on the differences between producers in a region. In the decade after the 1917 revolution debate about the nature of the peasantry was intense. Scholars of the Organization and Production School continued their empirical and theoretical work on the labour–consumption theory of peasant farms, but they were challenged by a new group, the agrarian-Marxists, who, working within a historical materialist framework, attempted to develop a methodology to measure the extent of class differentiation among the peasantry.[40] Using their methodology to analyse peasant class structure in the eastern commercial wheat-farming regions, the agrarian-Marxists showed that differentiation was less advanced than Lenin had argued in *The Development of Capitalism in Russia*. The importance of these findings was that they undermined the neo-populist thesis that peasant farms were largely impermeable to market influences and that demographic factors were of prime importance in farm evolution. By the same token, they challenged Chelintsev's theory of the evolution of farming systems and regions, but they could not wholly invalidate it.

One conclusion to emerge from the debate about the Russian peasantry was that contradictory processes were taking place in the countryside. The result was a complex regional picture of agriculture that was difficult to capture in its entirety. Contemporary theories of regionalization each threw a particular light on agrarian developments of the time, although this is not to say that every approach was equally valid. Any theory about an activity such as agriculture, which constitutes both an economic and an ecological system, must be affected by more general views about society–environment relations. These views informed the debate about the peasantry in the nineteenth century and in turn influenced work on the regionalization of agriculture.

# 6
# The Commune in the 1870s

A COMMON element in the agrarian history of Europe was the existence and persistence of the open-field system. Open-field villages were to be found in a broad band stretching across western and central Europe and extending north into Scandinavia and east into Siberia. During their long history open-field villages underwent transformations, and geographically they were far from uniform; but for all their variety they possessed common features that served to distinguish them from the enclosed farms which replaced them almost everywhere. By general consent, the four following features were the hallmarks of open-field villages: first, strips in the arable land; secondly, common pasture rights; thirdly, a common cropping-cycle; and fourthly, a local regulating body such as a manorial court.[1] All these elements were functionally interdependent, although they were not always simultaneously present in any one open-field village. In Russia the open-field system in the nineteenth century was associated with the land commune, called by the peasants by a variety of names such as *mir, sostav,* or *obshchestvo.* These names denoted a community of peasant households joined together by a web of economic and social relations. Territorially communes could correspond to a single village or to several villages or parts of settlements. Further, it was possible for communities of peasants constituting separate land communes to be linked to others by virtue of joint usership of a parcel of land. Superimposed upon communes were units of state and, before 1861, manorial administration which connected communes to the hierarchy of power that led ultimately to St Petersburg.

While sharing the features of open-field villages elsewhere in Europe, Russian communes are best known for the practice that became widespread in the nineteenth century of periodically redistributing, or repartitioning, land between their member households according to roughly egalitarian principles. There has been much debate, still unresolved, about the origin of the practice.[2] Nineteenth-century scholars of the statist school of thought, such as B. Chicherin, argued that repartitioning arose as a direct consequence of state fiscal policy, and that it was of relatively recent origin.[3] Other scholars, while agreeing that the onset of repartitioning was unlikely to have pre-dated the seventeenth century, believed that the practice arose as a result of 'natural processes' such as pressure of population on the land. The idea was that as population grew, competition for land resulted in the development of mechanisms for allocating the now scarce resource equitably between households; if the state had a role it was as modifier, not originator, of the mechanism.[4] After the 1917 revolution the debate was continued by Soviet historians, who, examining the commune's development in a historical-materialist context, attempted to show that repartitioning was of ancient origin and was in decline by the nineteenth century. This enabled them to argue that the commune was a relic institution which was out of phase with the developing capitalist order. Today few scholars, either in the Soviet Union or the West, subscribe to this view of the history of land repartition, and there is a general consensus in favour of its origin in the seventeenth and eighteenth centuries, although there is little agreement about why the practice came about.[5]

The debate about the origin of repartitioning has at times been heated and in the second half of the nineteenth century it came to dominate all discussion about the commune as a social institution. Gradually, scholars dropped the peasants' terms for their village community, such as *mir*, in favour of a new name, the *obshchina*.[6] The distinctiveness of the repartitional commune was recognized by the state in 1861, when the serfs gained freedom from their manorial lords. Under the terms of the Emancipation Statute peasants were granted their land in *obshchinnoe* or communal tenure. Later in the 1860s the allotment of state and apanage land to non-proprietal peasants was similarly in the form of communal tenure. Communal tenure gave groups of peasants joint ownership of land but individual user-rights, and it carried with it both an obligation on the part of member households to accept a parcel of land (along with the dues attached to it) and the right, should the whole community adjudge it necessary, to readjust the size and disposition of these land parcels at any time. Communal tenure was carefully distinguished in law from other forms of peasant land-ownership, such as private (*lichnoe*), household (*podvornoe*), family (*semeinoe podvornoe*), and common (*obshchee*)—all but the last being types of hereditary tenure. Official statistics for 1905 show that approximately 80 per cent of peasant land was held in such communal tenure. However, there was little uniformity in land-management practices in repartitional communes. The frequency and outcome of repartitioning, for example, showed considerable geographical variation, while not all communes possessing the right in law to repartition did, in fact, do so. The problem of explaining this variation came to be a major preoccupation of the commune's nineteenth-century theorists.

## Theories about the Commune

As has already been observed, one explanation given in the nineteenth century for the onset of repartitioning in the commune was pressure of population on the land. This idea was expanded by some scholars into fully fledged theories of the commune which accounted for its temporal and geographical variation. Despite differences in detail, all the theories attempted to define a series of stages through which communes must pass in their development. They also tended to see the commune's evolution as essentially unidirectional, reaching its apogee at the stage when repartitioning was introduced. K. R. Kachorovskii's work can stand as an example of this approach.[7] The stages in the commune's development Kachorovskii recognized were 'the commune in formation', 'the commune in being', and 'the commune in decay'. He believed that all these stages were represented in Russia at the turn of the century. The 'commune in formation', for example, was to be found in the most recently settled areas of the empire, and the 'commune in decay' in the longer-settled areas of the western borderlands. It was in the middle stage historically, and in the central provinces of European Russia geographically, that repartitioning was most widespread and that, according to Kachorovskii's evolutionary view, the *obshchina* was at the stage of maturity. In order to establish his propositions, he drew extensively on local authority surveys of communes in different parts of the Russian empire and, more particularly, on the research of A. A. Kaufman in Siberia.[8] Kaufman, like Kachorovskii, saw the newly developing village communities in the east as communes 'in the making'. In a thesis reminiscent of Frederick Jackson Turner's on the American frontier he tried to show how the history of the commune in central Russia was being repeated in the new settlement frontier in Siberia.

In identifying population growth as the mechanism of change in the commune theorists like Kachorovskii anticipated the work of the rural sociologists who, later in the 1920s, came to constitute the Organization and Production School. As Chapter 5 showed, A. N. Chelintsev, one of their number, sought demographic explanations for regional differences in Russian agricultural systems. Demographic theory was not without difficulties, however. For one thing, it did not give a full account of the reasons for the decline in the amount of land peasants held. These lay as much in the realm of changing social relations between the peasantry and other classes of society as in demographic factors. A more fundamental problem was that by concentrating attention so much upon the genesis and development of the practice of repartitioning, Kacharov-skii and his colleagues overlooked other facets of life in the commune that might have given a clearer indication of the way in which it was changing. This point has been made in recent years by the Soviet historian A. L. Shapiro: 'The question of the genesis and character of the ancient Russian commune cannot be confused with the narrower question of the origins and character of the specific mechanism of the equalizing repartition which occurred in Russian communes in the eighteenth and nineteenth centuries.'[9] Similarly, V. A. Aleksandrov argues: 'Accounting for the origin of land repartitions theoretically and empirically has always been very difficult for investigators, and has resulted in the identification of a whole series of explanatory factors: demographic, agro-technical, fiscal, and others. Undoubtedly the problem deserves attention, but it does not define the basic question [about the commune].'[10] The question to which Aleksandrov refers is that of 'when, how, and under what circumstances' communes lost or were able to preserve rights and functions they had acquired over time.[11] What is required is an approach to the commune that takes into account the whole range of its management customs and functions.

The rapidly changing economic and political environment of the second half of the nineteenth century brought new influences to bear on Russia's communes as their members were exposed to market forces and to the influences of urban and industrial society. Furthermore, legislation from the 1860s abolished former legal distinctions between serfs, state and apanage peasants, foreign colonists, and other categories of rural dweller; thenceforth all peasants became subject to much the same laws. Despite the levelling effects of such changes, differences between communities of peasants persisted in the nineteenth century. One reason for this was that the physical environment continued to set limits on the activities of peasant farmers, while at the same time providing them with the raw materials to build specialized economies. Meanwhile, customary practices developed in previous centuries provided communes with a variety of means, often regionally differentiated, of dealing with the new resource-management tasks the nineteenth century brought. The major division in landholding practices was between communes in which there was a tradition of individual or household tenure, and those in which the tradition was of more communal forms of tenure. Apart from differences in tenure, peasants were differentiated from one another by whether they fulfilled their feudal dues in labour service (*barshchina*) or in money (*obrok*), and in the range of their economic activities, the latter determined by the resources available to them. History and geography therefore combined to create locally and regionally distinct types of commune.

## THE COMMUNE IN THE SECOND HALF OF THE NINETEENTH CENTURY IN ARCHANGEL AND KHARKOV PROVINCES

In this section the findings of an investigation made in the 1870s are used to compare communes in two provinces of European Russia, Archangel and Kharkov. These provinces occupied different physio-graphic regions and they had somewhat divergent histories. Archangel province was the most northerly in European Russia. Environmental conditions in the province were harsh, as would be expected at latitudes north of 61°, but nevertheless peasant communities had grown up that were able to eke out an existence through combining farming with fishing, forestry, fur-trapping, and handicraft industries.[12] The prob-lems for agriculture consisted of long, cold winters, a short 90- to 105-day growing season, and podzolic and poorly drained soils. Arable land accounted for just over 0.04 per cent of the total area of the province in the nineteenth century,[13] and it was used primarily for the hardier grains, such as winter varieties of rye and barley, spring oats, buckwheat, millet, hardy vegetables, and flax. The saving grace as far as cultivation was concerned was the length of the summer days (between twenty and twenty-three hours), which helped to offset the effects of the short growing season. Livestock husbandry was the principal branch of farming developed in the province in the nineteenth century, but the need to provide over-wintering for seven or more months put limits on herd size. Hayfields and meadows were at a premium. They were mainly situated along the watercourses, although sedges and moisture-loving grasses, which were found in the forests and marshes, could be used for winter fodder. However, the principal wealth of the province lay in its forest, riverine, and marine resources. The forests provided the peasants with both subsistence and exchange goods in the form of berries, mushrooms, firewood, building materials, tar, and fur-bearing animals. Numerous rivers and lakes were another source of livelihood, providing a range of fish which could be sold commercially.[14] For communes situated on the coast the same was true of the rich maritime stocks of herring, cod, turbot, and plaice.

In contrast to Archangel province, Kharkov province had a favourable environment for the development of arable farming. It lay in the feather-grass steppe in the eastern Ukraine. Soils there were fertile black earths, and the climatic regime was continental with January temperatures averaging - 4 to - 8 degrees centigrade and July temperatures averag-ing + 20 to + 24 degrees centigrade. Although low summer precipita-tion (the annual average was 500 to 600 mm.) and unseasonable frosts could be a problem, environmental conditions had enabled Kharkov to become one of Russia's main wheat-producing provinces and from here grain entered both the domestic and export markets. By the nineteenth century Kharkov province no longer had Archangel's grazing lands and meadows, nor did it have abundant non-agricultural resources that could be used by peasants to supplement farm income. The peasant economy was therefore less diversified than in Archangel province, and, although livestock husbandry was practised, its purpose was to provide subsistence goods and motive-power to pull ploughs. Dung was used for fuel rather than being put on the land.

In terms of anthropological classifications, peasants in Archangel province were 'lowland forest peasants' or 'peasant-fishermen' and those in Kharkov province were 'plains peasants'.[15] They inhabited different ecological niches and can be said to have constituted distinctive 'peasant ecotypes'. In neither province, however, had the peasants' livelihood developed in a vacuum; they were subject to the laws of the tsarist state and to developments in the national economy. The experiences of peasants in the two provinces were different, since in Archangel they were largely black-ploughing, later state, peasants and in Kharkov they were proprietal serfs or military residents. The two provinces thus lay either side of Aleksandrov's dividing-line between regions with a long-standing tradition of individual landholding and regions with a tradition of communal landholding. These different physical and historical contexts had an impact on many spheres of activities in communes in the two provinces, as the accounts below will show.

The survey used to compare communes in Archangel and Kharkov provinces was made by the Imperial Free Economic Society and the Imperial Geographical Society in the 1870s. It investigated a sample of communes in a variety of provinces using a 155-point questionnaire which was sent out for completion to the Societies' local correspondents. These could be literate peasants, priests, schoolteachers, or village scribes. The questionnaire was drawn up with care and revised in the light of pilot surveys. It included questions about the demographic composition of each commune investigated, patterns of land usership, the types and frequency of repartitions, the system and practices of tax apportionment, membership rights, and tenure changes. Findings for a small number of communes were published by the Imperial Free Economic Society in 1880[16] but plans to analyse the returns for further communes were not realized and the completed questionnaires were simply stored in the Society's archive, where they have remained since the 1870s. The archival materials have recently begun to be published by the Soviet Academy of Sciences. The first volume includes surveys of seven communes in Archangel province and is used here,[17] while the previously published volume is the source for materials on Kharkov province. The use of the Imperial Free Economic Society's survey is not without problems and, as Lyudmila Kuchumova, who was responsible for reproducing them for publication, has observed, the way questions were posed often reflected contemporary misconceptions about the nature and principal functions of the commune.[18] The survey materials are nevertheless one of the few sources in the nineteenth century that attempt to give a comprehensive picture of the commune's activities.

## SEVEN COMMUNES IN ARCHANGEL PROVINCE

The peasant population in Archangel province lived predominantly in small villages and hamlets, sparsely distributed throughout the province. These settlements were populated by state and court peasants who had received their land on the same terms as proprietal serfs in special settlements made in 1863 and 1866. Seven communes of such peasants are described here. They were located in four districts in Archangel province: Onega and Kem, which abutted on one of the gulfs in the White Sea coast; Shenkursk, which ran along the banks of the river Varga, a tributary of the Northern Dvina; and Kholmogory district, which extended into the marshy interior of the province. Their location is shown in Fig. 6.1. The composition of the communes investigated in these districts differed: they were 'simple' in Onega and Kem, i.e. each was coextensive with a single village settlement, and 'complex' in Shenkursk and Kholmogory, each consisting of several separate settlements. All but two were coextensive with rural societies. The number of households in the largest commune was 242, but these were distributed among very small villages or hamlets, only one of which exceeded twenty households in size. In the other complex communes the constituent settlements were similarly small. The number of households in the villages making up the three simple communes in the coastal district was greater, but with under seventy-five households apiece they had still to be numbered among the smaller settlements in European Russia.

Each commune had access to a variety of land, riverine, and forest resources which, used in combination, allowed them to practise a form of mixed husbandry as Table 6.2 shows. The exception was Sorotskii commune in Kem district, which had no arable land. Its members depended for their livelihood on livestock husbandry, fishing, and forestry. Officially every commune held its land in communal (*obshchinnoe*) tenure, but each type of land was clearly governed by different usership customs and rights. These are shown in Table 6.3. First, in all

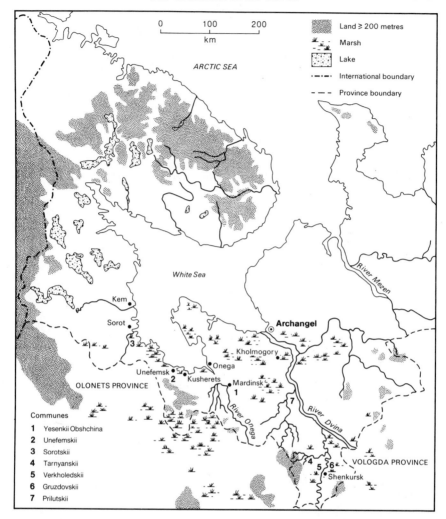

Fig. 6.1. The western region of Archangel province showing the location of communes investigated by the Imperial Free Economic Society in 1877–1880

communes some land was still subject to seizure (*zakhvatnoe*) tenure, whereby households were allowed to clear virgin land lying within the commune boundary and claim it for their personal use for a defined period of time, during which it might be put under either arable or meadow. That time ranged between twenty and forty-five, years and on its expiry the land had to pass into the joint ownership of the whole community of peasants. Alongside the land held in seizure tenure was land held in joint ownership. In all the communes this comprised arable,

TABLE 6.1. *Composition and population of seven communes in Archangel province, 1878*

| District (rural district) | Name of rural society | Name of land commune | Peasants' name for commune | No. of settlement[a] in commu... |
|---|---|---|---|---|
| ONEGA | | | | |
| (Mardinskoe) | Polskii | Yesenskii | *mir* | 1 |
| (Kusheretskoe) | Unefemskii | Unefemskii | *mir* | 1 |
| KEM | | | | |
| (Sorot) | Sorotskii | Sorotskii | *obshchestvo* | 1 |
| SHENKURSK | | | | |
| (Velikonikolaev) | Tarnyanskii | Tarnyanskii | *mir* | 19 |
| (Velikonikolaev) | Verkholedskii | Verkholedskii | *mir* | 10 |
| (Velikonikolaev) | Gruzdovskii | Gruzdovskii | *mir* | 21 |
| KHOLMOGORY | | | | |
| (Yemtsa) | Yemskii | Prilutskii | *sostav* | 11 |

[a] Figure is only for male revision souls.

TABLE 6.2. *Land use in seven communes in Archangel province, 1878 (des./sazheni)*

| Commune (District) | Household lots and gardens | | Separate vegetable patches | | Hemp plot | | Other speciality plots | | Arable | |
|---|---|---|---|---|---|---|---|---|---|---|
| | d | s | d | s | d | s | d | s | d | s |
| YESENSKII (Onega) | * | | * | | — | | — | | 96 | 6 |
| UNEFEMSKII (Onega) | 4 | 900 | 0 | 500 | — | | 0 | 30 | 20 | 2,0 |
| SOROTSKAYA (Kem) | * | | — | | — | | — | | | |
| TARNYANSKII (Shenkursk) | 13 | | 1 | 2,000 | 5 | 1,000 | 20 | 1,000 | 1,019 | |
| GRUZDOVSSKII (Shenkursk) | 4 | | 1 | 2,000 | 3 | | 3 | 2,000 | 483 | |
| VERKHOLEDSKII Shenkursk) | 5 | 2,000 | 1 | | 2 | 2,000 | 7 | | 297 | 2, |
| PRILUTSKII (Kholmogory) | * | | * | | * | | — | | 187 | |

*Note:* an asterisk indicates that the actual amount of land is unknown; 'state' indicates that the land was ... by the state.
*Key:* d = *desyatiny*; s = *sazheni*.

| Range of settlement size (in households) | Total population | No. of households in commune |
|---|---|---|
| — | 358 | n.d. |
| — | 430 | n.d. |
| — | 302 | n.d. |
| 4–27 | 780 | 242 |
| 1–19 | 293 | 95 |
| 1–15 | 489 | 145 |
| n.d. | 363[a] | 126 |

| ...atural ...yland | | Permanent pasture | | Calves enclosure | Wood and forest | Fish-runs | |
|---|---|---|---|---|---|---|---|
| | s | d | s | d | d | d | s |
| | | * | | — | (state) | — | |
| 3 | 600 | 33 | 1,800 | — | (state) | 10 | 600 |
| 5 | | * | | — | (state) | * | |
| 2 | | — | | — | 1,452 | — | |
| | | 140 | | 10 | 346 | — | |
| | 2,000 | — | | — | 184 | — | |
| | 1,300 | * | | * | (state) | — | |

TABLE 6.3. *Land tenure and usership in seven communes in Archangel province, 1878*[a]

| Usership | Existing land use | Category of land | Entitlement to use |
|---|---|---|---|
| Hereditary | Dwellings and vegetable plots and orchards | Household lots and kitchen gardens | All commune members |
| Individual (repartitional) or | Flax, cabbages, and other labour-intensive crops | Separate vegetable plots | ? All commune members |
| Individual (seizure tenure) | Cultivation of winter and spring grains | Arable | All commune members |
| Individual (repartitional) | | | Commune members minus *bobyli* and other 'landless' households |
| Communal | Livestock pasture | | All commune members |
| Individual (seizure tenure) | Hay cutting | Natural hayland | All commune members |
| Individual (repartitional) | | | Commune members minus bobyli and other 'landless' households |
| Communal | Livestock pasture | | All commune members |
| Communal | Livestock pasture | Natural Pasture land | All commune members |
| Rent or communal | Catching fish and retting flax | River resources | Either highest bidders or all commune members |
| Individual (repartitional) | Tar | Woodland and Forest | Commune members minus *bobyli* and other landless |
| Communal | Firewood, berries, mushrooms, and pasture | | All commune members |
| Communal | Pasture | | |

[a] A question mark indicates that the records are unclear.

hayland, pastureland, and dwelling lots and gardens. The general rule was that of these the pasture would be in the collective use of all commune members, the household lots, gardens, and yards in the permanent use of individual households, and the arable and hayland in the temporary use of a number but not all, of the households belonging to the commune. In the two communes in Onega district a third landholding regime applied to forest. Forest was owned by the crown and managed by the Ministry of State Properties, but customarily had been available for use by local peasants. The Ministry of State Properties strictly controlled the harvesting of timber and allocated parcels of woodland to the two communes by household. The distribution of user-rights in this land among members of the commune was then decided by each commune's *skhod*, or communal assembly. The use of forest by the peasants for other purposes, such as grazing livestock, gathering berries and mushrooms, and collecting firewood, was, in contrast, available to all commune members without interference from the commune.[19] These rights were extremely important to peasants in Archangel province because of the precarious nature of arable farming. Members of the former apanage communes in Shenkursk district had access to woodland of their own. This included pine stands, which could be used as a source of tar and were divided into separate lots for use by individual households, while the rest was in the use of all commune members. Finally, in a number of the communes there were parcels of land covered by special user-rights: in Prilutskii commune in Kholmogory district there were separate hemp, cabbage, and vegetable patches and dwelling lots that were in individual use but subject to periodic redistribution. Similarly, Yesenskii commune in Onega district had a separate cabbage patch, and in Tarnyanskii commune in Shenkursk district 100 *desyatiny* of unused woodland and scrub were allocated to individual households to clear for arable. To add one further complication to the picture of land-user rights in the communes, arable and hayland served for part of the year as pasture open for use by all members of the commune, whether or not they had an allotment in the arable.

The landholding arrangements in the seven communes in Archangel province indicate the importance of user-rights, as opposed to formal tenures, in determining how and by whom any given piece of land was used. The allocation of these rights was the business of the communal assembly and participation in its deliberations was therefore an important issue for individual households. In the communes invest-igated by the Imperial Free Economic Society participation in the deliberations of the assembly was not open to all members of the commune, let alone those resident in the village. The core membership of the communal assembly consisted of the heads of landholding households in all the communes, excluding outsiders who rented land in the commune. The landholding households had the right to vote on all matters brought before the communal assembly and were liable for service in any of the elected offices of the commune.

In addition to the landed, there were other households whose members, while belonging to the commune and resident in the village, did not hold land in either the arable or the hayland. Such households could include widows and orphans of former landholders, those whose male members had been conscripted to serve in the army, those whose land had been confiscated by the commune or voluntarily surrendered, newly formed households, or those consisting either of outsiders who had recently been admitted to membership of the commune or else of cottars (*bobyli*)—the traditional village landless. These landless households accounted for up to 15 per cent of the total membership in the communes in Shenkursk district. They were differently treated with regard to their participation in communal assemblies. Thus, in the communes in Onega district all landless households were excluded from any deliberations of the assembly that concerned the management of communal land, and they were debarred from holding office. This was in contrast to the communes in Shenkursk and Kem districts, where the landless took part in all meetings of the communal assemblies and were permitted to hold office in the commune. Usually only men were recognized as household heads for the purpose of attending and voting at communal gatherings, but in some circumstances widows and wives of temporarily absent male heads could attend. In none of the communes were village residents who were not members of the commune allowed to attend gatherings, although in Sorotskii commune, Kem district, they could be elected to serve as *desyatki*, the village police.

The terms 'landed' and 'landless' in the previous paragraph should not be taken too literally, since they refer only to the entitlement to a share in village land. Households which were nominally landless might rent land from a co-villager, while the nominally landed might have disposed of all or a portion of their land in rent. Just as formal tenures conveyed little about the pattern of land-user rights in Russian communes, so these rights were not necessarily a guide to actual patterns of use. One task communes had was to decide upon the entitlement of its members to a share of communally owned land and, within certain limits, the principles according to which such land was to be held or exchanged. Landholding was thus in the gift of the commune and it lay with the assemblies to develop the procedures and rules for bestowing or withholding the gift. Among these, repartitioning was potentially the most powerful mechanism available to the communal assembly. However, as has already been noted, there is some dispute among scholars about how far the practice had entered the culture of peasants in the north of Russia. The survey of the Imperial Free Economic Society does not, unfortunately, give many clues about how deeply embedded the practice of repartitioning was in the communes it investigated in Archangel province, although it does give details of procedures that had been developed for repartitioning in the recent past. The problem is that in the period before the settlement of land on the peasantry in the 1860s the majority of communes in Russia that repartitioned their land did so at the time of the national revision, the census which determined the number of taxable 'souls' in the country.[20] In the northern provinces of Russia, the re-allotment of land among member households of a commune at the time of these revisions was made on the basis of the number of souls each had. This was the case in the seven communes investigated in Archangel province. With the cessation of revisions in 1858 and the allocation of land to the peasants under the special statutes of 1863 and 1868, communes were in a position to decide for themselves whether or not to continue repartitioning. By the time of the 1877–80 investigation only two communes were recorded as having repartitioned their land since the last revision in 1858, but because the time-lapse since that revision was relatively short it cannot reliably be said that repartitioning had ceased in the others. The motive for land redistribution is recorded for only one of the two communes that repartitioned after 1858, but it is instructive. The commune was Unefemskii in Onega district and the repartition took place in 1878. It was brought about by the sudden death of a large portion of the village population in an epidemic which left some households unable to work their land. The land was redistributed among households according to the number of male members they had. Such catastrophic events in village life might well have been a rather important reason for repartitions in Russian communes in the second half of the nineteenth century, when migration, revolution, and famine could have dramatic effects upon village populations.

The 1870 survey indicates that at that time individual households held their arable and haylands in several separate parcels, although it does not record their exact sizes. Each commune had an established set of rules to guide them in determining the size, location, and number of strips to be held by each household, but it was only general repartitions that afforded the opportunity to apply these systematically. In all the communes the size of individual households' strips in the arable was determined by the number of their constituent souls. Thus, strips were of variable width depending upon whether a household had been allotted one, two, three, or more souls' worth of land, but the minimum everywhere seems to have been 2 *sazheni*. It was normal in all the communes for land to be classified according to some qualitative index, such as accessibility or fertility, and for each household's share to be made up of roughly the same mix of types. In none of the communes was the classification of land very precise. In Yesenskii commune, for example, no more than half a dozen simple categories were recorded, such as 'hilly', 'valley bottom', or 'new', and in Prilutskii commune physical and topographical attributes of land were dispensed with in favour of a fourfold division on the basis of distance from the dwelling areas. In theory, households' allotments need only have been divided into an equivalent small number of separate parcels, but in reality they were much more fragmented. This was because the communal arable from which households' strips were carved was itself scattered in a multitude of blocks associated with clearings made in the forests many years before. Wherever their size permitted the creation of strips of the minimum 2 *sazheni* in width, these blocks would be divided up among all the member households of a commune. Much the same regime existed for haylands, although the divisions were fewer. In Sorotskii commune, in which natural haylands constituted the main land resource, far and near fields were treated differently. In the near fields, termed the *polya*, households were each allocated three parcels, the location of which did not change between general repartitions, but in the distant haylands, the *pozhni*, the position of households' strips was rotated annually. The difference was related to the use of manure on the land: only the *polya* could be manured and so it was these that households strove not to lose. Similarly, in Prilutskii commune, Kholmogory district, it was recorded that households tried to retain their existing strips at repartitions, but this time these were of the arable. While one aim of allocating households strips in a variety of places was to achieve fairness in allocation, scattering was not the only way of achieving this. In Prilutskii commune, for example,

All the fields are divided into four divisions (*razryady*) according to their distance from the living-quarters . . . Then the number of *desyatiny* of each sort and the number of male souls in the commune are calculated and the figure for the latter is divided into the former. Using the number calculated every soul is allocated land in each of the four divisions . . . Differences in soil type are not taken into account, but if someone is given a very bad plot he is compensated either there or in another place with an extra allocation of land.[21]

This extract shows that one should view repartitioning as a rather flexible mechanism that could be used in various ways to achieve a given purpose. This flexibility is further demonstrated by the fact that when Prilutskii commune was compelled by circumstances to repartition in

1778, its members were able to retain some of the same strips they had previously worked, additions and subtractions being made to their allotments at the margins.[22]

The general repartitions that had been carried out in the Archangel communes before 1861, while serving to scatter peasants' land, did so in a systematic way. In the years following general repartitions other types of land exchange taking place between households must have broken down this pattern. According to the Imperial Free Economic Society survey, all the communes practised partial repartitioning, *svalka-navalka*, whereby portions of land were transferred between a small number of households. The tendency in the literature on the commune to refer to *svalka-navalka* as a 'lesser version' of the general repartition can be confusing, since it seems to have served rather different functions and to have been brought about for different reasons from those leading to repartitions involving all the land of all a commune's members. In Yesenskii commune, Onega district, for example, *svalka-navalka* referred to a process whereby the commune took over land for redistribution to other households in the event of a head of household's death, the death of male members of his family, or the economic failure of the farm. The survey noted that there was no standard form for such redistributions and that they could take place at intervals of anything between one year and five or more. The recipients were other revision households that were in need of land. The process seems to have been an extension of the practice that developed in the eighteenth century of communes building up a reserve of land that would be available for allocation to new households. To this end communes began to lay claim to the land allotments of 'dead souls' (*vymorochnye zemli*) and to abandoned land (*pustoshi*).[23] In Yesenskii commune the principle that governed the reallocation of land surrendered to the commune was not recorded, nor was it indicated how the competing claims of a number of households were treated. More details are available for the three communes in Shenkursk district. In all three, minor redistributions could be an annual event, and they involved taking land from households which had lost a revision soul. In none of the communes, however, was the full complement of land taken from such households; only an amount equivalent to one-half or one-quarter of a revision soul was redistributed, and the qualification for receiving land was an increase in the number of consumers in a revision household. Essentially, what these redistributions entailed was the surrender of a small portion of land to which one household had a claim by virtue of its labour-power to others whose claim was based on need. They did not involve the equalization of landholding between households in the way that general repartitions did, but seem instead to have been a way of assuring a minimum livelihood to all commune members, while special provision was made to protect households' rights to land that they had held for twenty years or more.

Before the eighteenth century one indication of the strength of individual property rights in northern communes was that peasants were able to make land transactions. According to I. V. Vlasova, the purchase and sale of land, mortgaging, and land seizure were widespread in the seventeenth century and early eighteenth, despite attempts by the state to supress them.[24] The General Survey instructions issued in 1754–6 categorically banned such transactions and after this they seem finally to have ceased.[25] The survey of the seven communes in Archangel province revealed that there was indeed a ban on the sale and mortgaging of commune land by individual households. Apart from this, the communes seem to have allowed substantial freedom in what households chose to do with their land. Thus, the survey recorded that peasant households in all seven communes were entitled to make independent arrangements to exchange their land parcels with one another. In some they could also enter into share-cropping or rental agreements, although this required the permission of the communal assembly. In Prilutskii commune a household disposing of its land was said to have 'sold' it, but such 'sales' were valid only until the next general repartition. This was the rule for any land exchanges in the other communes as well. As already observed, the combined effect of *svalka-navalka* and other land exchanges negotiated independently by the peasants must have been to break down the spatial order and the ratio of male family members to land set up at the time of general repartitions. Unfortunately, like so many of the nineteenth-century records, the Imperial Free Economic Society survey did not describe what the net effect of all these exchanges was in terms of where and in what amounts households held their land, but clearly general reparti-tions on the one hand, and *svalka-navalka* and independent land transactions on the other, had potentially contradictory outcomes. One interpretation of this is to argue, as Aleksandrov and Vlasova do for an earlier period, that it was a symptom of the historic struggle in the region between two competing principles of landholding, but there may well be other explanations.

Other issues relating to the management of communal resources falling within the competence of communal assemblies were numerous and no less important than the question of repartitioning. All the communes investigated had at one time or another devised rules covering the use by individuals of their strips, entitlements to communally used resources, land conveyance, and a host of other activities. In the communes in Onega and Shenkursk districts, for example, households were forbidden to fence their strips in the arable. The reason for this given in the survey report was that much land would be wasted in fences, but an additional explanation is that the absence of fences forced all households to offer their land up as pasture for part of the year. While unable to fence their own strips, all landholding households had to put up and maintain fences around the perimeter of the commune's fields. The length for which they were responsible was related to the size of their holding, and they were subject to fines in the event of animals breaking through. In the communes in both these districts there were also restrictions on the disposal of produce from communally owned resources. Thus, peasants in Onega district were forbidden to sell wood outside their commune, and in Shenkursk there was a similar ban on the sale of tar. There were even restrictions in some of the communes on the sale of privately owned resources. Members of Unefemskii commune, for example, were not allowed to sell manure as they were required by the commune to use it on their own land. Livestock could not be sold, although anyone who did sell would not be penalized. In Yesenskii commune no formal ban on such sales existed, but it was clear that the commune would take a poor view of any of its members who engaged in selling:

... it is not forbidden to sell manure, but if someone decides to do it the commune would consider such a peasant negligent. Similarly, it is not forbidden to sell off a farm's livestock, but this would be taken as a signal of the farm's demise and it would have unpleasant consequences, namely that the commune would pass a motion to take the land away from such a peasant in order to avoid the land's abandonment.[26]

Such restrictions, which were clearly designed to prevent the dissipation of rent on the commune's resources, were not matched by prescriptive measures. Thus, in none of the communes was there any obligation on peasants to follow a particular rotation or farming calendar. In reality, of course, the options open to individuals were limited not just because too marked a deviation from general practice carried the threat of trampling by livestock during the period of common pasturage, but also because the environment placed severe restraints on what could be sown and when. With one exception a three-field rotation was followed in all the communes with arable land. The exception was Unefemskii, where there was a simple two-field rotation.

The communes in Archangel province seem thus to have developed a number of rules aimed at managing the resources they owned communally but used individually. In matters concerning resources that were communally used as well being communally owned there was another set of rules. Such resources included permanent pasture, rivers, woodland, and scrub. In the six communes with mixed husbandry all member households, regardless of whether or not they had rights to an allotment in the arable and hayland, were entitled to use these resources. Thus, widows, orphans, *bobyli*, and other landless members had the same access to such common means of subsistence as their landholding neighbours. In none of the communes was there any limit placed on either the size of the fish catch or the amount of flax that could be retted in the rivers.

A characteristic feature of the open-field system in western Europe was that grazing on common pastures was regulated by establishing stints, or limits on the number of livestock that households could put with the village herd free of charge. In Archangel province no such system existed. Its absence can be explained partly by the relative abundance of pasture in the province, and secondly by the natural limits set upon the number of livestock any household or community could support by the need to provide fodder for the long period of over-wintering. Unless purchases could be made from outside, the annual harvest from the commune's hayland must have been an important determinant of how many livestock could be kept. In all the communes a source of summer pasture was state forest land, and two did not have their own permanent pasture. In Yesenskii commune such state land was used for pasture for three months beginning in May, before the herds were moved to the harvested hayland and arable. Summer grazing with the common herd was not free. Households putting their livestock with the herd had to pay towards the hire of cowherds and shepherds per head of livestock. In Yesenskii commune, for example, the charge on one cow was 25 copecks and on one horse 15 copecks. In addition to these fees, every household in the commune had to give the shepherd three pounds of grain in the autumn, and to provide one night's lodging for him and his assistants for each head of livestock. Much the same system of payments in cash and kind existed in the other communes and they must have acted as some limit on livestock numbers, but if the overcrowding of pastures had become a problem it had not yet reached levels requiring action. Meanwhile, the relatively free access for all commune members to such common resources helped to support the landless households for which the commune might otherwise have had to provide the means of subsistence.

Sorotskii commune was rather different from the others in that it had limits on the use of its communal fishing resources. Since fishing replaced arable cultivation as a major source of livelihood for households in the commune, the care shown over the management of fishing rights is not surprising. Fifteen years before the investigation by the Imperial Free Economic Society the commune had allocated fishing rights by lot, consisting of seventy-five river-bank positions and mid-stream fishing. Households were assigned to an *okol*, a small group, which drew lots annually for a stretch of the river. It was essentially the same system used in the commune for allocating plots of hayland in the *pozhni*. For the previous fifteen years, and presumably because of population expansion, this system had been superseded by another, under which the commune charged a rent for fishing rights and households bid for these against one another in an annual auction. A consequence of the new system was that not everyone entitled to bid for fishing rights gained a place on the river but all shared in the rent raised, which was distributed among households according to the number of revision souls. The market principle extended beyond the primary allocation of rights, however, as households were entitled to sell their place on the river-bank to co-villagers and were able to join together in co-operatives to bid for fishing rights. As if in recognition of the dangers of giving complete sway to the forces of the market, the commune prohibited the sale of rights to outsiders and would not allow members of a co-operative to do waged work for one another.[27]

A question that intrigued the nineteenth-century investigators of the commune was whether communal ownership of resources was matched by communal forms of production and labour organization. The answer to the question was important if the social character of the commune were to be understood. The communes investigated in Archangel province presented a contradictory picture of individualism in some spheres and collectivism in others. Apart from the pasturage of livestock in a common herd, farming was largely an individual pursuit, with households employing their own labour to cultivate the land, to bring in the harvest, and to perform other tasks about the farm. This applied to spheres in which, on comparison with the experience of open-field villages elsewhere, some co-operation between households might have been expected. Thus, for example, peasants did not contribute animals to a communal plough-team, nor was the hay crop normally harvested jointly. In Tarnyanskii commune, Shenkursk district, it was reported that at the behest of the Ministry of State Properties 55 *desyatiny* of communal land had at one time been cultivated jointly by all members of the commune. Some of the produce had gone into a communal grain store and some was sold to provide funds for the local hospital. This practice had ceased, evidently to the relief of the peasants: 'The peasants dodged doing their share of the communal cultivation and thereby risked punishment by the authorities. They viewed the communal labour as an unnecessary burden upon them. Cultivation ceased in 1861 on the order of the Ministry and the land was returned to the Imperial estates.'[28] While communal cultivation was eschewed in all the communes, there were circumstances when households helped each other in a system of mutual obligation, or *pomoch*.[29] In Shenkursk district this system was used by households which needed help carting manure to their fields, bringing in firewood from the forest, and harvesting grain. The first two tasks required the services of two men with a horse and cart; the third, women with their own sickles. In return for their help they would be given food and drink by the recipient household. In addition, in all the communes members were obliged to help out at neighbours' wedding celebrations, at births and deaths, and in cases of theft and fire.

While the control exercised by communal assemblies over the farming calendar of individual households was relatively light-handed, the same was not true when it came to fulfilling financial obligations to the state. In an attempt to ensure that taxes were paid, the state had instituted a system of joint payment, the *krugovaya poruka*, under which communes were entrusted with the task of collecting taxes from their members for the state. This system allowed communes to distribute the tax burden among members as they wished. The tax obligations of Russian peasants in the nineteenth century were numerous: in addition to the redemption payments there was the poll tax levied on males, and *zemstvo, volost*, and commune dues. In the majority of Russian communes the tax burden was apportioned among households according to the share of commun-ally owned land they held, but the precise details of this apportionment varied. In the communes in Archangel province landholding households were the principal taxpayers, picking up the burden of the poll tax levied against members of landless households. Some landholding households were also exempt from taxes, such as those with members conscripted to the army, and their burden was paid by the commune as a whole. Where the treatment of tax defaulters was concerned, it was usual to confiscate property in lieu of payments and to transfer land to another household. In Sorotskii commune it was the stronger households that received such land, and they kept it until the original holder was in a position to be able to take his tax burden on again. Unefemskii commune, in contrast to the others, did not have a system for coping with non-payment of taxes, since all its households were in arrears so that nothing was to be gained from removing any of their land. When any exchanges of land took place in the communes, whether they were informal transactions or reparti-tions, the total tax burden passed to the new occupier, poll tax included. Essentially, therefore, taxes on the peasants in Archangel province resolved themselves into a tax on land.

Taken together the various rights and obligations of members of the seven communes in Archangel province present a rather confused picture. Some practices seem to have been contradictory and their rationale obscure. However, the contradictions may often have been more apparent than real. An example is the way in which every commune had the right to redistribute its members' land, yet allowed individual households to organize their own exchanges, the latter having the potential to undo the results of the former. But this is a contradiction only if the purpose of repartitioning is assumed to have been to achieve quantitative and qualitative equality between households in the land they held. If by the 1870s it had a more modest purpose of delivering a minimum amount of land to households, the two types of land transaction are less in conflict with one another. If nothing else, therefore, the data from Archangel province demonstrate the importance of the framework within which the details of the internal operations of communes are assessed. It becomes even more crucial when comparisons are attempted with communes in different economic and physical environments.

## Two Communes in Kharkov Province

Two communes in Kharkov province were investigated in 1878 and appeared in an Imperial Free Economic Society publication of 1880.[30] They were located in the east of the province on the River Krasnaya, a tributary of the Severskii Donets, and they made up a single rural society. Both were populated by former proprietal serfs and on emancipation in 1861 they had been made into a single rural society. The larger commune, Morakhovka, had a total population of 260 and 165 *desyatiny* of land; the smaller, its 'daughter' commune of Khutor Dolya, had a population of 52 and 114 *desyatiny* of land. This land was held in communal tenure in both communes but, as in Archangel province, different types of land were covered by different user-rights. These are straightforward: household lots and a small area of former woodland (now scrub) were in the inalienable use of individual households, permanent pasture was in communal use, and the arable and meadow were in individual use but subject to repartition. All available land had been brought into use in the two communes and seizure tenure was thus a thing of the past. However, a version of it was practised on land which the communes rented from their former serf-owner, which involved each household claiming the right to the exclusive use of the furrows it ploughed in the rented land.

Although the Emancipation Statute had resulted in their liberation, peasants in both communes fifteen years on were still locked in a network of economic relations with the local large landowner that were reminiscent of the former era. Payment for the 15 *desyatiny* of land rented by Morakhovka commune, for example, was in labour service, which was used by the landowner to harvest 30 *desyatiny* of his estate. For a further parcel of land, which it had been forced to acquire because of the failure of its winter crops, Morakhovka had entered into a share-cropping agreement with the same landowner. Under its terms the commune received two-fifths of the harvest, or one-third of the harvest and one haycock per *desyatina*, from the rented land; the rest of the product was retained by the estate. The commune entered the land market as leasor itself, renting out 3 *desyatiny* of meadowland to landless residents of the village. The mosaic of landholding arrangements in the two communes was subject to further complication by the entitlement of individual households to rent out their land, to enter into share-cropping agreements with one another, and to mortgage their harvest.

The arable allotted to peasants in Morakhovka on emancipation had been divided into seven large fields, *lavy*, in each of which individual households received part of their entitlement. Later an eighth field was added as a result of the conversion of part of the commune's meadows to arable. The division into *lavy* had been made partly to take account of difference in land fertility and partly for ease of subdivision into strips. The best fields were numbers 2, 3, and 5, as shown in Fig. 6.2, considered so because of a valley-bottom site, proximity to the settled area, and a site protected from livestock and the threat of trampling; and the poorest were numbers 1 and 4, the former on an exposed site and the latter running parallel to the road. In 1861 plots of land in each of the fields had been allocated to households on the basis of their number of revision souls, which, as has already been explained, was the number of male household members alive at the time of the tenth national census enumerated in 1857–8. In the years following this initial land distribution there had been frequent repartitions, the last two in 1872 and 1874. Households carried out repartitions by drawing lots for new strips, but none was forced to accept an increase or decrease in the size of its allotment unless it wished.[31] Households which had received allotments in 1861 but which had lost revision souls in subsequent years were not compelled to surrender any of their land. The convention was for the allotment attached to such 'absent souls', *ubylye dushi*, to be passed to the nearest relatives. It was only if these kin were unwilling or unable to meet the payments due on the land that it would be passed to another household in the commune. The qualification for receipt was recorded in the survey report no more precisely than a 'willingness' to take on such land and its obligations. Another circumstance under which land might be transferred between households was the inability, for whatever reasons, of a household to work its full share of land, but here again the surrender was voluntary.

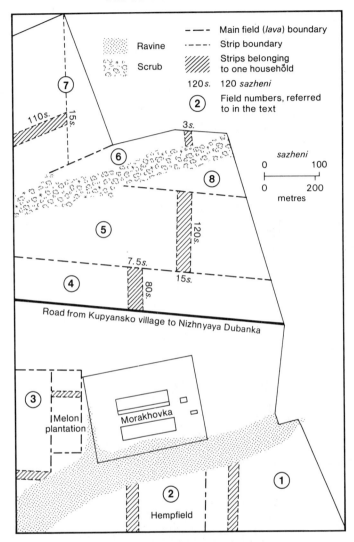

FIG. 6.2. The ground plan of Morakhovka commune,
Kharkov province (*c.* 1878)

In Archangel province direct transfers of land between households, such as those described above, were accomplished without involving other members of the commune by the process called *svalka-navalka*. In Morakhoka commune, by contrast, everyone became involved. Thus, whenever a repartition was called, all households holding land in the arable had to offer their strips for reallocation regardless of whether the total amount of land they held was to change. Households drew lots to determine the order in which land was to be parcelled out between them. Working systematically round the *lavy* that were to be repartitioned, households received the land due to them in a single strip. Repartitioning in Morakhovka therefore resulted in every household holding the same number of unequally sized strips in the arable. What changed from one repartition to the next was the location of each household's strips and, for some but not all households, the size of these. In addition, if any new households had been created or old ones dissolved, the total number of strips in the commune's fields would change. One final point to note is that not all *lavy* would necessarily be included in a repartition. In Morakhovka one field had been excluded from repartitions for the ten years preceding the investigation, and in neighbouring Khutor Dolya repartitions of only the fallow field were made at three-year intervals. As in the communes in Archangel province, households in Morakhovka and Khutor Dolya were not allowed to fence their strips, but they were not restricted in the crops they grew or in their cycle of work in the fields. A three-field rotation prevailed in both communes, with a selection of wheat, barley, buckwheat, oats, and rye grown in the spring fields and wheat in the winter fields. There was a separate lot put aside for the cultivation of hemp (as shown in Fig. 6.2, above), and this was the only land in either village that was manured.

In addition to their land being periodically redistributed, the activities of households in Morakhovka and Khutor Dolya were subject to other controls, some features of which can be explained by the nature of the local environment. Thus, the free access to common pasture that peasants had for their livestock in Archangel province was replaced in the communes in Kharkov by strict stints, or limitations on the number of livestock that could be run, free of charge, with the common herd. This was a consequence of the shortage of natural pastureland. In Morakhovka every household was allowed to run the equivalent of one and a half head of large livestock with the common herd, and in Khutor Dolya two large livestock for each soul's worth of arable land it held, and five to six pigs. Households with more livestock than their entitlement had to pay the commune for each surplus animal they had, while the landless had to pay for all their animals. The communal herd was divided into two: working livestock, heifers, and pigs were pastured on the common land, fallow, and stubble of the villages, while cows, bullocks, and sheep were pastured on land belonging to the former serf-owner. The commune had to pay for access to the landowner's pasture at a price per animal fixed by the market, these same rates determining its own charge on members' surplus livestock. Another 'tax' on livestock ownership existed in the labour dues paid in return for the 30 *desyatiny* of land rented from the former serf-owner. The distribution of the labour service was made according to the number of livestock individual households owned. In an interesting aside which reveals the way in which egalitarian arrangements in open-field villages were subject to market erosion, the survey report noted that households with large numbers of livestock would sometimes pay their poorer neighbours to perform this labour service for them.

The heavy steppe soils may in part account for another difference in practice between the communes in Kharkov and Archangel provinces. It will be recalled that households in Archangel did not pool their horsepower to plough the arable collectively, since the forest soils could be turned with a single-horse plough. In Morakhovka and Khutor Dolya, in contrast, plough-teams six-oxen strong were used, and this necessitated some co-operation between households. The convention in the 1870s was for groups of three to join together, each contributing a pair of oxen to a team and taking responsibility in turn to oversee the team at night. The survey returns do not reveal whether the commune at large had a hand in organizing the plough-teams. Instances of collective labour were few and far between in the two communes, and were confined to work on the communally used resources. In Morakhovka collective labour was used to maintain the commune's well, located half a *versta* from the village, to repair the dikes containing the village pond, to harvest brushwood from the landowner's forest, and to mend wattle fences. In the past there had been an attempt to cultivate 3 *desyatiny* collectively, for which purpose part of the commune's natural hayland had been brought under the plough. The harvest was intended for a communal reserve to tide the community over years of harvest failure but, as with the similar experiment in one of the communes in Shenkursk district, Archangel province, the attempt in Morakhovka failed. The survey reports that this failure was due to the combined effects of a poor harvest and lack of enthusiasm on the part of commune members. The land involved was subsequently rented out by the commune. In addition to the various instances of commune-directed co-operation between households, the system of mutual assistance, *pomoch*, existed in Morakhovka. The tasks for which households would ask for help from their neighbours included bringing in haystacks from the steppe and moving fences and farm buildings, and they would be performed on holy days in return for vodka and snacks. In addition, every member of the commune was expected to help out others in times of distress, e.g. a fire, and to attend at births, weddings, and deaths in the village.

Like other communes in Russia, Morakhovka had a budget to

administer. A principal source of income was the annual tax paid by members, but in addition there was clearly scope to increase income by careful use of communal resources. The sources of such extra income noted in the survey of 1878 included renting out a portion of the communally owned land, charging for grazing rights in the common, and working as a whole community as wage labourers on a nearby farm. The reimbursement of commune officers, payment for grazing the common herd on the local landowner's estates, and, it must be assumed, various public works constituted the expenditures of the commune. Communal income was evidently not used for welfare purposes: landless households were not relieved of the poll tax in Morakhovka and, as has already been observed, they had to pay for pasture rights. As elsewhere in Russia, members of the commune were jointly responsible for payment of state and local authority taxes. The tax burden, as in the communes in Archangel province, was apportioned among households roughly in accordance with their share of commune land, but with the difference that when land was transferred from one household to another it only took with it the redemption and local authority dues, leaving the poll tax to be paid by the original user. This may well have been an attempt on the part of the commune to encourage its members to take on 'liberated' allotments, rather than renting any additional land they might need from outside the commune; this seems likely since the local long-term rental price for land (3 roubles per *desyatina*) was lower than the sum due in taxes from a *desyatina* of peasant allotment land if poll taxes were included (4.5 roubles per *desyatina*).

In all the important decisions covering the disposition of their land and the apportionment of the tax burden members of both communes had to bow to the authority of their respective assemblies. Participation in the assembly was open to all landholding members, whatever the issue being discussed, but was limited for non-landholding members. The landless were not allowed to attend meetings concerned with the redistribution of land, but they could be present at discussions about taxes. Widows were permitted to attend and vote, but other women were debarred unless they were standing in for an absent husband. Residents of the two villages who were not members of the commune were also debarred from attending assemblies. Entry of new members both to the commune and to a share in the communally owned resources was covered by strict rules. Applicants for membership had to be vetted by the assembly and, if successful, would be given a household lot in return for a fee and vodka, but they could not expect any arable land. Member households had to ask their assembly's permission if they wanted to divide their property and, although the transmission of communally owned land from one generation to the next was allowed, this had to be between blood relatives who were members of the commune. The commune was thus an 'exclusive club' which guarded its membership and its resources. Within the club there was a pecking order. For Morakhovka it is described thus: 'influence at the *skhod* is exercised by the richer and cleverer farmers. Furthermore, brothers tend to stick together even though they might have separate farms.'[32] Such patterns of influence and control, which must have taken on specific forms all over Russia, would clearly have been important in determining the inner workings and objectives of individual communes. They remain perhaps the most elusive element in the history of the institution.

## CONCLUSION

There were marked differences in the way that communes in Archangel and Kharkov provinces managed their resources. These reflected both the specific nature of the physical environment in which they were situated and their customs of landholding developed over centuries. Whether the differences were such that the communes should be regarded as separate 'species' is not very important, but the contrasts observed underscore the necessity of examining communes in their geographical and historical settings. Hitherto much scholarship on the commune has dealt in generalizations, but materials are available which allow the description and analysis of the internal workings of the commune, and which reveal how they attempted to adapt to local environments and cope with new internal and external pressures. Nineteenth-century theorists understood the necessity of examining regional variations in the commune, but by concentrating upon differences in repartitioning practices they could achieve only a partial view. A re-examination of the nineteenth-century materials is now required to fill out the picture, and to reveal the full complexity of the commune as an economic and social institution.

# 7
# Modernization from Above

## *The Stolypin Land Reform*

THE village community described in the previous chapter served the needs of many generations of Russian peasants and, in its legal and administrative functions, it was also of service to the state. Until the latter part of the nineteenth century few people would have questioned the centrality of the village community in the daily round of life in rural Russia, or denied for it a role in the future. By the turn of the century, however, opinions in government began to change, and earlier support for the institution waned. The explanation of this change of heart lay in the village community's seeming inability to secure subsistence for its members, and a fear of the consequences of this for Russia's political and social stability. Long before the wave of peasant unrest that marked the opening years of the twentieth century, the government began to limit the autonomy of the commune, as, for example, in a law of 1893 which forbade repartitioning at intervals of less than twelve years, and to set in motion the processes that were to lead to the later decision to phase the commune out of rural Russia. In the aftermath of the 1905 revolution that decision was finally taken, and, in a series of laws and Imperial edicts, the complex of economic and legal structures that had been erected at the time of the serfs' emancipation to protect and safeguard the land commune were removed, and in their place a new structure was erected, the purpose of which was to promote the modernization of rural Russia.

The modernization measures were known as the Stolypin Land Reform, after the then prime minister, P. A. Stolypin. The Reform was enacted in the period 1906–11 and it operated for eleven years before war and revolution brought it to an end. Its principal purpose was to promote smallholder individualized farming among the peasantry at the expense of the existing communal system. This was to be achieved by giving peasants the right to withdraw from their communes and to consolidate their scattered strips into single unified holdings. The vision the legislators had was of a landscape of dispersed farmsteads worked by a class of yeoman farmers who, by virtue of their economic prosperity, would also be politically loyal. The landscape of consolidated farms would be very different from the existing one of nuclear villages and open fields, but the revolution of 1905 provided the government with a compelling reason for choosing radical measures. The advantages of farming on owner-occupied, consolidated farms had, in fact, been discussed on and off since Catherine II's reign, when they had been used in settling state peasants and foreigners in the southern and eastern peripheries of the empire.[1] The time seemed right for extending the experiment into the heart of old Russia. The case for individualized farms rested upon their apparent amenability to the introduction of intensive farming techniques. Thus, as scholarly and popular publications pointed out, whereas the commune stifled individual initiative and inhibited innovation, individualized farms provided just the right conditions for the transition to more intensive farming.[2] Thus, by the twentieth century agricultural progress in Russia had become linked to changes in the tenure and disposition of peasants' land, indeed the former was seen as conditional upon the latter. In essence, therefore, the Stolypin Land Reform sought to effect the transition from extensive to intensive systems of production in agriculture.[3]

The legislation which made up the Stolypin Reform contained a variety of provisions. In addition to allowing households to withdraw from their communes and to consolidate their land, it provided for them to take out title to their land without consolidating it, for land parcels to be exchanged between communities, for disputed boundaries to be fixed, and for jointly held commons to be divided between villages. In order to direct the changes involving the physical rearrangement of land—a process termed 'land settlement' (*zemleustroistvo*)—a new administration was set up in the Ministry of Agriculture, which had provincial and district-level organizations.[4] The administration's purpose was to process requests made by peasants under the provisions of the Stolypin legislation and to draw up and supervise the implementation of land-settlement projects. Where the task was simply to effect a change in tenure and no physical reorganization of land was involved, the law was implemented by the Ministry of Internal Affairs through its local offices. Supplementary legislation ended the system of joint responsibility for the payment of taxes.

The ideal type of farm that, as a result of the Reform's implementation, was to replace the commune was termed the *khutor*, a private farm owned by a household head, and consisting of a single parcel of land surrounding the farm buildings and appurtenances. A. A. Kofod, a naturalized Dane who worked for the Ministry of Land Settlement, was largely responsible for popularizing the idea of the *khutor* and for working out the details of its structure.[5] In the early months of the Stolypin Reform's implementation, in late 1906 and 1907, the task of forming *khutora* was vested in the Peasant Land Bank, an institution which had been established in 1883 to sell nobles' land to the peasantry. Rapidly, however, the task passed to the local Land Settlement Committees, which, in a series of circulars between 1907 and 1909, were given instructions about how to remould open fields into *khutora* and were urged to concentrate their efforts on this work.[6] At about the same time an initial assumption made by the legislators—that only a minority of households, the innovation leaders, would at first want to quit their communes and set up on individual farms—gave way to a belief in the possibility of the immediate dissolution of whole communes. Within a short time of its establishment, therefore, the land-settlement organization's priority had become the creation of whole systems of consolidated farms on the land of former village communities.

In the instructions that local officials in the land-settlement organization received, precise specifications were given of the shape *khutora* should be, e.g. that the ratio of the length of the landholding to its breadth should not exceed 3 : 1, and of the disposition of fields and buildings. Circulars also made clear that what existed in open-field villages as fixed categories of land use, such as the arable, hayland, pasture, and scrub, should be combined in the single land parcels of the *khutora*; common resources should thus disappear and, along with them, most of the business that had justified the existence of the village assembly. It is interesting that in the correspondence and reports of top officials returning from inspection tours of the provinces especially critical comments were reserved for local Land Settlement Committees that failed to include all possible land in whole village consolidations, leaving, for example, pasture or scrub in common use. The fear was that the retention of commons would perpetuate 'communal sympathies' among the peasants.[7] Sentiments such as these, coupled with the detail in which instructions about the technical side of land settlement were cast, reveal the deep faith some in the Reform organization had in the power of particular landholding systems to influence the farming practice and social life of the peasantry. To this extent the thinking underlying the reform was deeply deterministic and, in its attitude towards the peasantry, paternalistic; henceforth, the state would show the peasants the way forward, leading them out of poverty and backwardness. The role the peasants had to play as the future unfolded would be essentially passive, as the grateful objects of change, while the state, through the agency of the Land Settlement Committees, was to be the directing force of that change.

### LAND SETTLEMENT: POSSIBILITIES AND PROSPECTS

While the formation of *khutora* constituted the ultimate goal of the Stolypin Reform, the legislation passed between 1906 and 1911 in fact enabled peasant households to change their landholding in a variety of ways. The changes that could be effected were not entirely novel—most had been possible, and indeed in some cases had been carried out, within the framework of the existing landholding arrangements. For example, communes had always had the right to petition for land held in common to be divided between them, or for scattered blocks of land to be consolidated, while, where the internal disposition of their members' land was concerned, they could use repartitioning as the means to consolidate it if they wished. For individual households the provisions of the Emancipation legislation had established the right of legal withdrawal with their due share of allotment land so long as all redemption payments had been settled. With the cancellation of the redemption debt in 1905 this potential obstacle to secession was removed.[8] It was even possible for a peasant to set up on a *khutor*, either on a consolidated allotment or, after 1880, on land newly purchased from the Peasant Land Bank. Patterns of landholding in the Russian countryside were rather more fluid and diverse than is usually supposed.[9] The contribution the Stolypin legislation made to this changing picture was particularly important in two areas: first, the new bureaucracy expedited changes which had previously been subject to delays and frustrations, and secondly, the Reform legislation provided for the state to bear some of the costs associated with the reorganization of landholding that would previously have fallen to the peasants themselves. Assistance was given with the costs of legal work and of the services of land surveyors, and the provision of loans at low interest rates to help defray the cost to individual peasants of moving their farm buildings or bringing their land into a state suitable for cultivation. As a result of these measures previously expensive changes were now made to appear more attractive. But the government did miss some opportunities—notably in their failure to require nobles to co-operate in projects involving the rearrangement of peasant land. Before 1906 the inviolability of private ownership had served as a major impediment to the disentangling of communally owned land from noble land, and it continued to do so after that date.

Adoption of any of the provisions of the Stolypin Reform was voluntary. On learning of the possibilities offered under the law peasants were able to apply individually or collectively for land settlement. Alternatively, they could decide to do nothing. Should a peasant or a whole community decide to take advantage of the law to expedite change in some aspect of the way in which they held their land, the range of possibilities offered was considerable. Under the heading of 'simple tenure change' single households or whole peasant communities could ask for a transfer of title to their land into private ownership. Such a change meant that land was now protected from repartitions and became the inalienable property of the householder and his heirs. However, the protection was not absolute in that such land could still be subject to physical reorganization under an official land-settlement project. Full inalienable ownership of land was achieved for the peasant only if he settled on a *khutor* duly formed by the Land Settlement Committee or purchased from the Peasant Land Bank.[10]

The provisions of the Reform covered two types of land reorganization: unitary land settlement (*uchastkovoe zemleustroistvo*) and group land settlement (*gruppovoe zemleustroistvo*). The first referred to the consolidation of the land belonging to individual households, and the second to the resolution of problems concerned with disputed boundaries, shared commons, and fragmentation at the village level. These two basic types of land-settlement activity were divided into eight separate procedures:

1. The consolidation into individual farms of land formerly in the use of all the members of a commune (*razverstanie*).

2. The consolidation into individual farms of land formerly in the use of one or a small number of the members of a commune (*vydelenie*).

3. The physical disentanglement of land in the use of non-communal and independent peasant farmers and its consolidation. (This applied, for example, to the *odnodvortsy* with hereditary tenure.)

4. The legal and physical division of land used and administered in common by separate communities, the so-called 'single-plan settlements' (*odno-plannye seleniya*).

5. The creation of new settlements as a result of the physical separation of part of a commune's population and the legal separation from parent communes of existing subsidiary settlements.

6. The redistribution between communes of spatially intermixed parcels of land.

7. The legal and physical disentanglement of land used in common by peasant communes and private landowners.

8. The demarcation of boundaries between different peasant and non-peasant landowners.

Scope for the formation of *khutora* existed primarily with the first three of these types of work and to a lesser extent with the fourth, when resettling peasants in new villages. Consolidations that fell short of the formation of *khutora* were also possible. For example, consolidation could be confined to a single category of land in a village, such as the arable or hayland, and rights in communally used resources, such as pastures and wastes, could be left untouched by land settlement; different categories of land could be consolidated separately so that the resultant farms consisted of not one but several parcels; and farm buildings and dwellings could remain at their original site rather than being moved on to the consolidated land. *Otrub* was the name given to the type of farm where it proved impossible to unite the plot of land on which the peasant farmer's dwelling was located with his farmland. But even when farm buildings were moved *otruba* could still come into being, as, for example, in projects involving the resettlement of part of a commune's population at a new village site. The resultant villages were called *otrubnye poselki*. Figs. 7.1–7.3 show the differences between *khutora*, *otruba*, and *otrubnye poselki*. In the technical instructions sent to local land-settlement officials from St Petersburg there was a clear ordering of priorities for consolidation which reflected the official preference for *khutora*. Thus, any type of *khutor*, including those which consisted of more than one land parcel or whose owners continued to

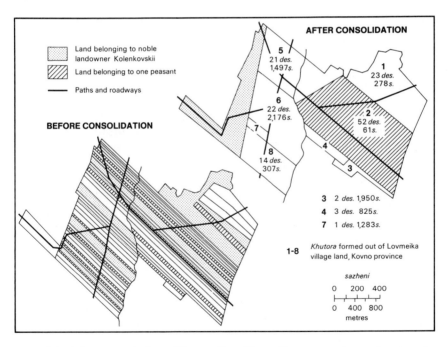

FIG. 7.1. The ground plan of Lovmeika village, Kovno province, showing the layout of *khutora*

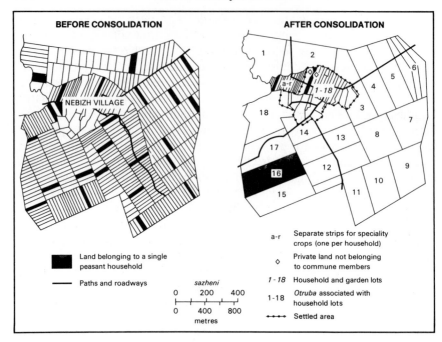

BEFORE CONSOLIDATION

NEBIZH VILLAGE

AFTER CONSOLIDATION

a-r    Separate strips for speciality crops (one per household)

◊    Private land not belonging to commune members

▇    Land belonging to a single peasant household

*1 - 18*    Household and garden lots

1-18    *Otruba* associated with household lots

——    Paths and roadways

◆◆◆◆    Settled area

*sazheni*

0   200   400

0   400   800

metres

FIG. 7.2. The ground plan of Nebizh village, Volyn province, showing the layout of *otruba*

use communally owned resources, was accorded a higher priority than any type of *otrub* or *otrubnoi poselok*.[11]

Procedures were developed to carry the Stolypin Reform to the local level by using traditional village decision-making institutions to debate and help implement land settlement. The decision to consolidate the land of all their members, for example, was entirely in the authority of communal assemblies, which had to pass a resolution stating the commune's desires. A two-thirds majority was needed for land consolidation to be effected in repartitional communes, and a simple majority in communes in which land was already held in hereditary tenure. For certain types of action the authority of assemblies was circumscribed, however. For example, they were powerless to prevent individual members from changing title to their land or, under certain circumstances, from consolidating it. In the first years of the reform a two-thirds majority vote at the village assembly was required for a whole community to change to hereditary tenure without consolidating, but from June 1910 communes which had not exercised their right to repartition during the previous twenty-five years were automatically transferred to hereditary tenure. Apart from the communes affected by

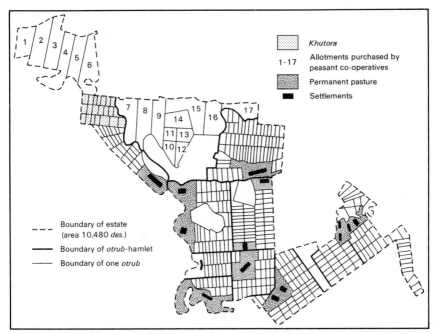

FIG 7.3. The ground plan of an estate divided for sale into *otrobny-poselki* by the Peasant Land Bank, Samara province

the 1910 law, all actions under the Stolypin legislation had to be initiated by the peasants themselves, whether they acted as individuals, whole communes, or groups of communes. Once in the hands of the local Committee, requests for land settlement would be processed, and successful applications would proceed to implementation through a number of planning stages. At any stage in this process the peasants who had put in the request for land settlement could withdraw from the project. This effectively enabled the peasants to veto projects they did not like.[12] This right, together with the range of options possible under the Reform, meant that the peasants had some opportunity to impose their own ideas on the Reform and to influence its course.

## THE STOLYPIN REFORM AND LANDSCAPE CHANGE

It is difficult to reconstruct a picture of the changes in the pattern of settlement and the disposition of land that took place in all provinces of Russia in the years after the promulgation of the Stolypin Reform. Although statistics about the course of the Reform were collected by the government after 1906, the bulk of them recorded various quantitative aspects of land settlement or measured the efficiency of the Reform's bureaucracy in processing the peasants' applications.[13] Fortunately, for a small number of places there are further records that allow a picture to be reconstructed of the sequence of land settlement and its impact on the physical structure of landholding. One such source was a survey conducted by the government in 1914 to check the accuracy of its statistics on land settlement.[14] Ten districts drawn from provinces in all the major economic regions of European Russia were surveyed, and for each the name of every village in which land-settlement work had been carried through to its completion, the types of projects undertaken, the year of their completion, and the numbers of households and area of land involved in each project were recorded. Two districts have been selected for presentation here, Ustyuzhna district in Novgorod province and Pokrov district in Vladimir province. They provide an interesting comparison because, although similar in their physical environments and settlement histories, they were affected rather differently by the Stolypin Land Reform. Ustyuzhna district lay in the broad band of provinces stretching from the Baltic in the north-west to a line linking Tver, Smolensk, and Mogilev to the east. This was a region in which, according to the government, most headway was made in introducing new types of farm, and especially *khutora*, to the peasant population. The only other region of Russia equalling the north-west for the numbers of new farms formed lay in the southern and eastern peripheries of the arable farming belt in the Volga valley and along the Black Sea littoral. Before the enactment of the Stolypin Reform, consolidated farms already existed in parts of the north-west of Russia, and in the Baltic provinces independent farming had always been the norm. These areas were now viewed by the Reform's enactors as an important innovation centre from which the *khutor* idea was expected to spread. By the time of the government's 1914 survey Ustyuzhna district was part of this centre, and an investigation carried out by the statistical department of the provincial *zemstvo* declared in 1913 that in Ustyuzhna and its two neighbouring districts 'there is every reason to conclude that the communal system of landholding has been completely obliterated . . .'.[15] Pokrov district, by contrast, was in a province noted for its relative failure to produce consolidated farms, and because of this it did not attract the attention of writers on the Reform and was not the subject of any special investigations until 1914. In this respect it was more typical of districts all over Russia than was Ustyuzhna. The Reform did not leave Pokrov untouched, however, as a number of its villages and individual households in them became involved in a variety of land-settlement projects.

According to official records there were 1,294 separate villages and hamlets in Ustyuzhna district in 1911. The majority of these (817) belonged to 290 rural societies (*selskie obshchestva*), and the others consisted of settlements such as small towns and the clusters of habitations around railway stations and wood-processing plants.[16] It was the villages and hamlets belonging to rural societies that were covered by the provisions of the Stolypin legislation. The majority of these were populated by former serfs who held their land in communal tenure. Fragmentation of peasant land was prevalent at both commune and household level throughout Novgorod province, where the number of strips held by individual households rose to between 80 and 100 in some communes.[17] Accordingly, it seemed as if landholding in the district was ripe for reform. Rapid change seems to have been a feature of the recent settlement history of Ustyuzhna district, however, the total number of settlements almost doubling during the period 1884–1910 from 736 to 1,296. Much of this growth had come about as a result of small groups of households hiving off from parent villages to form subsidiary settlements. The Stolypin Reform measures were thus added into a situation of dynamic settlement evolution.

The first land-settlement projects under the provisions of the Stolypin Land Reform were carried out in Ustyuzhna in four rural districts in 1909. In subsequent years work spread to other rural districts, and at the time of the government's 1914 survey there remained only two, out of a total of twenty-one, that had seen no land-settlement work of any kind. Statistics showing the results of the Reform are given in Table 7.1. The majority of projects involved the consolidation of the land of all the households in a village into separate parcels, and a minority involved individual consolidations and group land settlement. The same village, the survey revealed, was often the scene of more than one type of land-settlement work: group land settlement, in particular, preceded or accompanied whole village consolidations. Taking these cases into account, the final total for the number of villages that had been involved in some aspect of land settlement by the end of 1913 was 184. This represented 14.2 per cent of all the settlements in the district or 22.5 per cent of all the villages that, by virtue of being members of rural societies, came within the remit of the Reform. As a result of this activity 2,992 consolidated farms were formed in the district, but few of these were *khutora*. The figures show that 3 villages were broken up into *khutora*, 129 consolidated all their land into *otruba*, and 19 consolidated their arable into *otruba* while leaving some land in common use.[18] The impact of the Reform on the disposition of settlements in the district was therefore slight. Field patterns around villages did change, however, with the consolidation of strips in the arable and the dividing up of common pasture and woodland. The government survey did not record the number of separate parcels of land that households in these villages held after consolidation, but this was recorded for a smaller number of villages in the earlier *zemstvo* investigation. According to this, 38.7 per cent of all the new farms consisted of a single parcel of land, 32.7 per cent of two parcels, and the remainder of three or four.[19]

While compared with other districts in Russia the scale of land consolidation in Ustyuzhna was great, it is clear that in 1913 a large number of villages had yet to be touched by any change in landholding. The geographical distribution of the villages that were involved in some form of land-settlement, shown in Fig. 7.4, reveals that there were large areas where there had been no land-settlement activity. The villages that had consolidated the land of all or a number of their members were mainly concentrated in eight rural districts—Rastoropovo, Nikoforovo, Somino, Chernyanskoe, Verkhovsko-Volskaya, Vessko-Pyatnitskaya, Sobolevo, and Okhona—and within these they formed small clusters. In purely physical terms the impact of land settlement in the district was localized. The pattern of clustering suggests the operation of a weak diffusion process in the Reform's adoption, but it may in fact have been more closely related to the administrative organization of peasant villages in the district. Taking just the three rural districts in which the largest numbers of consolidated farms were formed, Rastoropovo, Nikiforovo, and Somino, Table 7.2 shows that the majority of villages which underwent consolidation were grouped in a relatively small number of rural societies. In Rastoropovo rural district, for example, sixteen villages out of twenty-six that consolidated their land belonged to just five rural societies out of a total of twenty; in Nikiforovo thirteen of the villages that consolidated belonged to six out of twenty-six rural societies; and in Somino twelve villages that consolidated belonged to six out of fifteen rural societies. It seems probable that a knock-on effect operated among villages in the same rural society, although it is difficult to determine precisely how it operated. It will be recalled that in 1861 rural societies became the lowest rung of the state administrative hierarchy and that there was little uniformity in their make-up and range of functions. O. A. Khauke, author of legal works on landholding at the beginning of this century, identified no fewer than nine different ways that rural societies could be constituted, from those that consisted of a single village that in turn corresponded to a single land commune (i.e. of a simple, or *prostaya*, commune), to those consisting of numerous villages which were grouped into a number of land communes sharing one or more resources.[20]

In the rural societies of Ustyuzhna district it was normal for the consolidation of member villages to take place sequentially rather than

TABLE 7.1. *Land-settlement projects completed 1909–1913 by volost in Ustyuzhna district*[a]

| Rural district (volost) | 1909 | | 1910 | | 1911 | | 1912 | | 1913 | | Total | | |
|---|---|---|---|---|---|---|---|---|---|---|---|---|---|
| | Whole | Other | Whole | Other | Whole | Other | Whole | Other | Whole | Other | Whole village consolidation | Individual consolidation | Group land settlement |
| Barsanikha | | | | | 1 | | | 2 | 1 | 2 | 2 | 5 | |
| Belye Kresty | | | | | | | 1 | | 1 | | 2 | 1 | |
| Cherenskoe | | | 1 | | | | | | | | 1 | | |
| Chernyanskoe | | | 6 | 3 | 1 | | 3 | 2 | | 1 | 11 | 6 | |
| Chirets | | | | | | | | | | | | | |
| Dubrovskoe | | | 5 | 1 | | | | | 2 | | 8 | 2 | |
| Khripelevo | | | | | | | 1 | | 1 | 1 | 2 | | |
| Kirvo-Klimovskaya | | | | | | | | | | | | | |
| Lentevo | | | | | | | | | | | | | 2[c] |
| Maloe Vosnoe | | | | 2 | 1 | 1 | | 1 | 2 | | 3 | 4 | |
| Megrino | | | | | | | 4 | 1 | 3 | | 7 | | |
| Mezga | | | 2 | | 4 | | 1 | | 1 | | 8 | | |
| Modensko-Plotichevskaya | | | 1 | | | | | | | | 1 | | |
| Nikiforovo | 1 | | 6 | 3 | 7 | | 5 | 1 | 7 | | 26 | 4 | |
| Okhona | | | 5 | 1 | 6 | 1 | 1 | | 1 | 1 | 13 | 3 | |
| Perya | | | | | | | | | | | | | |
| Rastoropovo | | 1 | 3 | 1 | 3 | 2 | 3 | 2 | 20 | 3 | 29 | 9 | |
| Sobolevo | | | 1 | 5 | 1 | 1 | 2 | 2 | 2 | | 6 | 6 | 4[b] |
| Somino | | | | | 9 | | 8 | | 3 | | 20 | | |
| Verkhovsko-Volskaya | | | 5 | 2 | 7 | | 3 | | 1 | | 16 | 1 | |
| Vessko-Pyatnitskaya | | | | | 1 | 1 | | | 4 | | 6 | | |
| TOTAL | 1 | 1 | 35 | 19 | 41 | 6 | 32 | 12 | 49 | 8 | 169 | 40 | 9 |

[a] The figures refer to projects completed. Villages consisting of more than one commune that were consolidated at different times appear more than once.
[b] This figure includes two group land-settlement projects for which the year is unknown.
[c] This figure is for two group land-settlement projects for which the year is unknown.

*Source:* Calculated from data in TsGIA, f. 408, op. 1, no. 894.

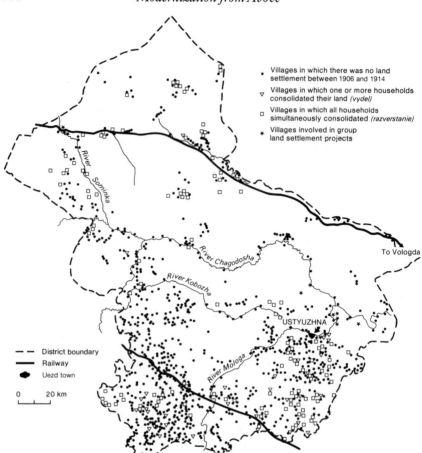

Fig. 7.4. The geographical pattern of adoption of the Stolypin Reform in Ustyuzhna district, Novgorod province, 1907–1914

simultaneously. Typically, one commune would consolidate the land of its members, to be followed by other communes in subsequent years. The sequence of whole village consolidations was often punctuated by individual separations by single households or small numbers of households from remaining unconsolidated villages, and also by group land settlement. Land surveyors must have had to make repeated visits to the same locale over a number of years. The consolidation of Vypolzov rural society in Nikiforovo rural district is illustrative. The rural society consisted of two villages, Vypolzov and Konyukhovo, the first of which was composed of two land communes and the second of which was coextensive with a single commune. The history of consolidation was as follows: in 1909 the first Vypolzov commune, numbering twenty-two households, consolidated its land into *khutora*, in 1910 one household from the second Vypolzov commune withdrew individually and consolidated its land, in 1911 the remaining eight households in Vypolzov consolidated, and in 1913 the peasants in Konuyukhovo consolidated. Each project required its own paperwork, plans to be drawn up, and visits made to the locale. Precisely because of the duplication in effort that could result from sequential consolidation, local Land Settlement Committees were encouraged by St Petersburg to concentrate on projects involving large numbers of peasants, and by 1909 they were being actively discouraged from spending time on individual separations and consolidations.

Just as the decision of one commune in a rural society to consolidate the land of its members may have triggered similar decisions in connected communes, the line dividing the decisions of individuals and of whole communities to consolidate also seems to have been a fine one. In the *zemstvo* investigation of land settlement in Novgorod province it was admitted that it was rare for a commune as a whole to initiate a request to the Land Settlement Committee for the formation of *otruba* and *khutora*. It was usually reported that the consolidation of whole villages began with one or two households petitioning the Land Settlement Committee for their land to be consolidated, but, because of the fear that the best land would be allocated to these separators, or because of a wish to avoid repeated repartitions, such requests could precipitate the joint decision to consolidate. Following a description of fifteen cases of whole village consolidation in Novgorod province that were hotly disputed at the communal assemblies, the final agreement to form *otruba* was explained in a local authority report thus:

... the majority of households were afraid that they would lose the best land in the village to the individual separators ... the Land Settlement Committee played a not insignificant role in resolving the situation by putting to the quarrelling parties the following proposition: 'either everyone agrees to consolidate their land or the best land will be given to the households which first asked for *otruba*'.[21]

An early report by the government minister Lykoshin, quoting from his observations in Kharkov and Voronezh provinces, noted the tendency for whole village consolidations to grow out of applications on the part of a few households for unilateral separation,[22] while a report from Yaroslavl attributed the slow formation of *otruba* in that province to the conversion of projects initially concerned with individual separations into whole village consolidations.[23] There were too few surveys of peasants involved in the Stolypin Land Reform for a reliable analysis to

TABLE 7.2. *Pattern of whole village consolidations in three rural districts in Ustyuzhna district, 1910–1913*

**Rastoropovo rural district**

| Rural societies | 1 | 2 | 3 | 4 | 5 | 6 |
|---|---|---|---|---|---|---|
| No. of villages and subsidiary settlements | 5 | 4 | 1 | 1 | 4 | 3 |
| No. of consolidations | 5 | 4 | 1[a] | 1 | 3 | 2(1) |
| Date(s) of consolidation(s) | 1910–13 | 1911–13 | 1913 | 1911 | 1910–13 | 1911, 19' |

**Nikiforovo rural district**

| Rural societies | 1 | 2 | 3 | 4 | 5 |
|---|---|---|---|---|---|
| No. of villages and subsidiary settlements | 3 | 2 | 2 | 2 | 2 |
| No. of consolidations | 3[a] | 2[b] | 2[a] | 2 | 2 |
| Date(s) of consolidation(s) | 1910–13 | 1909, 1913 | 1910, 1913 | 1910, 1913 | 1911, |

**Somino rural district**

| Rural societies | 1 | 2 | 3 | 4 | 5 |
|---|---|---|---|---|---|
| No. of villages and subsidiary settlements | 3 | 5 | 5 | 4 | 3 |
| No. of consolidations | 2 | 2 | 2[b] | 2 | 2 |
| Date(s) of consolidation(s) | 1911 | 1911, 1912 | 1911, 1912 | 1911, 1912 | 1912, 19' |

[a] Whole village consolidations preceded by withdrawal and consolidation by individual member(s).
[b] Consolidation involving two or more separate communes in village.

*Note*: Figures given in brackets are projects which had been approved but not yet carried out.

*Source*: Calculated from data in TsGIA, f. 408, op. 1, no. 894.

| 7 | 8 | 9 | 10 | 11 | 12 | 13 | 14 | 15 | 16–20 | Total |
|---|---|---|---|---|---|---|---|---|---|---|
| 4 | 2 | 2 | 2 | 3 | 4 | 4 | 4 | 6 | 11 | 60 |
| 2(1) | 1 | 1[a] | 1 | 1 | 1[a] | 1 | 1 | 1 | 0 | 26(2) |
| 1913 | 1912 | 1913 | 1910 | 1913 | 1911 | 1913 | 1913 | 1913 | | |

| 6 | 7 | 8 | 9 | 10 | 11 | 12–26 | Total |
|---|---|---|---|---|---|---|---|
| 3 | 2 | 2 | 2 | 3 | 3 | 31 | 57 |
| 2 | 1[b] | 1[b] | 1[b] | 1 | 1 | 0 | 18 |
| 1911 | 1910–13 | 1910–13 | 1911 | 1912 | 1912 | | |

| 7 | 8 | 9 | 10–15 | Total |
|---|---|---|---|---|
| 2 | 1 | 1 | 11 | 38 |
| 1[b] | 1 | 1 | 0 | 15 |
| 11, 1912 | 1911, 1912 | 1911 | 1911 | |

be made of how the decisions to consolidate came about, but confirmation that concern over the 'creeping depletion' of village land could be a factor comes from an investigation of Bogoroditsk district in the black-earth region of Tula province.[24] In the Bogoroditsk investigation many peasant households questioned about their decision to consolidate admitted that they had been influenced by a fear of losing land to their neighbours if they did not join in land settlement from the beginning.[25] An additional factor that may have operated derived from the provision in the law that individual separators could continue to run their livestock with the commune's common herd. Households staying in the commune faced the prospect of their fallow and stubble being used by separators to graze their livestock with the possibility, as the area of communal arable declined, of trampling and over-grazing. Households which had consolidated, meanwhile, did not have to open their fields to the common herd.

It is not possible to say how many whole village consolidations in Ustyuzhna district started with requests by individual members for separation from the commune. But, even if the direct evidence is fragmentary, there do seem to be grounds for believing that existing patterns of landholding meant that the decision on the part of a few peasants to consolidate their land could set off a chain reaction that extended far beyond their immediate sphere, and possibly to neighbouring villages. In such an account of the spread of the Reform, the decision of whole villages to consolidate does not require all, or even a majority, of the member households to be in favour of farming on *otruba* or *khutora*. Rather, it suggests that households could be pulled into land settlement as a consequence of others' consolidation. If this is an explanation for the consolidation of peasant land in some parts of Russia after 1906, it challenges the view put forward by the government that the increasing preponderance of whole village consolidations over individual separations as the Reform developed signalled a growing acceptance by the peasantry of the idea of consolidation. The original analysis by the government, suggesting that only a small minority of peasants would recognize the advantages of individualized farms, may in fact have remained the correct one; the missing element in the analysis was an appreciation of the repercussions any withdrawals could have on landholding arrangements in the countryside, given the complicated web of administrative and economic ties that bound peasant households together.

The course of the Stolypin Reform in Pokrov district, Vladimir province, presents a different picture from the one just described for Ustyuzhna district. While in Ustyuzhna district the focus of land-settlement work was on the formation of consolidated farms, in Pokrov district much effort was directed towards group land settlement, i.e. the reorganization of landholding at the macro or inter-commune level. As Table 7.3 shows, during the seven-year period from 1907 to 1913 covered by the government survey, group land-settlement work of some kind was carried out in 89 of the district's 423 villages, and it affected 3,561 peasant households. All but half a dozen of the twenty rural districts in Pokrov were the scene of group land settlement. By contrast only nineteen villages, all concentrated in four rural districts, consolidated the land of their members. As a result of these whole village consolidations 978 *khutora* and *otruba* were formed, to which must be added a further 161 new farms which came into being as a result of individual separations and which were scattered throughout the district. Apart from the few areas where relatively large numbers of these consolidated farms were to be found, there must have been few obvious signs in the landscape of Pokrov district of the changes brought about by the adoption of the Stolypin Reform, and, for the majority of peasants involved, there would have been few changes in their existing landholding practices: land settlement here mainly resulted in the rearrangement of peasant land within the existing framework of the commune. Given the government's priorities, it is not surprising that the district did not attract the attention of contemporary commentators, but, even if unacknowledged at the time, the changes brought about by group land settlement represented an important phase in the history of landholding in peasant communities, helping them to resolve longstanding problems associated with the use and disposition of their land.

The pattern of the land-ownership in Pokrov district made it a particularly likely candidate for group land settlement. As in many of the long-settled regions of central Russia, land belonging to different owners in the district was often held in several parcels and was covered by rights of co-usership with other landowners. The result was a mosaic of fields, pastures, and forests whose physical complexity was matched by an equally complex set of legal ties. Noble landowners in the district were numerous, and the majority of the peasant population, 62.7 per cent, was made up of their former serfs. In the rural district of Zavalino, for example, the former serf population of 633 households had belonged to twelve different landowners, in Zhary there had been seven noble serf-owners, in Kopnino again seven, and so on. In addition to these former serfs, the district had a sizeable population of former state peasants, 33.4 per cent in total, and small numbers of crown peasants and various types of independent peasant. These were scattered among all the rural districts, although in three—Kudykino, Lukyantsevo, and Filipovskoe—state peasants were dominant. As a general rule peasant households formerly belonging to different landowners constituted

*Modernization from Above*

TABLE 7.3. *Land settlement in Pokrov district, 1907–1913*

| Rural district | No. of villages in rural district | No. of communes in rural district | No. of communes involved in land settlement | Type of land settlement | | |
| --- | --- | --- | --- | --- | --- | --- |
| | | | | Whole village consolidation | Individual separation | Group land settlement |
| Anino | 15 | 16 | 11 | 5 | 4 | 9 |
| Argunovskaya | 9 | 11 | 11 | 4 | 5 | 6 |
| Dubki | 27 | 30 | 9 | — | — | 9 |
| Filipovskoe | 20 | 21 | — | — | — | — |
| Fineevo | 17 | 24 | 4 | — | 2 | 2 |
| Funikovo–Gora | 25 | 29 | 8 | — | 8 | — |
| Kopnino | 19 | 32 | 5 | — | 3 | 2 |
| Korobovshchino | 21 | 23 | 12 | 1 | 8 | 6 |
| Korovaevo | 20 | 27 | 8 | 1 | 3 | 5 |
| Kudykino | 16 | 18 | 2 | — | 2 | — |
| Lipnya | 15 | 37 | 2 | — | 2 | — |
| Lukyantsevo | 21 | 27 | 7 | — | 4 | 3 |
| Ovchinino | 22 | 26 | 11 | — | 4 | 8 |
| Pokrov–Sloboda | 40 | 64 | 23 | 5 | 11 | 8 |
| Selishchi | 13 | 22 | 3 | — | — | 3 |
| Voronovo | 32 | 32 | 7 | — | 7 | — |
| Yakovlevo | 8 | 15 | 2 | — | 2 | — |
| Zavalino | 31 | 35 | 9 | — | 1 | 8 |
| Zhary | 24 | 27 | 14 | 3 | 7 | 6 |
| Zherdevo | 28 | 28 | 16 | — | 4 | 14 |
| TOTAL | 423 | 544 | 164 | 19 | 77 | 89 |

*Source:* Calculated from data in TsGIA, f. 408, op. 1, no. 894.

discrete land communes, even if they occupied the same village. This explains why in Pokrov district there were 544 land communes but only 423 peasant villages. One example of a village with such multiple communes was Zhorovo, Kopnino rural district, which was made up of no fewer than five separate land communes. Where households formerly belonging to a single owner were numerous, or where they were physically separated from one another—living, for example, in different rural districts—they would normally be grouped into a number of separate land communes. In contrast, it was rare for peasants formerly belonging to different landowners to be members of the same land commune.

Complications in the use of peasant land in Pokrov district arose from the fact that, although members of separate land communes, peasants often had to use certain resources in common. As Khauke observed, such patterns of joint use were often the product of the settlements made at the time of the Emancipation.[26] At that time deeds to the land were handed either to villages or to whole estates of peasants. Much the same happened in the case of the state peasants, who, under the law of 31 March 1867, were allocated their land on the basis of physical proximity in so-called 'common settlement areas' (*obshchie dachi razverstaniya*). In the literature of the day, groups of villages sharing the ownership of land were given the name 'single-plan settlements' (*odnoplannye seleniya*). One outcome of the Emancipation settlement was predictable: communes which had previously formed discrete land-management units now had to manage land jointly with other communes. To complicate matters further, rural societies did not necessarily correspond to the new amalgams of land communes. Group land settlement offered peasant communes one way of sorting out anomalies created by the land allocations made in 1861, and longer-term disputes about land as well. In Pokrov district the first project was carried out soon after the Reform organization had been set up in 1907, two seasons before the first consolidations in Ustyuzhna district. It consisted of dividing jointly owned land between four villages in Zavalino rural district and, through exchanges with other landowners, consolidating into a reduced number of parcels the land of one of these. In subsequent years similar projects were carried out elsewhere in the district. By the end of 1913 thirteen projects had been completed which, like the first one in Zavalino, split up single planned settlements. The most complicated, involving the largest number of villages, were in Zherdevo rural district, where in 1911 and 1913 two single planned settlements, consisting of six separate villages apiece, were split up. This involved reallocating arable in discrete blocks to each village and subdividing land in common use. The other group land-settlement projects carried out in the district either effected exchanges of land between communes or withdrew from common use portions of hayland or forest for use by individual communes. These types of project were carried out in thirty-eight villages in the district.

In one of his later works on land settlement A. A. Kofod wrote of group land settlement that,

this work constitutes for the most part a preparation for the future consolidation of commune land into individual consolidated farms, although it is true that group land settlement can in its own right make a substantial improvement for a significant number of households. Work such as the breaking up of single-plan settlements, the formation of subsidiary hamlets, and the division of villages is absolutely necessary in order to satisfy the different interests of the population.[27]

The creation of consolidated farms, he continued, was, in contrast to group work, 'fully realized land settlement'. Kofod was not alone among central officials in viewing group work as subsidiary to the creation of *khutora* and *otruba*. The first consolidations of peasants' land into independent farms in the Pokrov district took place in 1909, when two villages, one of ninety households and the other of thirty-two, were consolidated, and there were eight individual separations from other villages. The number of consolidations rose through 1910 and 1911 but fell off thereafter, as Table 7.4 shows. There seems to have been an association of individual consolidations with group land-settlement projects: thirty-four of the households that withdrew unilaterally from their communes to consolidate did so either at the same time as, or soon after, group work in their commune, and the same applied to nine of the nineteen instances of whole village consolidations in the district. The greatest success in converting group work into consolidations was in

TABLE 7.4. *Number of households undertaking communal and individual consolidation in Pokrov district, 1907–1913*

|      | Whole village consolidation | Individual separation | Total |
|------|------|------|------|
| 1907 | —   | —  | —   |
| 1908 | —   | —  | —   |
| 1909 | 126 | 8  | 134 |
| 1910 | 214 | 58 | 272 |
| 1911 | 365 | 27 | 392 |
| 1912 | 145 | 22 | 167 |
| 1913 | 128 | 46 | 174 |
| TOTAL | 978 | 161 | 1,139 |

*Source*: Calculated from data in TsGIA, f. 408, op. 1, no. 894.

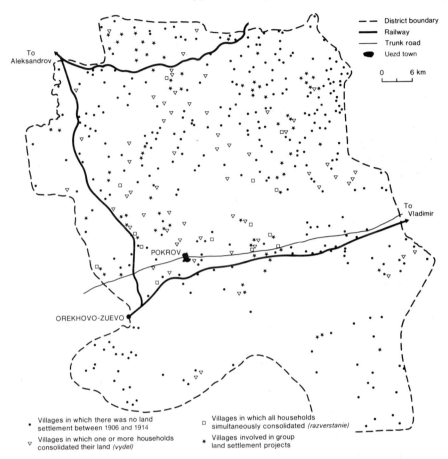

FIG. 7.5. The geographical pattern of adoption of the Stolypin Reform in Pokrov district, Vladimir province, 1907–1914

Anino rural district, where projects involving the division of forests and exchanges of land between eight villages in 1910 led on to the consolidation by 1913 of three whole villages. Elsewhere, if associated at all with the creation of consolidated farms, group land-settlement projects were more likely to spawn individual withdrawals from the commune than whole village consolidations.

A notable feature of land settlement in Pokrov district was that individual separation from the commune did not decline in importance as a means of consolidating land as the Reform developed. In 1913 over one-quarter of all new farms originated in this way. The individual separations could involve small groups of households, up to half a dozen withdrawing at a time, and there were cases of repeated withdrawals from the same village, e.g. in Zhorovo village, Kopnino rural district, where eight households out of a total of thirty withdrew and consolidated their land in three separate episodes. For the land-settlement officials cases like this resulted in the type of duplication of effort that was frowned upon by the Reform organization after 1909. Funikovo-Gora rural district must have given the local officials the greatest amount of work for the least reward: over the period 1910 to 1913, nineteen *otruba* and *khutora* were created in eleven separate projects. The continuing formation of consolidated farms through individual separations, together with the execution of projects exclusively concerned with group land settlement, meant that in Pokrov district the Stolypin Reform was following a different path from the one marked out in St Petersburg. There were also deviations in respect of the nature of the farms that were created. As elsewhere, *otruba* dominated over *khutora* in the district but, more exceptionally, over one-third (36.2 per cent) of the land belonging to the communes that opted for whole village consolidations was left in communal ownership and undivided. The extreme case was Petyshiki village, Anino rural district, which left 54.7 per cent of its land in communal ownership, some 939 *desyatiny*, and in three other villages the figure exceeded 42 per cent. In only two villages was all the available land distributed among the new farms. The nature of land left unconsolidated is unknown, but, since the district was forested it seems likely that a large portion of it was woodland.

## THE COMMUNE TRANSFORMED

It is clear that the course of the Stolypin Land Reform was different in Ustyuzhna and Pokrov districts. Such differences were reproduced all over European Russia, with the result that the regional pattern of the Reform's adoption was complex and is difficult to analyse. Given the wide scope of changes made possible by the Reform legislation, this was inevitable. One generalization that can be made, however, is that nowhere did peasants appear to share the government's enthusiasm for *khutora*. As the Reform developed it became clear that peasants preferred changes that fell short of the physical dispersal of their villages and retained at least some elements of their traditional landholding regimes. Thus, they opted for group land settlement or for *otruba*, they kept land in common ownership even after village-wide consolidations, or they took out title to their land without consolidating it, and if they moved from their former village, they did so in groups to form new communities at a new site. Table 7.5 summarizes data about the nature

TABLE 7.5. *Consolidated farms on former communally owned land, 1912*

| District (Province) | % of households with allotment land in three or more parcels after consolidation | Nature of tenure of consolidated farmers' land (%) | | | % of households retaining rights in common land |
|---|---|---|---|---|---|
| | | Individual ownership | Communal ownership | Other (unconsolidated) | |
| BERDYANSK (Tavrida) | 28.2 | 82.3 | 9.4 | 8.4 | 61.6 |
| BOGODUKHOV (Kharkov) | 41.9 | 75.8 | 11.3 | 13.0 | 81.8 |
| KRASNOUFIMSK (Perm) | 21.9 | 71.7 | 16.8 | 11.5 | 88.8 |
| KREMENCHUG (Poltava) | 23.4 | 55.7 | 0.1 | 44.2 | 0.6 |
| MOLOGA (Yaroslavl) | 36.4 | 68.6 | 18.9 | 12.5 | 40.6 |
| NIKOLAEV (Samara) | 42.0 | 74.4 | 18.0 | 7.7 | 75.9 |
| OREL (Orel) | 10.2 | 77.3 | 9.3 | 13.4 | 49.3 |
| OSTROV (Pskov) | 2.9 | 68.9 | 2.2 | 28.8 | 12.6 |
| RZHEV (Tver) | 16.1 | 63.1 | 3.9 | 33.0 | 25.5 |
| SYCHEVKA (Smolensk) | 27.5 | 78.1 | 3.0 | 18.9 | 31.1 |
| TROITSKOE (Vilno) | 6.6 | 92.7 | 2.2 | 5.1 | 14.1 |
| OVERALL | 24.6 | 76.9 | 9.8 | 13.3 | 45.4 |

*Source*: Zemleustroenie khozyaistva. Svodnye dannye sploshnogo po 12 uezdam podvornogo obsledovaniya khozyaistvennogo izmeneniya v perve gody posle zemleustroistva (St. Petersburg, 1915).

of consolidated farms collected in another government survey of districts in a range of provinces in European Russia. In terms of tangible changes to the rural landscape the impact of the measures was variable. In some places fenced enclosures did replace open fields, but elsewhere the only visible sign of a visitation by the Land Settlement Committee's surveyors was the new markers delineating the boundaries of a village. The ultimate test of the Reform, however, lay in the economic and political spheres; in its ability to promote the formation of the independent peasant producers who were believed to be necessary to secure agrarian modernization and political calm in the countryside. As, by the time of the outbreak of the First World War, the majority of peasants who had adopted the Reform had experienced barely half a decade of farming under the new conditions, it is impossible to say whether in the long term the government would have passed or failed the test it had set itself. But one aspect of the Reform's results about which comment is possible is the extent to which the measures adopted in the years between 1906 and 1917 succeeded in undermining traditional landholding practices in the countryside.

Two generations of commentators on the Stolypin Reform have taken the statistics relating to the numbers of tenure changes and consolidations as reliable indicators of the extent to which the peasant commune was eroded during the eleven-year period of the Reform's operation. But in recent years these have come under the critical scrutiny of Western and Soviet historians, and the more extravagant claims about the commune's destruction have been undermined. The criticism has focused mainly upon computational errors, but questions have also begun to be asked about the significance of the changes that the figures record. The main computational problem, to which the Soviet historian V. P. Danilov in particular draws attention, is that a government survey of land-ownership in Russia for 1905 is commonly used as a base against which the numbers of households adopting the Stolypin Reform are compared.[28] Since, in the years after 1905, the total number of household subdivisions would have increased, estimates for the percentage of households that either changed the tenure of their land or consolidated it must have been exaggerated. This exaggeration was likely to have been further compounded by land sales and the possibility that some of the recorded withdrawals from the commune were family partitions. Danilov concludes that the number of peasant farms formally held in hereditary tenure on the eve of the 1917 revolution was somewhere between 27 and 33 per cent of the total, including those already in existence before 1906.[29] Communal tenure still dominated the Russian countryside in 1917, despite two decades of the Stolypin Reform's operation.

The extent to which adoption of the Reform resulted in a real alteration in peasant landholding practices is the other aspect of the Reform's results that must be considered. Among the various measures that could be adopted under the legislation, group land settlement constituted the least threat to current practices and could in fact lead to a strengthening of existing communities. The main point of vulnerability of such communities was that their members would be exposed to pressure from officials to take land settlement further. By contrast the transfer of title from communal to hereditary ownership which was possible under the legislation did result in an immediate change in the principles by which peasant households held their land. For individual households taking out title, the change meant that their strips in the arable were exempt from any future repartitions the commune might undertake, although they were not exempt from a future decision on the part of their co-villagers to consolidate all the commune's land into *khutora* or *otruba*. As to other aspects of commune life, the position of such individual secessionists did not change very much, as Khauke explains:

the households . . . taking out title to their land [in the commune] did not leave their rural society as an administrative entity. They did not even truly leave their land commune, remaining tied to their neighbours by a whole host of conditions—first, by the fact that their land remained in strips, which necessitated the continuation of common farming practices with their neighbours; secondly, by virtue of their share in the jointly owned 'supplementary' resources of the commune; and thirdly, by the right the society retained to reabsorb land taken into title on the death of the owner, if he had no heirs.[30]

In some cases, Khauke noted, contrary to the provisions of the law, households which had withdrawn from communes continued to expect, and were sometimes allotted, land for their new offspring.[31] From the commune's point of view the emergence of hereditary strips in the arable could be a nuisance, but this depended upon whether it repartitioned frequently. Where whole communes changed title to their land, the break with repartitioning was complete, although Khauke's observations about the continued existence of elements of communality deriving from land fragmentation and the joint use of resources, such as pastures, was as true for whole communities that took out title as for individuals. The effect of such community changes of title on the communities themselves was to render them similar to the peasant communities with hereditary ownership which existed in the southern provinces of Russia and which at the time of the 1905 land-ownership survey accounted for some 2.8 million of Russia's 12.3 million peasant households. Whether such communities, holding their land in hereditary

tenure but retaining common pastures, strips, open fields, and con-
vening assemblies to transact community business, constituted a
species apart from the repartitional commune is open to debate. For
peasants in communes which had already ceased to repartition, the
change in the legal status of their land must have had little perceptible
impact upon current practices. This must have been especially true in
communes which, by virtue of the fact that they had not had a repartition
for twenty-five years, were automatically transferred to hereditary tenure
under the law of 14 June 1910. Some authors have included the
estimated 3.5 million households covered by this provision in their
analyses of the Stolypin Reform's performance despite the fact that the
majority were passive participants in change.[32]

If changing title to the land without altering its disposition moved
peasant households one step away from the repartitional commune, the
consolidation of land into *khutora* and *otruba* moved them one step
further. The principal importance of consolidation was that the
untangling of the strips of different households removed one condition
for communal action on the part of communities of peasants. Since their
land was no longer intermixed, peasant households no longer needed to
meet to decide upon cropping cycles or to devise other management
rules. Nevertheless, the transition to fully fledged private farming
remained incomplete in the majority of consolidations, whether these
took place on an individual or village-wide basis. The reason for this was
the failure of many land-settlement projects to achieve the consolidation
of all available land. The continued existence of unconsolidated land
meant in practical terms that peasant households continued to make
joint decisions about its management, and if the land were pasture, for
example, it usually meant the continuation of communal use. In legal
terms the position regarding such land and rights in it was ambiguous.[33]
Further problems were created for the legal definition of the new farms
when some of their land was left unconsolidated, as it was not clear what
proportion of all the land in the use of a peasant household had to have
been consolidated for that farm to constitute an *otrub*, or how many
separate land parcels were to be permitted before an *otrub* ceased to be
just that. As Khauke commented, 'whichever way the problem is
approached, if the supplementary land resources are left unconsolidated, then the surrounding *otruba* are not fully individualized and they
retain the same elements of communality which are characteristic of
such categories of land in the commune'.[34] As the authors of the Land
Reform were aware, links with the former communal order were finally
severed only when all the land of a commune, including its household
lots, was divided up and incorporated into single farms. For this reason
land settlement was said to be complete only when *khutora* were formed,

and it was these farms alone, out of all the types it was possible to form under the provisions of the legislation, that were protected from future land-settlement changes. In communes that had recently transferred the title of their land or in existing hereditary communes, individual households were secure in the ownership of their strips only so long as a majority of their co-villagers did not decide to approach the Land Settlement Committee with a request to consolidate. *Otrub* owners were more secure in that a redistribution of their land was forbidden unless there was unanimous agreement to disperse on to *khutora* or, in cases where more than one-third of all land remained unconsolidated, unless it was decided by a majority vote to rework the original land-settlement project. Since inviolability of ownership was conferred on only a small number of the peasant households that adopted the Stolypin Reform, the majority of new farms formed under its provisions remained legally distinct from other private farms in Russia. One important new right peasants acquired was that land transferred into hereditary ownership could be sold, and the years after the enactment of the 1906 edict saw a rising number of peasant land sales. Such transactions could only take place between members of the peasant estate; peasant land could not be sold to members of other social classes.[35] Hereditary peasant land, whether in the form of strips, *otruba*, or *khutora*, thus remained separate from other types of private land in Russia.

The discussion above has suggested that for the majority of peasants the adoption of the Stolypin Land Reform represented less of a break with the past than might have been expected given the government's priorities. In an article in *Sovremennik* Viktor Chernov, leader of the Social Revolutionary Party, summarized the situation thus:

Everywhere the picture is the same. A full break with the *mir* is extremely rare. The overwhelming majority [of households adopting the Reform] retain the household lot their *obshchina* gave them, they run their livestock in a common herd, every year they distribute common meadows along with the whole *mir*, and they receive their due share in common woodlands and so on.[36]

Even allowing for an overstatement of the case, Chernov's 'picture' raises the question that lies at the heart of any assessment of the Stolypin Reform's success in eliminating the commune from rural Russia, namely how the commune is first defined. It seems clear from the information available about the nature of the changes effected under the provisions of the legislation that the only way in which statements about the commune's disappearance after 1906 can make sense is if that institution is defined solely in terms of its legal right to repartition land. As mentioned above, Danilov's calculations suggest that, counting all types of households legally holding their land in hereditary tenure, some

27 to 33 per cent of the total number of peasant households in Russia were no longer in communes by the time of the revolution, and the figure rises to 50 per cent or more if we include communes covered by the 1910 act that failed formally to register the change in tenure.[37] However, there was much more to the commune than repartitioning. The existence of elective institutions of self-government, the joint use of resources, and social obligations and rights joined peasants together in a network of economic and social relationships regardless of formal land-tenure arrangements.

To argue that the peasant community continued to exist even for many of those peasants who adopted the more radical measures of the Stolypin Land Reform is not to deny that the period after 1906 saw important changes in peasant landholding in Russia. Village communities touched by the Reform could be strengthened as a result of improvements in access to their land, or they could be weakened, as must have been the case in almost all village-wide consolidations into *khutora*, but all were in some way transformed. As things stood at the time of the 1917 revolution, it was unclear how rapid the pace of change would be in the future or, indeed, the direction of that change. If the government's teleology is accepted, the changes imposed on the peasant commune during the course of the Reform were bound in time to lead to its elimination, but as events were to show, the commune's future was influenced by revolution in the wider society of which it was a part. Bolshevik policy at first constituted a threat to the commune, but from the beginning of the 1920s the New Economic Policy launched by Lenin witnessed a flowering of the peasant community in Russia.[38] It was finally abolished by force during the collectivization drive in the 1930s, although some elements were to survive even this traumatic period.[39] If one conclusion about the commune stands out from the experience of the Stolypin years it is its relative resilience to attack and its flexibility in the face of change.

# 8
# Regional Specialization and Trade in the Late Eighteenth and Early Nineteenth Centuries

SOVIET studies of the early modern period in Russian history have been principally concerned to trace the origins and development of capitalism in the economy. In this respect Lenin's much-quoted remark of 1894 has exercised a major influence. 'Only the recent period in Russian history (approximately since the seventeenth century)', he wrote, 'is marked by a real merger of all provinces, lands, and principalities into one entity.' This merger, he continued, was evoked by 'the ever-growing exchange among the provinces, the gradually rising turnover of goods, and the concentration of small local markets into a single all-Russian market. Since the leaders and masters of this process were capitalist merchants, the creation of these national ties was nothing more or less than the creation of bourgeois ties.'[1]

Lenin's remarks were made in the context of a heated political argument with the populists, and yet, in spite of their controversial character, they have had to be treated with considerable respect by Soviet historians. However, it would be a mistake to infer from this that Soviet historians agree about the significance of Lenin's statement. Quite the opposite is in fact the case. Neither the nature of the 'all-Russian market' nor the relationship between that market and capitalism itself has been understood in any agreed fashion. As a consequence, of course, the 'seventeenth century' part of Lenin's statement has also been interpreted in different ways. Some historians have seen an 'all-Russian market' appearing as early as the sixteenth century, others as late as the 1880s. Likewise, the origins of capitalist development in Russia have been detected in the sixteenth and seventeenth centuries, or alternatively very much later.[2]

Western historians, on the other hand, have been much less interested in the notion of the 'all-Russian market' than have their Soviet counterparts. In general they have agreed about Russia's relative backwardness in the period before the 1917 revolution, and there has been considerable discussion about the nature of Russian society at that time.[3] Some historians, however, have been impressed by the evidence for 'capitalistic' or 'modernizing' tendencies in Russia at a fairly early stage in her development.[4]

Lenin's observation concerning the 'all-Russian market' embraces the eminently geographical notion of growing regional specialization and inter-regional trade accompanying national economic development. Indeed, a good deal of Soviet scholarship concerning the concept has been devoted to empirical investigations of these very issues. Not all Soviet writers have agreed with this emphasis,[5] but it does at least accord with recent Western interest in spatial developments in the early modern period.[6] However, such work has yet to acquire a solid theoretical foundation.[7]

A serious failing of much of the scholarship concerned with the development of regional specialization and inter-regional trade has been the lack of attempts to investigate regional economies on a comparative basis.[8] Obviously it would be unrealistic to do this for Russia as a whole, but an analysis of economic activity in several provinces may well provide the basis for a geographical examination of the emergence of the 'all-Russian market'. This is what is attempted below. The choice of period for the investigation—the late eighteenth and early nineteenth centuries—has three essential justifications. Firstly, this is an era in which Russian economic development appears, from the available evidence, to have gathered real strength. Secondly, it is the period immediately before the introduction of railways, which obviously had a profound impact on the development of regional specialization as well as on much else. The detailed comparative study of selected regions before the coming of the railways (and before the abolition of serfdom) should permit a number of the peculiarities and problems of Russian regional development to be examined at a fairly important point in European and Russian history. Finally, in the late eighteenth century the historical sources become systematic and relatively comprehensive for the first time, although the accuracy of these sources cannot always be accepted at face value. The central question, therefore, is to what extent an 'all-Russian market' in the spatial sense had been established before the railway age. I intend to show that, while the term 'all-Russian market' can be accepted as a valid one to describe what was beginning to happen in many central and accessible parts of European Russia, several outlying regions were as yet only partially connected with this market. Moreover, the geography of trade was such that even many areas quite centrally located within Russia must have missed out on the new market developments.

## RUSSIAN ECONOMIC DEVELOPMENT IN THE EARLY ROMANOV PERIOD

In order to put the study of Russian regions into context, something needs to be said about the nature of the national economy at this time, and about the way it had developed during the early Romanov era.

Considerable work has been done to show the depressed state of the Russian economy in the early seventeenth century and how this had links with the sufferings of Ivan the Terrible's reign and of the Time of Troubles.[9] In addition to these problems, many other barriers hindered economic development. Lack of capital, internal customs barriers, legalized monopolies, the corruption of local officials, traditional social attitudes, and the pressures of service and taxation were all significant. Backward technology, labour shortages, and circumscribed markets hindered industrial growth. Everywhere long-distance trade was hampered by poor roads, bad weather, and frequent social disorder. The winter freeze on the rivers and the many transhipments that goods in transit underwent were obvious difficulties, because the rivers were the main trading arteries, despite the fact that they were unusable for up to half the year. At the same time winter ice encouraged the use of the sledge, a most efficient form of overland transport.[10] However, winter roads were often rendered impassable by snow-drifts.

In spite of such problems, the signs of economic growth became apparent early in the Romanov period. Different historians have accorded varying degrees of emphasis to the several factors at work. The greater measure of social order associated with the establishment of the Romanov dynasty was certainly one significant element in the situation. The state itself, both through its own economic activity and through the material and financial burdens it laid on its subjects, stimulated the economy. Even though obligations to the state often left little surplus beyond subsistence to encourage investment or enterprise on the part of the populace, a market was nevertheless provided by the state, the privileged classes, foreign merchants, townsmen, and other groups in society. Private trade and handicrafts were thus encouraged and some large-scale manufacturing made its appearance. Other factors stimulating economic activity seem to have included a growing population (reaching 14 million or more by 1722) and the expanding frontier of settlement.[11] In contrast to the situation in Western Europe, which appears to have suffered a period of depression in the seventeenth century, the Russian economy apparently grew more or less continuously at this time. A suggested explanation for this contrast is that Russia was in fact only in the process of recovery from the low points reached under Ivan the Terrible and during the Time of Troubles.[12]

The hub of Russian commerce and production in the seventeenth century was the city of Moscow. Its economy was closely associated with the court and government. Moscow was by far the largest city in the realm and had about a third of the total commercial turnover in Russia.[13] The city contained numerous small manufactories, and to north and south, reaching as far out as Tver, Yaroslavl, and Tula, an incipient industrial region was gradually forming. Commercial routes radiated from the city in all directions and, where these intersected river routes, or where river routes themselves conjoined, stood some of the important trading centres of the realm. Among these Kazan, Nizhnii Novgorod, Yaroslavl, and Vologda were particularly prominent.[14]

Historians have long argued about the impact of Peter the Great upon the Russian economy. Motivated by his military needs, Peter both founded and stimulated the development of many new industrial enterprises. Of particular importance were the new metallurgical and armaments plants, shipbuilding, sail-making, and rope-making concerns, and also the manufacture of woollen and linen cloth, leather, and other items. In addition, under the influence of mercantilism, Peter encouraged other types of manufacture to reduce dependence on imports. Both state and private industries were set up, the latter being especially stimulated by a liberalization of economic life after 1711. But the tsar's policies also had their negative side for the population. Many new financial and military burdens were laid upon them, the use of forced labour on public works was greatly increased, and the bonds of serfdom tightened. The number of state peasants compulsorily 'assigned' to work in industry grew greatly, and from 1721 merchants were permitted to buy serfs for their factories. Some historians have asserted that Peter's economic achievements were essentially transitory, but others have noted their long-term repercussions in industries such as metallurgy, shipbuilding, and military engineering.[15] His impact upon the development of St Petersburg and the Urals iron industry was particularly enduring.[16]

Those historians who have doubted the stability of the industries developed under Peter the Great have tended to regard the period after his death as one of relative economic depression. Others, however, have described it as an era of steady if unspectacular economic growth, which became especially apparent from the mid-eighteenth century. Population was rising throughout the period, from some 14 million in 1722 to about 35.5 millions at the end of the century.[18] A proportion of this increase came from territorial acquisitions. Continued settlement resulted in an expansion of cultivation. There is, however, little evidence of much technical advance in agriculture. Handicrafts and industry also continued to develop. By the end of the century there were 1,200 large-scale industrial enterprises, compared with only 663 in 1767.[17] Yet technology remained backward. There were perhaps 200,000 people in the industrial labour force.[19] Exports more than tripled in value between 1762 and 1793, and by the latter date 45 per cent of exports were in metals and textiles. Internal trade also seems to have grown remarkably. There were about 608 trading centres in the 1750s but 2,571 by the end of the 1790s.[20] The accent was upon periodic fairs and markets. Internal trade turnover at the time of the abolition of internal customs barriers in 1753–4 has been put at more than 22 million roubles. It may be assumed that this grew considerably thereafter.[21]

Official policy in the period encouraged greater participation in the market-place. In the reign of Elizabeth (1741–61) trade was liberalized, while Catherine II (1762–96) moved away from mercantilism, abolished monopolies, encouraged enterprise, and fostered the settlement and cultivation of the steppe lands to the south. Both government policy— emancipating the nobility from compulsory service in 1762 and strengthening the bonds of serfdom—and also rising prices encouraged the nobles' enterprise. They obtained a virtual monopoly on distilling in 1754, and in the following year they were freely permitted to trade wholesale and retail in the products of their estates and those of their peasants.[22] The availability of serf labour for their factories and for transport purposes, and of materials from their estates, led many lords to take to manufacture. Flax, hemp, wool, hides, and grains became the staples of their enterprise, with some involvement in metallurgy in places. Of course it would be easy to exaggerate this trend, since many nobles remained poor with few serfs, and many took little real interest in their estates. None the less, the growing incidence of fairs held in estate villages is probably an indicator of the developing economic enterprise of the class as a whole. In 1760 36 per cent of all fairs were held in villages on noble estates; in 1800 it was 51 per cent.[23]

Noble enterprise naturally meant competition for the urban-based merchants, who failed to establish any kind of monopoly over trade and industry, despite their loudly voiced wish to do so.[24] In 1762 they also lost the right to buy serfs for their factories and thenceforth had to rely mainly on hired labour, especially on state peasants and on those serfs whose obligations to their masters were in cash or kind (*obrok*) rather than labour-based (*barshchina*). This may not have been to the disadvantage of the merchants in view of the increasing incidence of peasant seasonal migration (*otkhodnichestvo*) and of the notoriously low productivity of serf labour. Merchants were active in both wholesale and retail trade and in those manufactures, such as metallurgy and certain textiles, where the nobility had no particular advantage or which required special technical skills.

Further competition for the merchants came from the peasantry, who constituted 90 per cent of the population. Spurred on by taxes, impositions, rising prices, and the more liberal trading laws of the period, many peasants traded the products of their fields and handicrafts and sold their labour to merchants and entrepreneurs. A few even became traders, middlemen, or manufacturers themselves. Subsistence agriculture remained predominant but market relations seem to have affected the countryside more widely as the century wore on. According to figures published by B. N. Mironov, in the 1750s 356 rural settlements had fairs or other forms of periodic trade; by the 1790s the figure was 1,975. Mironov also records 383 village fairs for the 1750s, rising to 3,180 by the last decade of the century.[25]

As a result of these developments, a growing tendency towards regional specialization has been noted by several scholars. N. L. Rubinshtein, for example, proposed the thesis of an economic regionalization of European Russia based upon his studies of the late eighteenth-century topographical descriptions, General Survey materials, and other sources.[26] Using Rubinshtein's regionalization as a framework, the remainder of this chapter will be devoted to a study of selected regional economies at this period.

## REGIONAL ECONOMIES IN THE LATE EIGHTEENTH AND EARLY NINETEENTH CENTURIES

### Vologda Province

The province of Vologda, falling within Rubinshtein's Northern Industrial Region, was formed in 1780 and consisted of nineteen districts. However, Archangel, with its seven districts, broke away to form a separate province in 1784. In spite of this diminution, Vologda province still occupied an extensive area, including considerable tracts of coniferous forest and the rather more fertile valleys of the Sukhona, Vychegda, Vaga, and upper Northern Dvina. Human settlement was almost entirely confined to these valleys and their tributaries, the major exception being the region to the south-west around Vologda itself. The broad, low-lying, or occasionally hilly interfluves were characterized by marshland and fir forest, the wide valleys by pine and birch interspersed by dry or flood-plain meadows. The province's population (278,000 males in 1782; 293,000 males in 1795)[27] was concentrated in small or medium-sized rural settlements coinciding with meadowlands, clearings, high river-banks, and well-drained terraces. Chains of settlement were sometimes found along the river-valleys.[28] Towns stood at strategic points and had a total population of at least 25,000 men and women in the latter part of the century.[29]

The most significant factor in the economy of Vologda province in the late eighteenth century was its position astride the major routes between central Russia and the port of Archangel. Although by that time easily outrivalled by St Petersburg and Riga, Archangel still accounted for 38.7 per cent of Russia's exports and 16.8 per cent of her imports in 1783, and was a town of some 15,000 people and a noted industrial centre.[30] The main overland route from Moscow and Yaroslavl ran by way of Vologda to Velsk on the Vaga, and thence by river or road to Archangel. Alternatively, goods could be transferred at Vologda on to the Sukhona River, and thence sent downstream along the Sukhona and Northern Dvina. At Velikii Ustyug and near Solvychegodsk the Sukhona–Dvina waterway was joined by rivers and roads allowing access to and from the east, the north-east, and ultimately Siberia.[31] The major routes are illustrated in Fig. 8.1.

The towns of Vologda province all participated to some extent in the long-distance and local trade generated by this system of waterways. With under 5 per cent of its population living in the twelve official towns, however, the province was somewhat less urbanized than many other parts of Russia.[32] There is also evidence that at least some of the townsfolk engaged in agriculture.[33] Of the seven new towns instituted in the 1780 administrative reform, six were very small, with under 2,000 inhabitants. Of the old towns, Velsk and Yarensk were also tiny.[34]

The two dominant towns of the province were Vologda and Velikii Ustyug. Vologda, with between 7,500 and 10,000 people, was the larger and its strategic location has been noted above. In the early 1780s, according to Klokman, it had 447 male merchants and 2,061 townsmen (*meshchane*).[35] The 1784 topographical description[36] records 16 stone and 1,485 wooden dwelling-houses, 106 factories and mills, 36 smithies, a trading-hall, and a total of 311 wooden shops. Velikii Ustyug, situated to the north-east of Vologda and further down the Sukhona River, was somewhat smaller, with a population of just under 7,000, and had 4 stone and 1,080 wooden dwelling-houses, and 15 stone and 104 wooden shops. In 1784 265 male merchants and 2,027 townsmen were recorded in the town.[37]

Both these towns had a commercial and manufacturing profile with the broad, unspecialized character typical of early modern urbanism.[38] Vologda, however, had a slightly more developed industrial base with two silk-, two cotton-, and several dye-works being recorded by the economist, M. D. Chulkov in the 1780s.[39] Unfortunately, the sources give no real indication of the numbers employed in these concerns, but they were almost certainly quite small. Otherwise the economies of the

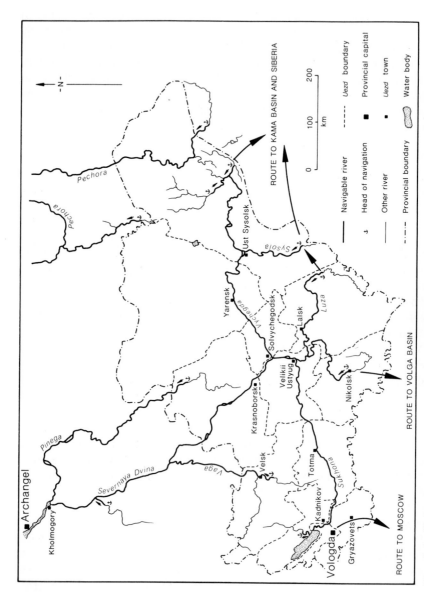

FIG. 8.1. Vologda province: towns, waterways, and trading routes in the late eighteenth century

two towns were in effect extensions of the rural economy: tanners, maltsters, tallow-, soap-, and chandlery workers, spinners, weavers of linen cloth, flour- and saw-millers, and metalworkers are recorded.[40] Velikii Ustyug was noted for its fine metalwork and especially for its silver-work.

Some of these activities reflect not only the close connections between the two towns and their immediate surroundings, but also their trading links much further afield. Here the towns' locations on the Sukhona–Northern Dvina waterway and their links southwards towards Moscow and the Volga, and eastwards towards the Urals and Siberia, were especially important. Thus, the wealthier merchants of Vologda were engaged in the selling of hemp, candles, tallow, flax, canvas, bristles, grain, and wooden goods as far afield as Archangel and St Petersburg. Such goods were obtained not merely locally, but from a considerable variety of northern, central, and eastern locations.[41] Merchants from both towns were also involved in the trade with Siberia and even China, and some of the wealthiest from Vologda had trading links with Western Europe.[42]

The other towns of the province had economies which were much more oriented to the local, rural milieu, although, in the case of Totma and Solvychegodsk at least, some long-distance trading links were in evidence. Seven of the twelve towns of Vologda province had permanent shops, according to the 1784 topographical description, but in only Vologda and Velikii Ustyug was there a significant number of these. According to data assembled by Mironov, in the last three decades of the century the northern region, which included Vologda province, had more fairs and bazaars relative to its population than the average for Russia as a whole. However, only a handful of these northern fairs had a large trade turnover.[43] This may have reflected a dispersed population, among other factors. Winter fairs at Nikolsk, Krasnoborsk, and in Yarensk district attracted merchants from well outside the province, including Moscow.

Unfortunately, the sources do not permit a detailed enquiry into the rural and peasant economies of Vologda province, although indirect indications can be obtained from the topographical descriptions. The northerly environment and poor podzolic soils were detrimental to agriculture, but grain was grown wherever possible and especially on the more favoured soils, such as the glacial clay loams and Permian marls. Rye, barley, oats, and flax were the principal crops. On the whole, the peasants of Vologda province were not well provided with arable land, but those in the south-western districts around Vologda were better off than most. These districts also formed the main area where noble estates predominated.[44] The 1784 topographical description notes that regular manuring was responsible for good yields of grain and that this was rendered possible by grazing on pasture and haylands released by the General Survey.[45] Some of the surplus found its way north to the Archangel market or even to St Petersburg, but much was also sold locally within the province.[46] A market for surpluses was provided by the population in the poorer northern and north-eastern districts. Thus, at Ust Sysolsk to the north-east the inhabitants are recorded as having abandoned the poor land entirely for the sake of trade.[47]

Those who were not fully occupied in arable farming engaged in a variety of other occupations. The raising and keeping of livestock was an important occupation sustained by the relative abundance of pasture and hayland.[48] It helped to supply the many urban-based manufacturing activities mentioned above. Hunting was another northern occupation. Thus, the inhabitants of Vologda district traded in pelts and hides, particularly those of the wolf, bear, marten, fox, lynx, deer, ermine, rabbit, and squirrel.[49] Pelts, hides, and leather were exchanged at an annual fair on 6 December near Nikolsk, and also in Ust Sysolsk and its district.[50] Handicrafts were widespread in rural areas. The 'black' peasants of Krasnoborsk district were noted for the making of candles and soap, silver- and copper-work, locksmithing, and other skilled metalwork.[51] Handicrafts were often tied to local needs or, as at Gryazovets and Kadnikov on the main road north through Vologda to Archangel, to the needs of overland travellers.[52]

Mention should also be made of the timber trade. In Nikolsk district, for example, some peasants were engaged in boatbuilding, lumbering in the state forests, and the rafting of timber downriver to Archangel.[53] Near Vologda, on the other hand, the depletion of the nearby forests was responsible for local shortages of good construction timber.[54] One reason for this was the continuing practice of slash-and-burn cultivation (*podseka*), which had ruined the forests close to many settlements. The temporarily cultivated patches could sometimes be 30 or more *versty* from the nearest settlement.[55]

An important phenomenon in the northern provinces was the tendency for peasants to hire themselves out to work in local trades and transport, or to travel to distant locations in search of work. The state peasants (black, economic, and others), who were particularly numerous in this province, found this to be a significant means of supplementing their livelihood, whereas privately owned peasants on *obrok* were regularly granted permission to work by lords anxious to increase the quit-rent they derived from their estates. Thus, the peasants of Vologda district are described as being engaged mainly in grain farming and handicrafts, but some travelled by permission (*po pashportam*) to various Russian towns and hired themselves out as carriers and boatmen for the trade to Archangel.[56] In Velsk district the peasants were principally engaged in farming, making resin for the Vologda or Archangel markets, and fishing, while some sought temporary work in Moscow and St Petersburg.[57] The sources mention Archangel, Moscow, St Petersburg, Riga, and even Siberia as favourite work destinations for the northern peasants.[58]

Only a handful of peasants in Vologda province were employed in the few rural factories and plants, most of which were in the hands of merchants or belonged to the state. Several salt-works are recorded, dependent upon hired or assigned labour, and there were three ironworks in Ust Sysolsk district hiring local labour. The province also had a number of small sawmills.

In conclusion, then, it is evident that, apart from districts in the south-west of the province, inadequate resources of arable land encouraged handicrafts, hunting, fishing, and seasonal labour migration among the peasantry. The numbers occupied in each went unrecorded. Only the cities of Vologda and Velikii Ustyug constituted major centres of trade and manufacture. The provincial economy was dominated by the Sukhona–Northern Dvina waterway and the Archangel market, with links to the Central Industrial Region, the North-West, and provinces to the east. Livestock-based products (leather, tallow, and wool), the products of hunting, woodwork (timber, resin, and wooden implements), and skilled metalwork were in demand outside the province. Some grain and flax were also exported from the province, especially to the north. Textiles and cloth-making existed in both urban and rural settings. On the whole it was a diversified rather than specialized regional economy characterized by the export, or the processing and export, of primary and animal-based products. Subsistence agriculture remained important. The incidence of peasant seasonal migration suggests the limited local opportunities for supplementing income in this part of northern Russia.

## Tula and Other Provinces

Although the province of Tula is assigned by Rubinshtein and other scholars to the Central Black Earth Agricultural Region, its northern portion was similar in character to that of the Central Industrial Region. Thus, Vasilii Levshin, who composed his topographical description in the first decade of the nineteenth century for the Imperial Free Economic Society,[59] contrasts the inhabitants of the northern districts of the province, who frequently suffered from a lack of good-quality land and were therefore obliged to engage in handicrafts or to travel about in search of work, with the residents of the south, who were utterly dependent upon grain and livestock farming. In the autumn and winter, says Levshin, as soon as the agricultural season was over large numbers of northerners visited the steppe districts of the province, as well as Orel and neighbouring regions, with cartloads of tubs, barrels, sledges, shovels, crockery, bast shoes, and other items. These they bartered for grain, pork, sheepskins, and similar agricultural products. Levshin praises the superior 'education, agility, and ingenuity' of the northerners as compared with the 'idleness' of the better-favoured southerners, whose agricultural resources allowed them to spend the winter apparently doing nothing. No doubt Levshin's moralizing is exaggerated, but there is nevertheless some truth in his comments. Thus, a lack of livestock, associated with inadequate pasture, reduced the availability of manure, especially in the north, where it was most needed. Grain yields were naturally better in the south, and the province's rye surplus came mostly from southern steppe districts such as Novosil and Yefremov. The surplus was used in distilling or exported to Kaluga and Moscow provinces.[60]

Tula province was located in the north-eastern portion of the Central Russian Uplands, where the gently undulating uplands were frequently split up by a network of river-valleys, gulleys, and ravines. The valley floors were usually occupied by meadowland, above which there were terraces characterized by loess-like clay-loams, that could be readily cultivated. Apart from some areas of fluvioglacial sand in the north-west, the province's soils were fertile, trending from turfy podsols and grey forest soils in the northern and central parts to weakly podsolized black earths in the south. Following this transition was a vegetational change from the mixed forests of the north to a landscape of mainly deciduous woodland and natural grassland in the south. The expansion of arable farming, however, had greatly reduced the forests, and by the beginning of the nineteenth century timber shortages were common even in the formerly well-wooded north.[61] Some woodlands were still conserved for industrial purposes. At the time of the General Survey 69.6 per cent of the province's land was arable, 7.8 per cent hayland, and 18.2 per cent forest.[62]

The economic geography of Tula province was governed by the disposition of navigable rivers, namely the Oka in the north and west, as shown in Fig. 8.2. Towns such as Belev, Odoev, Aleksin, and Kashira, situated on or close to the Oka waterway, benefited from the trade between the southern agricultural provinces and the Central Industrial Region. Those in less accessible places, such as Venev in the east, or Krapivna, Bogoroditsk, Yepifan, Chern, Novosil, and Yefremov in the centre and south, were more unfortunate in a commercial sense. These towns therefore tended to be small agricultural trading centres without

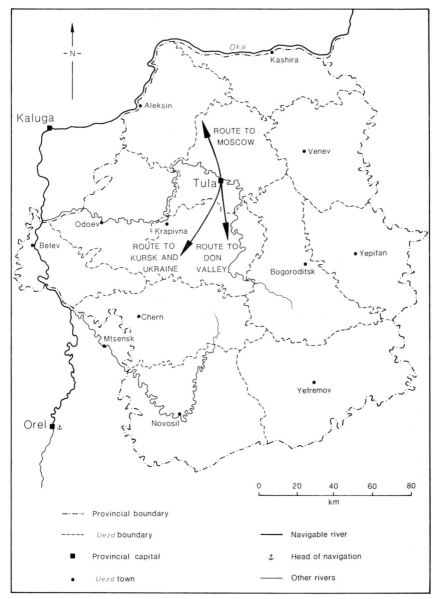

FIG. 8.2. Tula province: towns, waterways, and trading routes in the late eighteenth century

significant processing of agricultural products, and surpluses were sold in the many small markets and fairs in the region. Surpluses found their way either to local distilleries or to the Oka trading towns (Belev,

Mtsensk, and Orel) and Tula. The province was noted for its production of rye, oats, and hemp, and also for its livestock and livestock products (tallow, hides, bristles, and meat). Hemp was grown universally and especially in the northern districts, where it was an important item of trade. Grain was more significant for trade in the south. The southern districts, in some parts of which there were still areas of open steppe grassland, constituted the main livestock region. However, there were also considerable livestock imports from Little Russia and neighbouring provinces.

Unlike Vologda, Tula province was a region of private peasants, who made up 80 per cent of the rural population. The overwhelming majority were on *barshchina*. Most estates were small and this fact, together with the fertility of the soil, is the main reason for the predominance of *barshchina*, according to V. I. Semevskii.[63] In any case relatively few peasants left the province in search of work (only 16,000 in 1803, according to Levshin[64]), and even at the industrial centre of Tula only 1,085 people from outside the town were employed.[65] Local handicrafts were important, as already noted, as well as other forms of employment, such as carting. In distilling and in the other mills and factories of the province, employment opportunities were limited.[66] Complaints about poverty and the consequences of land shortage were occasionally heard among both peasantry and lords.[67]

Of the 50,000 people living in the towns of the province in the early nineteenth century (5.6 per cent of the total population), about half lived in Tula, which at the time was one of the major industrial towns in Russia. Tula had been a centre of metallurgy and armaments manufacture since the seventeenth century. In the 1780s it had a population of 25,000 with 4,600 houses, few of which were of stone.[68] At the time Levshin wrote, Tula had 300 stone and 446 wooden shops, 192 mills, 62 factories, and 432 smithies.[69]

The most important enterprise in Tula was the armaments-works established by Peter the Great's decree of 1712. The first specialist weapon-smiths (*oruzheiniki*) settled in Tula in 1595, but the effect of Peter's decree was to introduce an element of large-scale technology into the manufacturing process. Even so, in 1781 only 213 people out of a total of 3,128 master weapon-makers actually worked at the enterprise; the others laboured at home, usually on specialized tasks according to order.[70] Moreover, many of the weapon-makers also worked on their own account, and Tula was particularly noted for the manufacture of small arms, tools, instruments, and ornamental metalwork. *Oruzheiniki*, together with merchants and townspeople, controlled or worked in many other enterprises in the town.[71] In Levshin's day these included eight samovar-makers (employing an average of 23 workers each), five resin-makers, ten hatters, four glove-makers, five brush-makers, four makers of sheepskin coats, one silk-weaver (employing peasants from Moscow and Ryazan provinces), three rope-makers, and silversmiths, black-smiths, and tilers. In the smaller workshops were found soap-boilers, wax- and tallow-chandlers, tanners, curriers, leather-dressers, brewers, copper-smiths, a vinegar-maker, a felt-maker, potters, and brick-makers. Most of these concerns were small, employing only a handful of workers.[72] Tula's economy, therefore, showed the diversified pattern typical of early modern towns, although the metallurgical bias certainly had something of a proto-industrial character. As with Vologda and other towns, Tula had a significant local trade. The city's wealthier merchants often employed agents in other towns to undertake trade transactions on their behalf. Such wealthy merchants were also noted for their national and even international trading links.[73]

Of the other towns in the province, Odoev, Aleksin, Kashira, and Belev benefited from the grain, hemp, and timber trades of the Oka waterway and from the St Petersburg market via Gzhatsk. Aleksin (population 1,330), for example, had a sail-making factory owned by two local merchants which employed 159 workers, of whom eighty were factory-owned peasants.[74] Together with Kaluga, Serpukhov, Kolomna, and certain other towns it formed an incipient textile zone south of Moscow. The sailcloth was intended for the St Petersburg market. Odoev (population 2,000) was a centre for the purchase of grain for distilling[75] and also sent 15,000 *chetverti* of grain annually to Kaluga and Vyazma, mainly overland in the autumn and winter. Hemp was sold to the wholesale merchants of Belev and Kaluga, while some Odoev merchants dealt in cattle from the Ukraine and fish from the Black Sea. The town had leather-, soap-, and tallow-works.[76] Belev (population 5,800) had a similar profile and, like Tula and Odoev, many of its merchants and townspeople were Old Believers, a group who were often noted for their commercial enterprise. According to Levshin, up to 500,000 *pudy* of hemp, 150,000 *pudy* of tallow, 20,000 *vedra* of hemp oil, and 10,000 Russian leathers were sent every year from Belev to Petersburg via Gzhatsk.[77] Belev also had a thriving grain trade and the Thursday and Friday bazaars were especially noted for the crowds of *odnodvortsy* from Orel province who would arrive with considerable quantities of grain to sell. The grain was resold by the merchants to peasants and merchants from Kaluga province.[78]

In economic terms, therefore, Tula province was intermediate between the Central Industrial Region, with its incipient specialization in textiles and metallurgy, and the black-earth provinces to the south. Tula dominated as the most important industrial centre, and there was a handful of smaller commercial centres in the north and west oriented towards the Oka waterway and the Central Industrial Region. The province was, of course, essentially agricultural, but the agriculture of the northern districts had much in common with that of neighbouring parts of the industrial centre, being characterized by shortages of pasture, hayland, and timber and the importance of hemp, while the southern districts were more akin to the other black-earth provinces, with an orientation towards grain-marketing and livestock. The lack of good waterway links acted to the disadvantage of the grain market in several southern districts.[79]

In view of the detailed analysis given of Voronezh province in Chapter 3, only a word need be said here about two other provinces of the Central Agricultural Region, Tambov and Kursk. By the late eighteenth century both were developing as important agricultural provinces based on the rich black earths, and both were noted for the export of many food and associated products. Tambov province, however, looked north-east towards the Volga, exporting its grain, hemp, tallow, wax, and honey via Morshansk pier on the Tsna to the middle and upper Volga towns, St Petersburg, and Moscow, and also either south to the lower Volga, Astrakhan, and the Don cossacks, or north up the Kama towards the Urals. The peasants of the province were much given to participating in the boatbuilding, trades, and industries of the Volga towns.[80] Kursk province, by contrast, looked south, and sold its grain to the Ukrainian districts, particularly for distilling. However, some grain, as well as many other products, did travel overland to Orel and hence towards the centre and St Petersburg. Kursk, with over 12,000 people in the 1780s, was a centre for small-scale agricultural processing. There were numerous factories and mills in the rural districts especially devoted to the making of woollen cloth, canvas, and rope.[81] The Kursk Korennaya Fair, held every June, was one of Russia's major fairs at this period.[82]

## The South-West and South

Rubinshtein considered such provinces as Kharkov and Yekaterinoslav to be peripheral territories, not yet fully integrated into the 'all-Russian market'. Only a few words will be said here to illustrate their distinctive character. Kharkov province, part of the Ukrainian territories, was cut off from direct waterway linkages with central Russia, and its grain production was largely oriented towards distilling, a cossack privilege. Grain was also imported from elsewhere for this purpose.[83] This part of Russia was noted for its many small fairs and markets, reflecting a lively rural trade, and also for the number of *official* towns, mainly the result of cossack traditions of settlement and administration. The province, which was still being settled, had considerable areas of hayland and so livestock farming was important. Handicrafts based upon livestock products such as wool, leather, and tallow were common, as were the working of metal, wood, and clay. Specialized products such as fruit, vegetables, tobacco, and hemp were regularly marketed. Livestock and livestock products, liquor, and some handicrafts were the province's main exports northwards.

A feature of Kharkov and other Ukrainian provinces was the seasonal migration of large numbers of peasants south to the still lightly populated New Russian and Don cossack territories for agricultural work, fishing, and the livestock trade. These were territories whose steppe grasslands provided an indispensable basis for livestock farming, constituting their main contribution to the 'all-Russian market'. Thus, many fairs and markets in the Central Agricultural Region and even in some central provinces were noted for the sale of horses, cattle, and sheep from the south. Livestock-based handicrafts were also common in the south (except in the Don cossack lands),[84] although many of these were for local consumption only. The southern territories are illustrated in Fig. 8.3.

## THE 'ALL-RUSSIAN MARKET' IN EARLY MODERN RUSSIA

Soviet and Western historians alike have described the growing regional specialization to be found in societies undergoing economic development in the early modern period. However, it is a complex process with many contradictions, and it is also a difficult feature to measure. In the case of the Russian provinces described above, the reader may well be impressed by a lack of regional orientation in many of the activities described. Indeed, given the reluctance of the sources to describe in any detail the character of the household economy (especially in rural areas), this impression of sameness may actually be underemphasized. This should come as no surprise in view of the relative 'underdevelopment' of the Russian economy at the time. Most of the population were still peasants, labouring in the fields for scant reward in terms of grain, meat, and other necessities. The grains grown and the domestic animals reared did not vary greatly. Each regional economy was a complex of interlinked activities designed for the most part to support life in the local area. Everywhere tools had to be fashioned, cloth woven, household implements wrought, buildings constructed. The materials for these things—wood, iron, clay, wool, bristles, and so on—were often locally acquired. The agricultural life was the basic life, and even many town dwellers were involved in it. Only where the environment

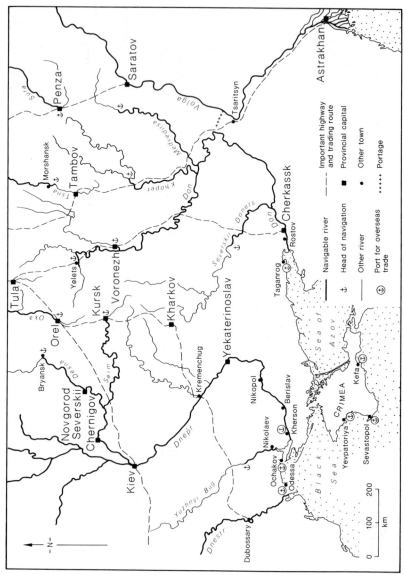

FIG. 8.3. South European Russia: towns, waterways, and trading routes in the late eighteenth century

presented special opportunities—hunting in the coniferous forests, herding on northern pastures or southern steppes, fishing (and, as exchange developed, salt-boiling, iron-smelting, potash-making)—was the agricultural economy diversified.

Russia was therefore mainly rural and agricultural with limited opportunities for specialization. However, earlier sections of this chapter have described a growing propensity to trade by the late eighteenth century, and have also noted the sources of some of this expanding activity: the growing population, the territorial expansion, the overseas market, the demands of the state, the towns, the nobility, and even the peasants. Capital, both material and human, was being invested in production and trade; communications were being improved, if only slowly; inter-regional commerce was growing: hence the beginnings of regional specialization alluded to by Rubinshtein. Several facets of this specialization have been noted in this chapter. The grain trade was obviously fundamental, since it permitted those without the resources and land, and those with the calling, to engage in non-agricultural activities. Its lineaments were complex, and surpluses and deficits in grain supply varied greatly through both space and time. But the most obvious pattern to emerge was the growing dependence of the Central Industrial and North-West Regions upon grain from the black-earth provinces. Moscow and St Petersburg were the dominant markets, but a number of other industrial towns, as well as some rural districts with land shortages, added to the demand.

The livestock trade was another important consequence of regional specialization. Here, however, as with the timber trade, it was the unavailability of the right type of land which was the governing factor. The lack of hayland and pasture in many well-populated central districts forced livestock farming for the market out towards the periphery—to the north and on to the steppe grasslands to the south. Although there was some correspondence between regions with extensive livestock farming and livestock-based handicrafts and industry (tanning, tallow, soap, woollen cloth, etc.), the correlation was by no means exact. Thus, the wool of the south and Left Bank Ukraine supplied both local weavers of rough cloth and the more specialized textile industries of the centre.[85] Tanning, tallow-making, and similar trades were, of course, very common in Russia, and the cattle and horse drives from the Ukraine, the Don, and the steppe provinces were a well-established feature of the period.

Further elements of regional specialization in agriculture have also been mentioned above. The hemp trade of the lands along the Oka, feeding the canvas and sail-making factories of towns south of Moscow, and ultimately the shipbuilding and associated industries of St

Petersburg, is widely alluded to in the sources. According to Rubinshtein, the cultivation of hemp for this purpose was especially characteristic of those provinces in the southern part of the Central Industrial Region and the northern part of the black-earth provinces, such as Tula and Kaluga, which lacked the fertile agricultural land further south and the alternative employment opportunities available near Moscow.[86] Similar patterns of specialization were to be found in the case of flax-growing, for example in Vologda province. Here the Archangel market was particularly important. Distilling was yet another agriculturally based activity, to be found especially in the Ukraine, where it was a cossack privilege, and also in provinces such as Tula and Kaluga, where the nobles had a monopoly. Such factors as the ready availability of grain, fuel, and labour, and in some cases the absence of waterways for disposing of grain surpluses, were evidently important.

The pre-industrial character of the Russian economy in the late eighteenth and early nineteenth centuries is apparent from the low levels of urbanization and the predominance of very small towns, the incidence of periodic forms of trade, the underdevelopment of manufacturing industry, and the common occurrence of handicrafts. The unspecialized nature of urban economies, processing local raw materials for local markets, and the frequency with which industries, handicrafts, and periodic trade located in the countryside fully accord with our understanding of the early modern period in other European societies.[87] In the Russian provinces described above, the possibilities for peasant employment in the towns, in industry, or in other regions were obviously still limited. The opportunities in local handicrafts or in servicing trade and transport were evidently greater, but hard statistics are difficult to come by. Labour surpluses rather than shortages must have characterized many areas, and thus serfdom may not have acted as a major brake upon industrialization at this period, as has sometimes been asserted.

In spite of the traditional character of much of the Russian economy, however, enough has been said in this chapter to indicate the beginnings of a very different type of economic structure. Industrial centres, such as Kaluga, with its cloth- and sail-making, and Tula, with its specialized metallurgy, formed part of an incipient manufacturing and industrial region centred upon Moscow. As yet this region bore all the hallmarks of what some historians have referred to as a 'proto-industrial' rather than a fully industrial character.[88] But the larger industries of Moscow, St Petersburg, the Urals, and other districts of Russia were already beginning to form specialized economies around themselves and to exercise their influence far and wide. The economic importance of St Petersburg and Moscow, based as much upon the activity of government, administration, and trade as upon industry, is especially apparent in the above discussion of Vologda, Tula, and adjacent provinces.

Russian towns in the early modern period were noted as much for their trading as for their industrial activity. Indeed, this chapter has already described several cities whose wealthier merchants' contracts embraced much of the nation, and who aspired even to dealings involving foreign commerce. Trade at this time was relatively disorganized by modern standards, goods in transit passed through numerous hands, and the propensity for towns to compete with one another in many different markets must surely raise questions about the meaning of the neat seven-tier urban hierarchy suggested for this period by G. Rozman.[89] Towns were the chief organizers of trade even when, as in the case of many fairs, trade took place in the countryside. But trade fluctuated, punctuated by weather, natural disaster, disorder, and many other factors. The picture was far from the well-ordered pattern of exchange and integration suggested by Central Place Theory.[90] Trade was also uneven through space. Goods circulated by means of an untidy network of trade routes, the major arteries of which were the waterways. Waterways in different river systems were interlinked by overland routes, and roads also acted as feeders to the waterways. In winter, or in times of low water, roads came to act in place of waterways, but overland travel was expensive. Hence it is hardly surprising that, in a period of growing inter-regional trade in bulk items such as grain, waterways should have played such a vital role.[91] Nor is it surprising that many of the major towns of Russia should have been situated on water routes, especially where these linked up with important highways. The towns of eighteenth-century Russia were thus nodes in the trade arteries formed by the waterways. This chapter has suggested, however, that this was not the complete picture. Tula, for example, was situated well south of the Oka, and owed its importance in part to its historic defensive role and to its long-standing tradition of metallurgical and armaments manufacture. But it also stood on the main overland route between Moscow and the Don basin, with its important agricultural economy. Similarly, Kursk and Kharkov lacked direct water communication with the Central Industrial Region, and yet both were becoming important commercial centres by the late eighteenth century. The explanation is that they too stood as gateway towns on the overland route between the agricultural frontier of the Ukraine and the basin of the Oka.

This pattern of trading links suggests an important and insufficiently appreciated aspect of economic development in the Russia of the pre-railway era. This is that inter-regional trade and regional specialization must have developed as they were fed and nourished by the chief arteries—i.e. the waterways. Those areas that were not so fed, even if they were in fairly central provinces, lagged behind the rest. Even the brief analysis presented in this chapter, derived from sources that leave much to be desired in terms of solid statistical evidence, has been

enough to suggest the importance of the friction of space. Proximity to
the Moscow and St Petersburg markets, and major arteries such as the
Volga and its many tributaries, was all-important. Comparatively
inaccessible areas, such as the southern districts of Tula province (to say
nothing of wide parts of neighbouring provinces such as Tambov and
Penza), lagged behind in terms of urbanization, industrial development,
and market penetration. Merchants and traders visited wide areas of
Russia in their perambulations, connecting even inaccessible places into
an intricate web of fairs, markets, and commerce. But where a lord had
his estate, or where a peasant laboured in the fields, must have had a
major influence on their ability to benefit from that commerce.

A chapter which focuses upon inter-regional trade will inevitably
emphasize the importance of lines of communication, and the problems
attending communications in the early modern period are widely known
and appreciated. This chapter has been unable to deal directly with the
many other factors which held back Russian development, such as the
lack of capital and enterprise, serfdom, the traditional economic order,
technological backwardness, and so on. Yet by tracing a portion of the
commercial map of the period, and by showing the way it was oriented
towards waterways, it has reinforced an old lesson—the extreme
importance of geography to both the mode and the speed of Russian
development. Regions were not uniform economic spaces, but rather
unevenly integrated areas only partially organized around rivers and
towns. Hence, in the early modern period, no really marked degree of
regional specialization is to be expected.

To return, finally, to the question of the 'all-Russian market', it is
clear that by the late eighteenth century Russia was embraced by a
partial and unevenly interconnected network of trade and commerce—
involving not only high value commodities but also numerous bulk
goods and manufactured items—particularly organized around Moscow
and St Petersburg. The influence of these two cities was apparent far
and wide, even in the more far-flung parts of the Empire, as goods
which had been handled by the merchants of these cities penetrated
distant markets, and as these markets in turn sent their own specialized
products in the reverse direction. Moreover, the growing interconnec-
tedness between the Central Industrial Region, the North-West, and
the black-earth provinces, linked in turn with the Volga basin and the
Urals, was a harbinger of what, later in the nineteenth century, was to
become a much closer interdependence. But in the period under study
some regions did not yet fall fully into this central net. The north's
commercial orientation, as illustrated by Volodga province, was more
towards Archangel and the export trade, and to a lesser extent towards
St Petersburg. The southern part of the Central Agricultural Region,

the Ukraine, and New Russia looked south, and were to be increasingly influenced by the Black Sea trade.[92] In other words, the 'all-Russian market' was still in process of formation, not yet fully embracing the periphery or even all the many interstices within the network of waterways. It was to fall to the railways to make the 'all-Russian market' a reality; in the pre-railway era it was still severely limited by geography.

# 9
## Peasant Manufacturing in Nineteenth-Century Moscow Province

THE nineteenth century saw the rapid development of industry in Russia, especially during its last two decades. One effect was to attract ever-increasing numbers of peasants into towns. Industrialization, however, also developed in the countryside, and factories owned by the nobility and by peasants appeared in and around villages, tapping the large rural labour force. In addition, peasant domestic industry expanded at this time. Production of manufactured goods by peasants was not a new phenomenon in Russia; for centuries demand for a variety of manufactures in the village had largely been met by local craftsmen and farmers' own families. The difference in the nineteenth century was that these manufactures were now more than ever destined for distant markets, far beyond the boundary of village and rural district. Such production for the market was most developed in those parts of European Russia where agricultural incomes were particularly low, where pressure on the land was intense, and where, in terms of markets, transport, and accumulated skills, the potential for industrial expansion was high. These conditions were met in the forested provinces clustered around St Petersburg and Moscow, the latter being the focus of the Central Industrial Region.

The term *kustar* or *kustarnyi promysel* was often given to peasant manufacturing in contemporary writings.[1] Although this is generally translated as 'domestic industry' or 'handicraft industry', a strict definition of *kustar* would limit it to the manufacture of goods for the market by peasant households using their own labour and materials. But the use of such a strict definition became difficult in the nineteenth century, and it was often broadened to include a wider range of activities, or was replaced by the more general term *krestyanskii promysel*, roughly 'peasant industry' or 'peasant business'. Investigators of peasant industries in fact described a variety of situations from pure handicraft to industrial outwork, and included among the peasant manufacturers entrepreneurs and wage-labourers, as well as independent producers using their own and their family's labour. Such catholic definitions did not please the Russian Marxists, chief among them Lenin, who delivered an attack on most existing studies of peasant industries in *The Development of Capitalism–Russia* (1899). He called for a strict distinction to be drawn between artisans, small commodity producers, and wage-workers in the village, observing that the term *kustar*, in its all-embracing meaning, was 'absolutely unsuitable for scientific investigation' and, in its narrow meaning, described a type of peasant rarely to be found in the Russian countryside.[2] For Lenin, the most important feature of peasant industry was its social character and the contribution this made to the historical transformation of the peasantry:

The combination of industry with agriculture plays an extremely important part in aggravating and accentuating the differentiation of the peasantry: the prosperous and the well-off peasants open workshops, hire workers from among the rural proletariat, and accumulate money for commercial and usurious transactions. The peasant poor, on the other hand, provide the wage-workers, the handicraftsmen who work for buyers up, and the bottom groups of petty-master handicraftsmen, those most crushed by the power of merchant's capital. Thus the combination of industry with agriculture consolidates and develops capitalist relations, spreading them from industry to agriculture and vice versa.[3]

As Lenin had argued in earlier chapters of *The Development of Capitalism in Russia*, one consequence of agricultural progress was the ultimate disappearance of the 'middle peasant'. A similar fate now awaited the small-commodity-producing peasant of Russia's industrializing provinces, and would be the outcome of the progressive separation of manufacturing from agricultural work in the peasant household:

The separation of industry from agriculture takes place in connection with the differentiation of the peasantry, and does so by different paths at the two poles in the countryside: the well-to-do minority open industrial establishments, enlarge them, improve their farming methods, hire farm labourers to till the land, devote an increasing part of the year to industry, and—at a certain stage of the development of industry—find it more convenient to separate their industrial from their agricultural undertakings, i.e. to hand over the farm to other members of the family, or to sell farm buildings, animals etc., and adopt the status of burghers or merchants. The separation of industry from agriculture is preceded in this case by the formation of entrepreneur relations in agriculture. At the other pole in the countryside, the separation of industry from agriculture consists in the fact that the poor peasants are being ruined and turned into wage-workers (industrial and agricultural). At this pole of the countryside it is not the profitableness of industry, but need and ruin, that compels the peasant to abandon the land, and not only the land but also independent industrial labour; here the process of the separation of industry from agriculture is one of the expropriation of the small producer.[4]

The materialist framework that made Lenin concentrate on the class differentiation of the peasantry also led him to counterpoise industry against agriculture and town against countryside; although at the time many peasants were engaged in farming as well as manufacturing in the industrial regions of Russia, industrial labour had to prevail, and, as a consequence, many villages had to lose their agricultural function. In Lenin's view, therefore, peasant manufacturing was a passing phenomenon destined to be displaced by factory production.

Temporary it might have been, but in nineteenth-century Russia domestic manufacturing was vital to the survival of many peasant households in the mixed and coniferous forest belts. Here a particular type of multi-occupational household existed which was different from the primarily agricultural households to the south, and within the forest belt there was a distinct geography of peasant manufacturing, reflecting differences in the resources available for manufacture, labour availability, and markets. Although the general tendency was towards the development of 'polar groups' among peasant manufacturers, and for industrial earnings to displace agriculture in the peasant economy, these differences meant that the speed of transformation and the form it took varied between communities and locales. Indeed, involvement in manufacturing could lead to a stability in household economies that helped stave off the abandonment of the land. In such cases, peasant manufacturing has to be seen as a successful adaptation to the changing conditions of the nineteenth century rather than as indicative of the demise of traditional rural society.

Moscow province, one of the most industrialized in Russia, has been taken in the pages that follow to illustrate the diverse forms peasant manufacturing could take. From the 1880s the city of Moscow witnessed the growth of modern factories that were among the largest in the country. These contrasted with the multitude of small-scale works that continued to flourish in the rural areas and in the city itself until the 1917 revolution and after. If peasant industries were to be extinguished by competition from the modern sector, the signs would be most clearly detected in Moscow province. For this reason it is a particularly good example to take to examine peasant manufacturing in the nineteenth century.

## THE GROWTH OF THE DOMESTIC SYSTEM IN MOSCOW PROVINCE

It has already been observed that peasant manufacturing was particularly strongly developed in the Central Industrial Region. The emphasis in the region was on textiles: cotton, linen, wool, and quality materials such as velvet, silk, and brocade. Their production constituted the single most important branch of factory-based and domestic industry in the province. Related industries, such as the manufacture of fringes, tassels, lace, braid, frills, and clothing, using the products of the spinning-mills or finished cloth, were also well developed. Vladimir and Moscow provinces were the principal centres of the textile industry, and they boasted growing numbers of factory villages, the best-known being Ivanovo in Vladimir.[5] The Central Industrial Region also had other industries. Some, requiring complex or mechanical processing (e.g. chemicals and machine-building), were confined to factories, but many were shared with the domestic sector or were exclusive to it. It was usual practice in the nineteenth century to classify domestic industries by the raw materials they used: animal products, wood products, stones and clay, metals and fibres. Counting all the forms a single product could take, one could identify hundreds. The following list, a mere sample from the censuses of peasant industries in Moscow province, indicates the range of goods entering the market from the peasants' huts: glass beads, cloth, guitars, toys, felt boots, leather boots and shoes, abacuses, bark, reed and straw baskets, hats, gold leaf, silver and gold ornaments and crosses, furniture, bobbins, pins, spectacle frames, gloves, cigarette filters, blotting paper, buttons, fishing nets, veils, scarves, clocks, silver foil, trays, and wooden utensils. In addition, peasants in Moscow province engaged in activities such as colouring wrapping paper and lithographs, painting icons, and recovering gold and silver from house fires, which were counted as domestic 'industries'.

Before the second half of the nineteenth century, although their existence was noted, such peasant industries were not the subject of any special investigations. However, starting from 1861, with the publication of A. Korsakh's *On the Forms of Industry*, interest mounted, leading in the 1870s to the establishment of the Special Commission for the Investigation of Handicraft Industries, which sat from 1872 to 1885.[6] The Commission published a series of essays of varying length and detail on individual industries taken from all provinces. At much the same time *zemstvo* statistical departments began their own investigations of peasant industries. One of the most complete was for Moscow province, where one of the leading figures of *zemstvo* statistics, Count V. I. Orlov, helped to direct a survey of over one hundred domestic industries, which was published in seven volumes between 1876 and 1882.[7] Two decades later there was a follow-up survey, this time directed by P. A. Vikhlaev, another well-known *zemstvo* statistician and economist.[8] The analysis that follows is taken from these publications of the Moscow *zemstvo*.

In the nineteenth century very few peasant households in Moscow province relied solely upon working their allotment land for their livelihood, and in 1899 more than 90 per cent of households which were still registered and resident in their native villages had at least one of their members employed outside farming for all or part of the year.[9]

Over one-half of the male peasant population and one-quarter of the female population was involved. The explanation for such high rates of participation in non-agricultural work is not difficult to find—poor physical conditions for farming combined with small landholdings meant that it was hard to eke out an existence from agriculture. There was accelerated population growth in the province in the nineteenth century which contributed to mounting pressure on land resources, so that by the end of the century peasant landholdings were among the smallest in European Russia. The encroachment of arable land on to natural pastures and meadows, which was a consequence of attempts to maintain grain output levels on minute holdings, threw the traditional peasant farming system out of ecological balance. Reduction in livestock numbers meant that manure, needed to improve the fertility of the province's acid soils, was in short supply, and that dairy and meat production, which, given the close proximity of Moscow, was the obvious basis for commercial farming in the province, was hindered in the initial stages of development. Over-felling of forest and woodland destroyed another vital resource in the peasants' subsistence economy. A level of taxation on the land that exceeded the income normally available from it compounded the peasants' difficulties. The effect of this was to force peasants to search for supplementary sources of income and, paradoxically, to try to avoid taking on extra allotment land.

In other parts of Russia where land shortage was a problem, many peasants earned supplementary income by making labour excursions to destinations far from their native village. This applied to peasants from the overcrowded central black-earth belt who travelled south, or north to the factories of Moscow and its region. In the Central Industrial Region such labour excursions also took place, but additional opportunities to find employment existed nearer home. According to the 1898–1900 census analysed by Vikhlaev, 41.3 per cent of all the people earning supplementary incomes in Moscow province found work in their native village. Among the remaining 59.7 per cent over one-third ventured no further than their district boundary.

Whether searching in their own village or district or further afield, peasants in Moscow province did not usually have difficulty finding work. As industry developed in the nineteenth century, opportunities for employment increased. Some of the new industrial recruits found their way immediately into urban and rural factories, while others entered the much more varied domestic sector. Unfortunately, it is impossible to make an assessment of the distribution of labour between factory and peasant hut, and of their relative contributions to Moscow province's industrial development. This is because of inaccuracies in contemporary industrial surveys, which often neglected the household sector or were selective in their inclusion of domestic enterprises. In an 1872 atlas of industry in Moscow province, for example, the author wrote apologetically in the introduction.

> It is true that, in order to give a full picture of manufacturing in the province, our atlas should include information on artisans and on small peasant handicraft industries, which, as is well known, play such an important role in the province, but the statistics on these enterprises are so scanty and so unreliable that we decided not to include them with our published data.[10]

Other directories of industrial enterprises in Moscow province, compiled before and after 1872, adopted much the same policy, but each used different criteria to identify the cut-off point between factory and 'handicraft hut'; sometimes this was a particular size of workforce, sometimes the use of hired labour or machine-power. 'In the majority of cases', concluded the factory inspectorate at the end of the century, 'it is impossible to determine . . . what a particular investigator understands by the term 'factory', 'plant', or 'artisan workshop'; to date no theory or practice has been developed to categorize the various transitional forms of industry—the so-called domestic and handicraft industries.'[11] As the author of these words observed, the inconsistencies were so great as to cast doubt upon generalizations made about the nature of industrialization in Moscow province, and especially of the role played by the factory.

Whatever the contribution of domestic industry to overall industrial growth in Moscow province, contemporary accounts testify to an increase in output from the sector during the nineteenth century. Individual manufactures experienced different periods of expansion. Military campaigns—the 1812 war, the 1849 Hungarian campaign, and war with Turkey in 1877—were good years for industries supplying the army. The goods produced included boots, cloth for uniforms, brushes, tassels and braid, mementoes and charms for recruits, and, for relatives at home, pictures and lithographs depicting successful battles. Coronation years benefited other domestic industries: the accession of Alexander II, for example, was followed by two boom years in industries using precious metals, such as gold lace and silver-work. Traditional goods catering for peasant demand, such as felt boots, bast-work, wooden utensils, and toys, had a guaranteed market that expanded with population growth, although years of cholera and famine resulted in serious slumps.

The fortunes of some industries were, of course, tied to developments in factory production. Domestic weaving, for example, expanded after the fire of Moscow in 1812 and the lifting of import restrictions in 1819, when many large textile factories failed. Domestic production was further encouraged by legislation in 1808 and 1818 allowing rural inhabitants to produce and sell textile goods 'without hindrance or obstacles'. For the industrial capitalists, rural weavers constituted a pool of skilled labour to be exploited, and it made sense in the pre-machine era to use this labour at home, thereby saving on overheads. The 1830s, 1840s, and 1850s thus saw the proliferation of domestic weavers attached through the outwork system to spinning factories and dye-works. As was observed by Tugan-Baranovskii, author of one classic account of industrial development in Russia, domestic looms sprang up 'like mushrooms after a storm' around any new rural textile plant.[12] At the end of the nineteenth century much weaving was still in peasant huts, but the end of the outwork system was already signalled by the introduction of mechanical looms in broadcloth and wool production, which had the effect of moving labour into the factory. In the industrial surges that took place in the 1890s and 1910s, mechanization similarly threatened the survival of many other domestic industries, while exhaustion of local raw materials or over-production threatened others. Although such developments suggested that the 'Golden Age' of domestic industries was drawing to a close by the end of the nineteenth century, they still had a role to play. Moscow province's population remained predominantly rural in the early twentieth century, and, although employment in factories could be expected to account for an ever-rising share of the supplementary earnings of peasant households, the advantages of finding work near home would not be diminished, at least in the short term. This latter point was borne out when, in the troubled conditions of revolution and in the 1920s, domestic industries again came into their own.

## THE GEOGRAPHY OF DOMESTIC PRODUCTION IN MOSCOW PROVINCE

Figs. 9.1–9.6 show the geographical distribution of domestic production in Moscow province. Fig. 9.1 shows the distribution of workers in the domestic sector of the textile industry. The remainder pick out the much more specialized industries, in which overall participation rates were low but which imparted a distinctive character to individual villages and occasionally whole rural districts.[13]

It is clear from Fig. 9.1 that the whole of the eastern part of the province constituted a single region devoted to the production of textiles and related goods. In the heart of the region, in Bogorodsk and Bronnitsy districts, the majority of households had a loom either in their own hut or workshop. Scattered throughout the region, on the perimeter

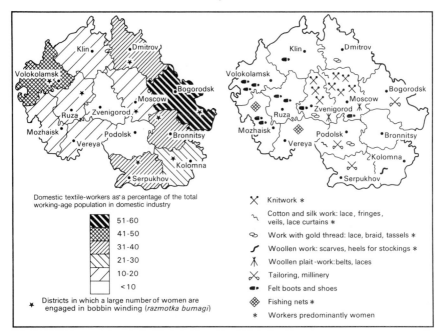

Domestic textile-workers as a percentage of the total working-age population in domestic industry

| | |
|---|---|
| ▨ | 51-60 |
| ▨ | 41-50 |
| ▨ | 31-40 |
| ▨ | 21-30 |
| ▨ | 10-20 |
| ☐ | < 10 |

✳  Districts in which a large number of women are engaged in bobbin winding (*razmotka bumagi*)

| | |
|---|---|
| ✕ | Knitwork ✳ |
| ⌐ | Cotton and silk work: lace, fringes, veils, lace curtains ✳ |
| ◇ | Work with gold thread: lace, braid, tassels ✳ |
| ⌠ | Woollen work: scarves, heels for stockings ✳ |
| 大 | Woollen plait-work: belts, laces |
| ✂ | Tailoring, millinery |
| ◖ | Felt boots and shoes |
| ❀ | Fishing nets ✳ |
| ✳ | Workers predominantly women |

FIG. 9.1. The distribution of workers in domestic textile production in Moscow province, 1898

FIG. 9.2. The principal areas of fibre-based domestic industries in Moscow province, 1870–1880

of villages, were the factories of the larger cloth manufacturers and the dye-works. It was in this part of Moscow province that the greatest number of 'factory villages' were to be found. Contemporary accounts paint unflattering pictures of conditions of work in the textile villages, as the following description of a peasant weaver's hut in Moscow district (taken from a 1900 newspaper) illustrates:

> The little hut is small, cramped, and dirty, with three windows along the façade. Inside there are two diminuitive, smoky, and dirty rooms and a kitchen. In one of the rooms, the smallest, three weavers are at work . . . the room has such miserable dimensions that it is impossible to turn round in it . . . the impression I got when I came in was very oppressive; dust thrown up by the work made a thick, black cloud and lay in flakes on the looms, the floor, and on the workers, getting up their noses and into their mouths and ears.[14]

Women constituted a large part of the labour force in the textile region. In Bogorodsk and Bronnitsy districts they worked at looms alongside men, weaving the lighter, lower-quality, and less valuable materials, or they wound cotton yarn onto bobbins. Elsewhere in the textile region,

FIG. 9.3. The principal areas of reed, bark and other woodland-based domestic industries in Moscow province, 1870–1880

FIG. 9.4. The principal areas of animal-product-based domestic industries in Moscow province, 1870–1880

women were involved in knitting, lace-making, and assembling card-board filters for the cigarette industry. Such women's manufactures made up a series of sub-regions within the textile belt, and could, as the following extract indicates, impart a special character to an area. It describes the roads leading into Moscow on market-days:

> Whole files of carts stretch out loaded up with hay, straw, firewood, potatoes, and other vegetables, and on them or walking alongside are women, who talk loudly to one another, all the time quickly moving their hands, their fingers darting backwards and forwards—they are knitting stockings. You see women carrying sack-loads of pitchers and bottles full of cream and milk on their backs with stockings in their hands, knitting.[15]

Outside the textile region it was rare to find one type of domestic production dominating over a wide area; in the western and north-western districts domestic production was more varied. Some industries, such as leather-work and sheepskins, were more or less ubiquitous, in that someone in every local area would be engaged in their production. Other more specialized manufactures were highly localized, and their

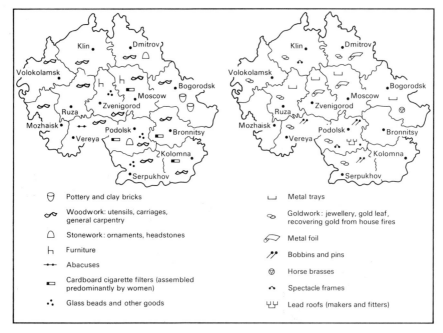

FIG. 9.5. Miscellaneous domestic industries in Moscow province, 1870–1880

FIG. 9.6. The principal areas of metal-based domestic industries in Moscow province, 1870–1880

production became associated with a single village or rural district. Thus, the brush industry was confined to twenty-seven villages on the borders of Podolsk and Zvenigorod districts, and all 'were closely interrelated and almost formed a complete brush-making region'.[16] Similarly, lithograph painters were concentrated in fourteen adjacent villages, glass bead-makers were concentrated in several neighbouring rural districts, silver lace-makers constituted 'one continuous region', and so on. Localization in some industries was such that a village or group of villages would produce only one particular variety or design of an item; in the textile industry rural districts would concentrate on the production of one type of cloth—silk, broadcloth, or cotton—while the design of scarves, wagons, wooden utensils, and trays bore in each instance the mark of the village or area in which they were made. This was not always an advantage when it came to marketing the products: in one group of villages around Vyshegorod in Vereya district the identically patterned red, orange, and indigo blue ribbons produced locally were apparently so unattractive that it was difficult to find buyers for them.[17]

The extent to which individual domestic industries made an impact on their local area varied. In the case of some, production was confined to a few households and the impact was slight, but in others, where a substantial number of local people were involved, villages could be given a distinctive character, like those in the textile region. Buildings were often a clue to the type of industries present in a village. If furnaces were used in production, for example, separate huts would be built away from settled areas to minimize the risk of fire. Such 'primitive zoning' was characteristic of villages where glass beads and metal toys were made. Lean-to sheds, outbuildings, and the design of workshops and small factories could each denote particular activities, and in some cases modification to the peasants' own dwellings—extra windows, raised ceilings, and widened doorways—could signify a loom or work-benches within. Peasant tailors, for example, lived in specially enlarged huts in which half the space was allocated to the cutting and sewing shop.

Involvement in *otkhozhie promysly*, jobs that took peasants away from their native villages, was most strongly developed in the western districts and least in the south and east of the province. In the eastern textile region, the majority of peasants included in the category of those going away to work were in fact employed in a spinning factory or dye-works in a neighbouring village. Elsewhere labour excursions could take peasants much further afield, co-villagers often pursuing the same trade and seeking out similar destinations. Among the 'going-away workers' included by *zemstvo* investigators in the volumes on domestic industry, picture-sellers, roofers, window-fitters, tailors, itinerant furriers, and cockroach-exterminators all originated from neighbouring villages and rural districts. Apparently such identification with co-villagers was developed among those who were destined for Moscow factories as well.[18] Men were more inclined to go on labour excursions than women, and when large numbers of them were involved whole village populations could lose their menfolk. This was the situation in neighbouring villages in Mozhaisk and Vereya districts, where, according to the *zemstvo* investigators, it was rare to encounter men other than at the hay harvest or on feast-days, since, on reaching the age of nine, all boys were sent off to Moscow to work as tailors. The women remaining in the villages did all the farm work and fetching and carrying of firewood and manure. Remarkably, they also pursued their own domestic industry, making fishing-nets out of flax they grew on their allotments.[19]

From the discussion above a rough division of Moscow province emerges. The main dividing-line was between a western region of varied domestic industries (many heirs to traditional peasant crafts) and high rates of participation in labour excursions, and an eastern region dominated by textile and related domestic industries. The division corresponded roughly to areas of more and less productive agriculture in the province. In the eastern district soils were poorer than in the west and landholdings smaller, so that by the latter decades of the nineteenth century farming could provide for the subsistence needs of households for no more than four months of the year. Households were forced to deploy all their family labour into supplementing farm income. This included female and child labour. Fig. 9.7 shows that the female participation rate in domestic industry was greatest in the east of the province. Later, it will be shown how such mobilization of women's labour into domestic manufacturing affected the long-term ability of households to continue as joint agricultural and industrial enterprises.

Outside the textile districts, many peasant industries in Moscow province owed their origin to chance factors. Their histories, which were compiled by *zemstvo* investigators from interviews with peasants, tell stories of peasant 'wiliness' and 'ingenuity'; a peasant boy working in a Moscow factory learnt the secret of a process and returned with it to his native village, a house-serf taught herself lace-making by examining her mistress's foreign gowns, carters happened upon a foreigner who taught them the skill of glass bead-making, and so on. By the second half of the nineteenth century such legends had become part of local folklore. Beneath the legend it is clear that many of the industries that appeared in the nineteenth century were indeed introduced into villages by peasants who had learnt skills in Moscow or in rural factories; but far from setting up in opposition to the Moscow or noble entrepreneurs, peasants frequently retained strong links with their former employers, relying upon them for raw materials and custom. In later years the more successful were able to break such dependence. Sometimes Moscow entrepreneurs provided the capital to set up a peasant in domestic industry. The history of hat-making in Klenovo village, Podolsk district, for example, was bound up with the name of A. T. Aleksandrov, who provided a peasant from the village with the capital to set up his own workshop. Within two decades, Semen Ivanov's hats were known, apparently, the world over, and not only had the workshop in Klenovo become a factory with a substantial labour force but there had also been expansion into a large number of new workshops in neighbouring villages.[20] In other traditional crafts, the nineteenth century witnessed modifications of technique and design that allowed goods to reach a wider market than previously; again, the impulse seems often to have come from experience gained in factories and Moscow workshops. Despite a tendency, once a new manufacture was introduced, towards repetitiveness and conservatism in production, the peasants seemed able to adapt to major market upsets. The list of illustrations is long and includes lithograph painters who, because of over-production in the 1870s, switched to making jar- and hat-labels; silk-weavers who changed to cap-making, lace-makers to brush-making, silversmiths and purse-makers to millinery, and so on. Thus, at any one time domestic industries were in a state of flux, some in decline and some expanding.

A number of domestic industries prospered because of local availability of a raw material. Clay brick-making in the environs of Gzhel, Bogorodsk district, where there were deposits of white clay, was an example of this. Gzhel was also the location of a large porcelain factory. A large number of domestic industries, from furniture-making to toys and bast-work to brush-making, relied upon woods and scrubland. By the second half of the nineteenth century the depletion of forests had caused a contraction of the wood-using industries in the northern and western parts of the province. This also happened to those industries using quantities of firewood, such as metal-working. However, at the end of the nineteenth century a large majority of peasant producers relied on non-local sources for their raw materials. Weavers, for example, were supplied with yarn by the factories for which they worked, while peasants engaged in other industries might travel to Moscow or as far afield as Nizhni Novgorod to buy their raw materials from trading rows and warehouses.

Moscow exerted an influence on the make-up of domestic industries within its immediate sphere of influence. Rural districts close to the capital were distinguished by the great variety of their industries, and by industries with a low value-to-bulk ratio, such as knitting, glove-making, and the making of cigarette filters. Other markets existed in the province which exerted an influence on the domestic industries in their immediate area. Ivashkovo, one of Russia's principal leather markets, stood at the centre of a region of domestic leather-work, boot- and shoe-making, and fur-dressing in the north-west of the province, while the Sergeev monastery, a destination for pilgrims from all over Russia, had become a major market for toys, icons, and religious paraphernalia produced in the local villages. Loosely defined and overlapping as they were, regions of domestic manufacturing existed in Moscow province, which together made up a distinctive micro-geography.

## THE ORGANIZATIONAL STRUCTURE OF DOMESTIC INDUSTRIES

It has already been observed that the organizational structure of domestic production was varied. Variation was in terms of the scale of operations, composition of the labour force, ownership of the means of production, and arrangements for the supply of raw materials and marketing. The example of the weaving industry can serve as an illustration. Weavers could work at home using their own labour, they could rent space for their loom in a neighbour's workshop, they could be entrepreneurs using hired labour alongside their family's, or they could themselves be hired by others. Where the means of production were concerned, weavers might have their own loom, or rent one, or they might have the use of one free. For marketing and obtaining the raw materials with which to work, some might be under contract to a factory, others to an agent or wholesaler.[21] The term used by contemporaries to describe organization in the textile industry, 'the domestic system of large-scale production', thus hides a variety of relationships between producers, their suppliers of raw materials, customers, and the means of production. Further, these relationships were far from static; it was not unknown for a weaver's status to shift from that of hired worker to independent producer and back again to hired worker in tune with fluctuations in the market. Such changes in status were particularly common in those industries where everyone controlled their own means of production. The traditional categories of capital and labour are clearly rather difficult to apply to domestic industries at this time.

All but a few of the peasants engaged in domestic industry in Russia were dependent in some measure either upon other producers or upon merchants and wholesalers for obtaining raw materials or for marketing their products. The slide into dependency is described in much Marxist literature as the 'capture' of peasant production by 'merchant capital'. In Moscow province such merchant capital was represented by a number of different types of intermediary standing between peasant producers and the market. Roughly classified according to the scale of their operations, the functions they performed, and where they were located, they included the following six groups: factories, subcontracting offices (*kontory*), master entrepreneurs, rural-based middlemen, urban-based middlemen, and retailers. The first three were found in the textile and other industries, such as hat-making, the cigarette industry, and the glass industry. The simplest situation was when a factory dealt directly with the peasant outworkers; in return for some security the factory would give out yarn and purchase back the finished products, but more and more frequently, as business boomed in the nineteenth century, enterprises specializing in the distribution of yarn were set up to deal with peasant outworkers. These were generally independent enterprises contracted by factories to handle yarn and finished cloth. As in the case of peasant households, subcontractors were expected to leave a security with the factory against the yarn taken and they would be paid upon delivery of the finished product. In addition to organizing weaving, such subcontractors could also take responsibility for getting yarn dyed before distribution to the peasants. A third way in which factories could deal with peasants was via a master entrepreneur, essentially a peasant agent who, in addition to weaving himself, organized the distribution of yarn among his neighbours. Quite often subcontractors would themselves subcontract to master entrepreneurs. Thus, it was possible for a normal peasant weaver to be separated from the factory making the yarn he used by two intermediaries.

Master entrepreneurs, who, in addition to manufacturing products, performed a marketing and supply role for other peasants, were found in several domestic industries in Moscow province. Sometimes their role as traders took over from production and they became full-time middlemen. The enterprises of these rural intermediaries differed from the subcontractors in the textile industry in the scale of their operations, which were generally smaller, and in the fact that they dealt predomin-antly with urban merchants and retailers rather than with factories. In the nineteenth century such rural intermediaries were known by their fellow villagers simply as the 'bosses', in women's industries *khozyaiki* and in men's industries *khozyainy*. It was rare for rural middlemen to have secured a monopoly of business in a village or rural district by the 1870s, but in some industries they succeeded in getting near to this; in the 1870s the 56 households making fringes in one *uezd* worked for just three 'bosses', for example.[22] In the lace industry it was difficult to persuade Moscow merchants to supply gold and silver thread to a new lace-maker without a recommendation from one of the established *khozyaiki*.[23] According to contemporary investigators, rural intermediar-ies did not necessarily earn more than the peasant manufacturers they 'serviced', and they were viewed in such cases as equal partners in a common endeavour to produce and market goods in as rational a way as possible, but there is no question that some industries provided fertile ground for exploitation. In the knitting industry, for example, it was reported that

The buyers-up and *khozyaiki* take enormous commissions on the materials they supply and these are of the very worst quality. As a commission, they force the knitters to work on their land and vegetable plots, and they get them to hand over food for nothing—also as a commission. They squeeze and drain the population. Who are these people?—nothing other than female kulaks.[24]

In an attempt to defend themselves against such exploitation, the knitters developed various ways of deceiving their 'bosses' about the amount of materials contained in finished stockings and socks, the most widespread of which was to wash them in salted water. As all accounting was done in weight, this practice allowed the knitters to keep back some of their work to sell privately. The advantage remained with the intermediaries however, and the investigator of the 1870s noted that every year saw more women knitters attempting to become the more lucrative intermediaries themselves.

Not everyone who supplied peasants in their villages with raw materials or bought their manufactured goods was a member of the local community; many were itinerant traders who, in the course of their travels, would buy from some peasants and sell to others. The roads of Moscow province must have been full of such people, and among the wares they traded the products of peasant domestic industries had a prominent place. Some traders developed a regular circuit, and producers came to rely upon them to service their industries. For example, fishing-net-makers of Mozhaisk and Vereya districts usually sold to itinerant traders as they rarely had time to get to the market themselves. It was the smaller households which tended to make most use of itinerant traders—although this was not a general rule.

Whether working through intermediaries or making the journey to market themselves, many manufacturers in Moscow province had dealings with urban merchants. The relationship between peasants and merchants was not always a happy one, as the following account of the annual excursion of braid-makers to Moscow indicates:

On the road to Moscow the peasants refrain from drinking vodka, only allowing themselves a small glass if they meet someone they know. Having deposited the goods they have brought with their 'boss' they beat a retreat . . . they wait an hour and a half or more and then approach the merchant's shop again . . . 'Will we be called in?' they ask themselves. Sometimes they are called quickly, sometimes they have to wait several hours . . . but as soon as they are summoned the 'boss' asks, 'What is this you have brought?' 'Braid,' they reply. 'This is what such goods are worth,' says the merchant, stamping the braid underfoot.[25]

Such treatment may not have been typical, but contemporary surveys do contain a catalogue of complaints about cheating and sharp practice in the dealings between merchant and peasant. Trusting relationships could exist, however. In the lace-making industry, for example, Moscow merchants would allocate gold and silver thread to peasants with whom they had regular dealings without demanding any securities. Where no permanent link existed with a merchant, peasants could seek out suppliers and market outlets among the retailers in Moscow's many trading rows (*ryady*): papier mâché boxes were sold directly to tobacconists, sweet-shops, and shoe-shops in *lapotnyi ryad*, leather for shoe- and boot-making could be bought in *mytnyi ryad*, silver in *serebryanyi ryad*, hooks and eyes and pins were sold in *igolnyi ryad*, and so on.[26] In the specialist markets outside Moscow, such as Kimry and the

Sergeev *posad*, peasants might have their own stalls or, as in the city, use established traders to sell their goods.

The precise organizational form of domestic industries in Moscow province seems to have been partially influenced by the distance of producers from their markets. The further away a peasant household was situated from the supply of raw materials and the market, the greater was the likelihood that it would have to employ the services of intermediaries. This relationship was observed in many of the industries in the province. In the textile industry weavers in the immediate vicinity of a factory would deal directly with it, further out they would deal with subcontractors, and further out still with local master entrepreneurs. Travel-time and cost were, of course, the reasons why peasants were forced into the hands of middlemen. In the conditions of travel in nineteenth-century Moscow province, forty kilometres was the greatest distance anyone could expect to travel on horseback in a day, and the distances that could be covered were considerably less in winter months. Peasants living beyond the twenty-kilometre radius of the market or their suppliers therefore stood to lose more than a day on the road if they made the journey themselves. Paradoxically, therefore, the hold of 'merchant capital' over the peasant producers may have been greatest in villages apparently most out of Moscow's reach and the ideal of the independent domestic producer was most often approached near to the city.

## THE AGRICULTURAL AND INDUSTRIAL YEAR

Like Lenin, *zemstvo* analysts were interested to discover the ways in which involvement in domestic industry affected the economy of peasant households. As might be expected, the picture they revealed was complicated. There were differences, some of them following geo-graphical contours, in the numbers of households that had withdrawn from the land to devote themselves full-time to domestic industries, in the relative contribution industry and agriculture made to household budgets, and in the way household labour was deployed between the sectors. On the crucial question of whether domestic industries were associated with high or low rates of withdrawal from the land, the 1899 census data provided some evidence to support Lenin's view that involvement in industry did undermine the farm economy, since in the eastern, most industrialized districts the percentage of households that had ceased to work the land they held reached its highest levels. As Fig. 9.7 shows, this was in the region of 15 to 25 per cent in Bogorodsk, Bronnitsy, Kolomna, and Serpukhov districts, but fell off sharply to the

Percentage number of households entitled to hold land
in the commune that were not working it in 1898-1900

| | |
|---|---|
| ▨ | 20.01-25.00 |
| ◨ | 15.01-20.00 |
| ◧ | 10.01-15.00 |
| ☐ | 0.00-10.00 |

FIG. 9.7. Percentage of peasant households in Moscow province not working their land at the beginning of the twentieth century

north and west. When added to the number of landless, the percentage of rural households that did not make a livelihood from the land reached one-third in some rural districts, as Table 9.1 shows. The number of landless included both those households that had left their commune for good (*otsutstvuyushchie dvory*) and the resident landless.[27] The figures tell of peasants' difficulties in making a living from their land, but whether it is right to attribute the withdrawal of peasants from the land to their involvement in industries is another matter. As Vikhlaev showed in his analysis of the 1899 census data, the size of landholding was a crucial determinant of households' ability to continue farming. The households with the smallest landholdings were able to command the least remunerative jobs in industry.[28] As a result of the combined misfortune of a small landholding and low-paid work in domestic industry, they were likely to fail.

For households which continued to combine farming and domestic industry (the majority in the province), the successful dovetailing of

TABLE 9.1. *Incidence of peasant withdrawal from land in Moscow province,*
*1899*

| District | % of all registered households | | % of resident households not working their land |
|---|---|---|---|
| | Not working their land | Absentee | |
| Bogorodsk | 34.9 | 17.1 | 24.6 |
| Bronnitsy | 35.5 | 25.7 | 21.9 |
| Dmitrov | 31.5 | 19.7 | 5.4 |
| Klin | 27.5 | 14.9 | 9.1 |
| Kolomna | 35.1 | 24.1 | 23.3 |
| Mozhaisk | 28.1 | 19.0 | 11.6 |
| Podolsk | 25.8 | 19.7 | 12.2 |
| Ruza | 30.9 | 20.5 | 9.6 |
| Serpukhov | 34.2 | 25.3 | 17.7 |
| Vereya | 28.2 | 18.1 | 12.9 |
| Volokolamsk | 27.1 | 18.4 | 5.6 |
| Zvenigorod | 26.2 | 16.1 | 9.1 |
| OVERALL | 30.3 | 19.2 | 14.7 |

*Source: P. A. Vikhlaev, Moskovskaya guberniya po mestnomu obsledovaniyu 1898–1900, i 1 (Moscow, 1903).*

work in the two sectors was important. In all but a few domestic industries, work was confined to the winter months, when the fields lay under snow, and dates of the commencement and termination of manufacturing were regulated by major events in the farming year, such as spring ploughing, the midsummer hay harvest, and autumn harvesting and ploughing. By the last decades of the nineteenth century changeover dates had become fixed by tradition in a number of industries. In the knitting, pin, and glass-bead industries, for example, the working year extended from 8 September to 8 July, in the braid industry it was from 1 October to 8 July, and in weaving from 15 September to 15 April in the western and northern parts of the province and from 15 September to 19 July in the east. Each of these dates corresponded to a major Orthodox church holiday. In some industries peasants would use rainy days during the summer months to take up work, but, even when the processes involved lent themselves to stopping and starting, this was the exception rather than the rule. Like their urban counterparts, rural employers had difficulties lengthening the working year to include the summer months. Therefore, despite increases in pay and the promise of bonuses, the output of small rural factories and of outworkers in the larger industries could be halved in the summer months. The withdrawal of labour was often graduated, men leaving first and minors, the elderly, and women following later.

The tradition of the summer changeover to agricultural work was so deeply ingrained in peasant custom that even landless peasants would give up work in industry on the prescribed date. This was reported as being usual in the textile industry. Such landless peasants would work as day-labourers in agriculture in the summer or would spend this time collecting berries and mushrooms from village common lands.[29] Equally ritualized were the beginning and end of the industrial year and the observance of church holidays during the work period. As with everything to do with domestic industries, the number of days taken off or on which holidays were observed differed according to industry, although the large festivals such as Easter and Christmas were universally kept. Glass bead-makers and knitting-workers took the greatest number of days off during the winter months, the total adding up to seven weeks. With the appropriate number of days subtracted for holidays, journeying to market, and different start-up and put-down dates, the length of time spent at work in Moscow province's domestic industries ranged from 192 days for braid-makers to 270 for milliners.

Wherever the length of the manufacturing season had not become fixed by tradition, other factors influenced households' patterns of work. Household size was among the most important. The general rule was that the greater the number of family members a household had, the more flexible its strategies of labour deployment could be. A study of households in the brush-making rural districts, for example, showed that whereas a household consisting of husband and wife alone could work from 22 October to 6 May at making brushes, larger families could work from 1 September to 8 July and, if their size exceeded the number needed at the height of the hay harvest, work could continue all year round. In a number of industries the most successful enterprises were those where the head of household had managed to avoid family partition. Among the rural silversmiths and goldsmiths the most successful was a household of thirty-two members which occupied three separate dwellings but operated as a single economic unit. Similarly, in the hat industry it was mainly the heads of 'multiple' households who were the master entrepreneurs. The relationship between household size and status was observed in a detailed survey of households in three districts, as shown in Table 9.2. Isaev, author of the two earliest volumes on domestic industries in Moscow province, charted for the wood-processing industry the growth and development of what he termed the 'family association': as successive sons 'came of age' (at twelve years in this industry) so the prosperity of woodworking households would increase until a point was reached when the family association would

TABLE 9.2. *Average size of households involved in domestic industry in three districts of Moscow province*

| District | Av. family members per household | | |
|----------|-----------------------------------|---|---|
| | Master entrepeneur households employing hired and family labour | Independent households using their own labour | Households of wage-workers in domestic industry |
| Bronnitsy | 7.53 | 6.38 | 6.38 |
| Vereya | 8.04 | 6.50 | 6.41 |
| Moscow | 7.16 | 6.11 | 5.70 |

*Source:* P. A. Vikhlaev, *Moskovskaya guberniya po mestnomu obsledovaniyu 1898–1900*, i 1 (Moscow, 1908)

split up, and the process would begin again for each newly formed household.[30] This model of demographically determined change was later developed for all peasant households, 'industrial' and 'non-industrial' alike, in the works of the Organization and Production scholars.[31]

Not all households relied on their family labour alone. As contemporary investigators observed, the use of hired labour was a common feature of many domestic industries. Depending upon their number, the position of these hired workers in the employer household could differ. When they were not numerous, they could be treated almost as part of the family, working alongside members of the household and eating and taking breaks with them, but in the larger enterprises this was less common. The use of hired labour could increase the options available to heads of households in deploying family labour. Another bonus of hired labour was that it could be used in agriculture, which was normal practice in a number of industries.

A similarly important decision every household had to make was whether its women should work in industries. Women's work on peasant farms was less seasonal, with fewer slack periods than men's: in addition to sowing and harvesting alongside the men, women were responsible for looking after the small livestock, cultivating the kitchen garden, carrying water, and preparing food. These tasks went on all year round and were labour-intensive. The withdrawal of women's labour and its transfer to industry were, therefore, potentially more difficult to accommodate than the withdrawal of men's labour. This might explain why in the east of the province, where, as Fig. 9.8 shows, women's participation rates in domestic industry were high, the number of households that had given up working their land was above average. It might also explain the tendency for women to opt for the low-paid domestic industry, such as knitting, which could be picked up and put

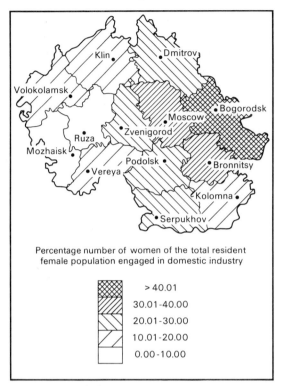

FIG. 9.8. The participation rate of women in domestic manufacture in Moscow province, 1898

down quite easily, and could thus be fitted into the domestic timetable. Whether to use their children's labour in domestic industry was another decision to be made by households. Child labour was in widespread use in Moscow province and took the form either of a supplement to the parents' activities or of a formal apprenticeship to a master. From the age of seven or eight, girls could be expected to spend almost as much time as women knitting, making lace, and sewing gloves. In the textile districts young boys fared no better, having to spend the working day loading bobbins for the weavers' looms. With such use of child labour there were good reasons for households to strive to increase family size and there must have been a premium on childbirth. In an interesting reversal of the usual state of affairs, *zemstvo* investigators noted that the birth of a girl was more welcome than a boy in the lace-making districts in the south.[32]

DOMESTIC INDUSTRY AND THE HOUSEHOLD ECONOMY

The *zemstvo* publications of the 1870s include a detailed breakdown of the annual budget of individual peasant households involved in a number of different domestic industries. These allow some conclusions to be drawn about the contribution of industrial earnings to household economies. Contemporary commentators usually concluded that such earnings met peasant households' need for cash income to pay taxes and to purchase goods in the market-place: Vikhlaev, for example, wrote in 1908 that 'in the majority of cases agriculture does not provide families with the wherewithal to cover their need for money income. The source of such money is found almost exclusively in non-agricultural employment.'[33] V. Vorontsov also referred to a subsistence agriculture combined with a market-oriented industrial production.[34] The *zemstvo* investigations revealed a more complicated reality for the 1870s. At this time households in Moscow province seem to have been occupationally plural with both subsistence and money income coming from a variety of agricultural and non-agricultural sources in differing proportions.

Table 9.3 is a summary of detailed budgets contained in the censuses of the 1870s for a sample of households involved in different domestic industries. The sources of income included in the tables were diverse. Apart from the principal domestic industry and agriculture, household members were involved in a variety of income-generating activities, such as haulage, agricultural wage-labour, carting firewood and manure, taking in washing, and renting rooms. As the table shows, the main domestic industry practised by each household was not always the principal source of household income: with the exception of the brush-making industry, no household derived more than half its income from the main manufacturing activity. Farming clearly continued to be important in the household economy of peasants in Moscow province in the 1870s, despite their outwardly industrial aspect. The *zemstvo* investigations indicated that the division of household income into 'subsistence agricultural' and 'money-earning non-agricultural' sectors did not hold good for all households in the 1870s. First, not all farm products were consumed at home but could enter the market as commodities. This applied in particular to produce from the kitchen garden, such as vegetables, and small livestock. Secondly, domestic industry was not always rewarded in cash. The methods of payment for domestic industry were varied and could include money wages, consumption goods (such as tea, sugar, soap, tobacco, vodka, grain, kerosene, and matches), or a combination of the two. Most commonly, it was peasants working for a local employer or middleman who had to

TABLE 9.3. *Summary budgets of six peasant manufacturing households in Moscow province*

| | % share of household income derived from | | |
| --- | --- | --- | --- |
| | allotment and kitchen garden | main domestic industry | other non-allotment sources[a] |
| Hat-maker[b] | 32.2 | 48.5 | 19.3 |
| Brush-maker[c] | 43.2 | 29.6 | 27.1 |
| Braid-maker[d] | 52.0 | 31.0 | 17.0 |
| Basket-maker[e] | 32.6 | 56.5 | 10.9 |
| Boot-maker[f] | 46.4 | 33.7 | 19.9 |
| Abacus-maker[g] | 43.3 | 49.6 | 7.1 |

[a] These included secondary domestic industries and services, such as haulage or domestic service.
[b] *Promysly Moskovskoi gubernii*, i (Moscow, 1879), 178–83.
[c] Ibid. 50.
[d] Ibid. ii (Moscow, 1880), 151.
[e] Ibid. iii (Moscow, 1882), 41–3.
[f] Ibid. 95.
[g] Ibid. 142–4.

accept the 'gift box' system of payment. Sometimes agreement to accept goods in place of money was used by a head of household to ensure that he received the earnings of a family member. For example, it was reported that in comb-making villages 'it is mainly young people who work in the workshops and their earnings are usually taken by the older members of the family in the form of various goods: the comb-makers are left with very small payments of 15 to 20 copecks'.[35] In other cases, peasants accepted goods as a form of credit in lieu of money wages. As pay-day was often only once in three months the economically weaker households had no alternative but to go along with this method of payment. It was generally unpopular among the peasants, but for the employer households and merchants payment in goods could provide an easy way of increasing profits: 'All these provisions taken by the peasants', continued the *zemstvo* investigator, 'are valued at 25 to 30 per cent, and sometimes more, above the local market price.'[36] Many of the larger local employers opened their own shops, where their workers could obtain goods on credit, which served much the same purpose.

When a peasant craftsman worked for another peasant or merchant he usually had to keep a work-book, in which all advances made to him were recorded, and which was used at intervals to reckon up his earnings. An extract from such a book belonging to a master button-maker included in a *zemstvo* survey showed that approximately 75 per cent of the advances the peasant received from his employer were in the form of goods such as tea, herrings, soap, and tobacco.[37] Earnings from industry, in whatever form they came, were not necessarily all destined for the satisfaction of basic needs. Some went towards meeting rising peasant aspirations for a higher standard of living. The nineteenth-century investigators of domestic industries charted the incidence and frequency of tea-drinking; the normal diet; clothing; and other habits of peasant households in different industries. In at least one case distance from Moscow seems to have been a factor accounting for variations. The domestic industry concerned was knitting, where the 'sober' and 'hard-working' attitudes of girl and women knitters in villages far from Moscow were contrasted with the 'disruptive' attitudes of those living within easy reach of the city. According to contemporary observers, women in villages near Moscow broke from their knitting to drink tea more frequently than their counterparts further out and, moreover, expected to be bought 'nice dresses' for their trips to market. By the 1870s such demands had already led to family breakdowns and were the cause of the departure of some young women to Moscow in search of work.[38] This example is included not as a comment on peasant women's expectations but to make the more simple point that, while the domestic system represented an adaptation to economic changes in the nineteenth century, its development at the same time brought new factors into play that set the stage for its own transformation.


CONCLUSION

The study of peasant industry in Moscow province underlines the great diversity of the peasant economy which followed regional lines. This is not to deny, however, the power of changes in the macro-economic environment to break down such regional differences. Where peasant domestic industries were concerned, the historical prediction by Lenin and others that they must in time be replaced by other forms of industrial organization does not seem to be unreasonable. Mechanization would destroy the rationale for much domestic industry from the point of view of the large employer, while the problem of keeping family labour at home would become increasingly difficult for household heads as the lure of city lights and regular wages grew. However, in the last decades of the nineteenth century domestic industries in all their various forms were still flourishing, and they were destined to survive through the years of revolution in the twentieth century and beyond. They were finally eliminated in the 1930s during the collectivization drive.

# 10

# City and Tsar

## *The State and the Towns in the Romanov Era*

THROUGHOUT the Romanov era Russia remained an overwhelmingly rural society. In the early seventeenth century Russia's towns had less than 5 per cent of the population, and still contained only 17 per cent at the end of the period.[1] Despite this apparent insignificance in demographic terms, the towns played an important role in Russian life. Not only were they of profound significance to the country's economy, but they were also pivotal points in politics and administration. To quote J. Michael Hittle:

Even in vast, rural, peasant Russia, political organization proceeded through the cities. Not only were major decisions of state made in the great cities, Moscow and St. Petersburg, but their execution depended to a significant degree on crown representatives who had final authority over cities great and small. And within the cities, the elite constituted a potent force—as representatives of the crown, as consumers, as holders of large blocks of urban real estate, and not infrequently as sponsors of the urban economic activities of their rural dependents.[2]

To enquire into the nature of cities and towns over the three-hundred-year Romanov era, therefore, is to enquire into a key feature of the society of the day.

That towns were central features of Romanov society, however, does not necessarily mean that they were the major engines of social change, or that they had the same dynamic role in Russian development that Western European towns are generally believed to have had in Western history. As is well known, many, especially pre-revolutionary, historians have described the apparent backwardness and sluggishness of Russian urban life, and have traced the causes to the effects of serfdom and the centralizing policies of the state.[3] Questions have been raised about the forces which hindered urban enterprise and the development of a commercially minded middle class on the Western model. There have even been suggestions that the Russian town failed to evolve a fully urban character in the way that Western European cities did.[4] Arguments about the backwardness of Russian urbanism lose some of their force at the end of the Romanov period, when industrialization was unleashed, but even then a comparative perspective reveals how unlike Western industrial cities they still were.[5]

The changing spatial distribution, relative importance, and physical form of Russian towns during the Romanov era is an important topic for historico-geographical investigation, not only because towns were an integral part of the landscape, but also because urban changes are symptomatic of, or caused by, other social changes, and themselves serve to promote change. The forces which influence urban development are complex, but a significant question which suggests itself in the Russian context during the Romanov period is the extent to which the absolutist state itself was an important agent of urban change. The geographical theory of the development of urban networks and hierarchies has tended to be based on essentially economic models, such as Christaller's Central Place Theory and the mercantile model suggested by Vance.[6] There is no developed body of theory on the historical role of the state in this area.[7] The central question here, therefore, concerns the role of the state in moulding the system of towns and cities and an assessment of how successful the state was in its policies towards towns. The broader question of the relationship between the towns and Russian development as a whole is also considered, although this is a topic which has already been tackled by a significant number of scholars and has yet to become the subject of overall agreement.

## THE TOWNS IN THE PRE-PETRINE PERIOD (1613–1682)

The sources describing the population and social characteristics of seventeenth-century towns are incomplete and of doubtful accuracy. However, a number of analyses have been undertaken and a list of the largest towns, based on Ya. Ye. Vodarskii's survey of the enumerations of *posad* dwellers of 1646, 1652 (i.e. after the incorporation of 'white places' or ecclesiastical and private suburbs into the *posady*), and 1678, will be found in Table 10.1.[8] This shows that the largest places, with over 1,000 *posad* households (Moscow, Vologda, Kazan, Kaluga, Kostroma, Nizhnii Novgorod, and Yaroslavl), were all old towns dating from before the sixteenth century and, with the exception of Kazan, all with a long history of connection with Muscovy. A common feature of these large towns was their strategic location on waterways, as shown in Fig. 10.1. As regards population trends between 1652 and 1678, it is not easy to discern any overall pattern, except to note some sharp decreases in some of the large towns, such as Vologda, Kazan, Kostroma, Nizhnii Novgorod, and Yaroslavl. H. L. Eaton has argued that urban growth was sluggish at best after the middle of the century, and has questioned the conclusions of Vodarskii and other Soviet scholars that urban expansion

TABLE 10.1. *Urban household totals in the seventeenth-century (towns with 500 or more households in the* posad)

| Town | 1646 | | | 1652 | 1687 | | |
|---|---|---|---|---|---|---|---|
| | 1 | 2 | 3 | 4 | 5 | 6 | 7 |
| Archangel[a] | 645 | 1,018 | 263 | 715 | 835 | 4 | 138 |
| Arzamas | 430 | 2 | 135 | 559 | 560 | — | 98 |
| Balakhna | 637 | — | 112 | 661 | 642 | 9 | 140 |
| Galich | 729 | (41) | 46 | 788 | 481 | 19 | 46 |
| Kazan | 1,191 | 1,600 | 200 | — | 310 | — | — |
| Kaluga | 588 | 339 | 105 | 694 | 1,015 | — | 45 |
| Kargopol[b] | 538 | 20 | 6 | — | 666 | — | — |
| Khlynov | 624 | 1 | 26 | 661 | 616 | 20 | 142 |
| Kolomna | 615 | 8 | 261 | 740 | 352 | — | 79 |
| Kostroma | 1,726 | 54 | 414 | 2,086 | 1,069 | — | 106 |
| Kursk | 270 | 396 | 20 | — | 538 | 104 | 11 |
| Moscow | 1,221[c] | (20,000)[c] | 8,000[c] | 3,615 | 7,043[d] | — | — |
| Nizhnii Novgorod | 1,107 | 500 | 666 | 1,874 | 1,270 | — | 600 |
| Novgorod | 640 | 1,050 | 145 | 770 | 862 | 153 | 344 |
| Olonets | — | — | 155 | 155 | 637 | — | — |
| Pereslavl-Zalesskii | 525 | (80) | 104 | 624 | 408 | — | 110 |
| Pskov | 940 | (1,306) | 51 | 997 | 912 | 372 | 1,043 |
| Rostov | 416 | (15) | 167 | 552 | 491 | — | 217 |
| Simbirsk | — | — | — | 19 | 504 | — | 114 |
| Sol Kamskaya | 549 | 9 | 146 | 686 | 831 | 25 | 20 |
| Suzdal | 360 | (14) | 495 | 435 | 519 | 7 | 596 |
| Torzhok | 486 | 8 | 58 | 508 | 659 | — | — |
| Tver | 345 | 53 | 250 | 497 | 524 | — | 110 |
| Uglich | 447 | — | 226 | 603 | 548 | — | 49 |
| Ustyug Velikii | 744 | 53 | 36 | — | 920 | — | 119 |
| Vladimir | 483 | 58 | 405 | 703 | 400 | — | 290 |
| Vologda | 1,234 | 175 | 363 | 1,674 | 1,196 | 13 | 284 |
| Yaroslavl | 2,871 | 174 | 564 | 3,042 | 2,310 | 57 | 468 |
| Zaraisk | 446 | (127) | 65 | 587 | 254 | — | 1 |

*Key:* 1. *Posad* households, 1646.
    2. Servitor households, 1650 (figures in parentheses—1632).
    3. Other households, 1646.
    4. *Posad* households, 1652.
    5. *Posad* households, 1678.
    6. Servitor households, 1670s (partial data).
    7. Other households, 1678 (partial data).

[a] With Kholmogory.
[b] With Turchasov.
[c] Data for 1638.
[d] Figure for 1700.

*Source:* Ya. Ye. Vodarskii, 'Chislennost i razmeshchenie posadskogo naseleniya v Rossii vo vtoroi polovine XVII v', in *Goroda feodalnoi Rossii* (Moscow, 1966), 271–97.

was a symptom of the development of the 'all-Russian market'.[9] The total number of *posad* inhabitants does seem to have continued to grow after the middle of the century, although probably not as quickly as

Fig. 10.1. Towns in seventeenth-century European Russia

Vodarskii suggests.[10] A fair proportion of this growth can be accounted for by the contribution made by new towns. It seems that some of the sluggishness of the old towns was offset by a more buoyant situation on some of the frontiers.[11]

In addition to the *posad* dwellers, who were responsible for much of the urban commerce and trading activities, Table 10.1 also gives data for the state servitors or military personnel living in the towns. For Russia as a whole Vodarskii estimates that servitors constituted 51 per cent of the urban population in 1646, and 45 per cent in 1678.[12] They were particularly numerous towards the frontiers. Indeed, Table 10.1 fails to indicate some of the really big urban garrisons. Thus, Belgorod, with only 44 *posad* households in 1646, had 459 servitor households in 1650. In Kursk the respective numbers were 270 and 396, in Sevsk 0 and 6,017, in Voronezh 85 and 1,135, and in Astrakhan 0 and 3,350.[13] The frontier towns have thus been widely recognized as a different type of town from those in the centre and north of the country.[14] Some Western scholars have described them as mere forts, on the grounds that the servitors were employed on military duties and supported themselves through agriculture or on subsistence provided by the state. The tendency has been to minimize their contribution to trade and commerce. However, this is to overlook much evidence to the contrary. Thus, at Voronezh in 1615, of sixty-three shops and half-shops, twenty-three were held by various types of servitor. Servitors held twenty-five shops in 1648, twenty-six in 1654, and thirty-three in 1658 (out of a total of 137 in this last year).[15] A similar picture is to be found elsewhere in the south, and later in the century a particularly important trading role was played by the Ukrainian cossacks. Far from the 1649 Law Code restraining their activities, it seems that, in Siberia at least, the commercial activities of the urban-based servitors may actually have expanded in the second half of the century.[16]

The primary significance of towns in the early Romanov period was twofold: in their role as military and administrative centres, and in their role as commercial nodes.[17] With respect to the first of these, a singularly important function of the early Romanov towns was the part they played in the consolidation of the new dynasty. It was during this period that the office of military governor, or *voevoda*, which had previously existed only on the frontiers, was extended to all the centrally located towns of the realm and largely displaced the previous semi-elected forms of local administration. Hence, the military governor became a major agent of Muscovite centralization, ruling over his town and its subsidiary district in the name of the tsar, and supervised by the latter's Moscow-based officials. State power thus became town-centred, and the towns themselves were closely controlled by the state. During the course of the seventeenth century that power was enhanced by the absorption of private or ecclesiastical suburbs and urban households (especially under the terms of the 1649 Law Code) and the gradual disappearance of private towns.[18]

Not only did the Romanov dynasty enhance its power over towns during the seventeenth century, but it also greatly added to the network of towns. R. A. French estimates an increase of more than a hundred towns in the Russian network during the century, taking the total to over three hundred.[19] In addition to the towns acquired in recently annexed western territories, many new towns were founded on the southern frontier and in Siberia. Generally speaking, the development of these towns, from their initial siting to their settlement and the subsequent regulation of their societies, was under the close supervision of the state. On the southern frontier the founding and settling of towns and their districts was undertaken as part of the policy on frontier defence, and was therefore the task of the Military Chancellery (*Razryadnyi prikaz*) in Moscow. In Siberia a rather similar process occurred, although here, on the swiftly moving frontier, it is less clear that the government was always as closely involved at the initial stage. Nevertheless, the organization of administration under a military governor sent by Moscow quickly followed more local initiatives.[20]

Towns, then, were pivotal points in tsarist administration, so much so that the word 'town' (*gorod*) was used in official documents to denote both the town and its subsidiary area. Although each military governor was individually responsible to his masters in Moscow, in certain frontier areas the needs of defence gradually led to the setting up of military districts, or *razryady*, in which the armed forces, and much of the civil life as well, came to be organized and co-ordinated by the military governor of one particular town. After certain attempts to organize such districts along the old Zasechnaya Line—the fortified line running to the south of the Oka—the first permanent military district was established in the 1640s and 1650s centred on Belgorod.[21] This was followed by other frontier *razryady* (Sevsk, Smolensk, Novgorod, and Kazan) and, in the last quarter of the century, by some less peripheral ones (Moscow, Vladimir, Tambov, and Ryazan). The centres of these districts seem to have been chosen for their strategic positions rather than for their size or previous importance, but in some cases the new administrative roles must have added to their potential for growth.

In addition to their administrative and military roles, many towns were also commercial centres. Through taxation and other means, such towns were major sources of government revenue, and therefore the state took a keen interest in their economies. On the frontiers, where new towns were developed, special provision was made for the laying out of market squares and trading rows, and *posady* were soon added to the original servitor suburbs. However, by no means all the new towns proved commercially successful. Elsewhere the state strove to control and regulate economic activity and to ensure that the urban residents paid their taxes and fulfilled all the many duties that were required of them.

An important effect of the 1649 Law Code was to bind the *posad* dwellers permanently to their *posady*. Thus, in exchange for a legal monopoly over urban trade, the *posad* dwellers were no longer free to migrate and thus to escape their many obligations. Such migration controls were, of course, not always effective. The same is true of the state's ability to control the recruitment and movement of its military servitors.

The state therefore dominated the economic and social life of the towns, and this has led numerous historians to lament the negative consequences of such control for commercial enterprise and activity. Important though many restraints no doubt were, it would be all too easy to exaggerate the extent of the state's power. The military governors, for example, though appointees of the tsar, were by no means always faithful servants, and were much given to corruption and arbitrary practices. Nor were they always successful in enforcing the tsar's will or in maintaining law and order. In the late 1640s a series of urban disturbances forced the tsar to summon an Assembly of the Land, which led in turn to the promulgation of the 1649 Law Code. The abolition of the 'white places' was a concession to the *posad* dwellers, who were resentful of the tax privileges of their neighbours. However, as noted in Chapter 1, the riots of the 1640s were only some among many to disturb the realm during the course of the century.

Just as the state was limited in its ability to control the social and economic life of the towns, so it was limited in the degree of its influence over the town's physical form. Only in the new frontier towns, and in certain instances when the 'white places' were incorporated into the *posady*, was effective planning possible.[22] Elsewhere the traditional urban pattern, based on the divisions between *gorod* (fortified core), *posad*, and 'free suburb' (*sloboda*), continued as before.[23] Public welfare matters, such as fire protection and road repair, were badly neglected. It was only where the physical form of the town bore on essential matters, such as defence and the building of fortifications, that priority was given to construction, and even then the record was patchy. The prevailing situation has been well described by Hittle:

Crowded homes, crudely constructed and modestly furnished, sat on narrow muddy streets. Amenities and refinements were few. And everything was subject to the periodic ravages of fire. Public buildings, where they existed, were equally humble, reflecting an indifference towards broader civic aims. Only the churches, marvellous in design and sumptuous in decoration, suggested the presence of accumulated wealth . . .[24]

Compared with towns in many parts of western Europe, the outstanding characteristic of Russia's towns in the seventeenth century was their lack of autonomy and unity. Indeed, the term *gorod* had no definite meaning at all in law.[25] For example, in the administrative or even social sense there was no clear demarcation between town and country, since the power of the military governor extended over both, and some social groups, such as certain servitors, overlapped between the two locations. Nor was the town or city itself a single entity, being subdivided into different social strata with different service functions. The *posad*, or merchant and trading quarter, was by no means the entire city, and neither the inhabitants of the *posad* nor the other urban residents enjoyed anything in the nature of urban citizenship.[26] In the words of I. I. Dityatin, the city was a 'conglomeration of taxable communes' having little in the way of common interests to bind them together.[27]

Yet, in spite of this diversity, there was one characteristic that almost all urban residents had in common—their obligation to serve the state. As J. Keep[28] and others have argued, the designation of such service was actually a device whereby a poor and undergoverned realm could ensure the stability and resources it needed. The burdens it implied added significantly to the problems of the townsmen, but in the circumstances of the day these seemed essential to the state, and possibly inevitable to its citizens. They certainly did little to encourage the self-confidence and animation which have characterized urban life in certain other times and places; but then the effects of environment, poor communications, and limited opportunity undoubtedly tended in the same direction.

## The Petrine Period and the First Half of the Eighteenth Century (1682–1762)

The Romanov tradition of adding to the urban network continued in this period. French estimates that some thirty-five towns were added by Peter the Great.[29] Some of these, like Revel and Riga, were pre-existing towns in territories gained by Peter from his enemies. Others, like St Petersburg and Petrozavodsk in the north-west, or Pavlovsk and Boguchar to the south, were new. Succeeding years further extended the list—towns were won in territories taken from the Swedes and the Turks, and yet others were built in Bashkiria, Siberia, and elsewhere. V. M. Kabuzan lists 335 towns (including forts) for the period 1730–67; however, some smaller places may have been omitted.[30] At the first revision in the 1720s 189 towns had *posady*.[31] The total *posad* population was 183,453 males in 1724–8, rising to 228,365 in 1769.[32] The official *posad* population in 1769, then, was only 3.1 per cent of the registered population of the country.[33]

Eighteenth-century urban population data are as difficult to use as those for the seventeenth century, and there has been little detailed analysis by scholars. In particular, the data on the non-*posad* populations are inadequate. An indication of the position in the larger *posady* is given in Table 10.2, revealing some interesting comparisons with Table 10.1. Thus, as in the earlier period, Moscow remained predominant, though its *posad* appears to have suffered in the intervening era, possibly because of the many exactions of Peter's reign. Other towns had similar problems down to 1710, but there seems to have been some recovery thereafter. Yaroslavl, with over 8,000 in the *posad* in 1722, remained a very important town, but Kostroma, Vologda, Nizhnii Novgorod, and Kazan had slipped in relative significance. Nevertheless, Kirilov records a military garrison of over 5,000 at Kazan in the late 1720s.[34] Remarkably consistent in its growth since the first half of the seventeenth century was the town of Kaluga, situated to the south-west of Moscow on the Oka waterway. This no doubt reflected the gradual settlement of the Central Black Earth Region and its development as a grain surplus area. Towns such as Orel, Kursk, Tula, Belev, Bolkhov, and Voronezh were thus growing quite quickly in this period. To the north, Archangel, Kholmogory, Vologda, and Velikii Ustyug were suffering as a result of trade restrictions imposed by Peter in favour of his new capital. Novgorod and Pskov also remained in the shadow of the new city.

St Petersburg was undoubtedly Peter's most spectacular contribution to the development of Russian urbanism. The city was constructed on territory conquered from the Swedes at the beginning of the eighteenth century, and became the capital in 1712. From 1705 Peter began a fixed policy of forced settlement in the city, and in 1719 decreed that all Russian landowners possessing more than forty serf-households were to build a house there, in which they were themselves to live.[35] In 1710 St Petersburg had a permanent population of at least 8,000, and this had grown to 40,000 by 1725. As noted already, the development of the city was important not merely for its own sake, but also because of the effects it had on other cities, to say nothing of the great burdens it imposed on the Russian population as a whole.

More than the other cities of Russia, the twin capitals of St Petersburg and Moscow came to be characterized by a mixed population of merchants, servitors, officials, and others. Hence Moscow, with a total (male and female) population of nearly 139,000 in the 1730s, had only 24,000 *posad* dwellers and 15,000 military personnel. There was a large population of nobles, dependants, and others.[36] Likewise St Petersburg, with 68,000 people in the 1730s, had only 4,800 *posad* dwellers but nearly 26,000 military personnel. Many nobles and their dependants lived in the capital.[37]

TABLE 10.2. *Urban population totals in the eighteenth and early nineteenth centuries (towns with over 10,000 people in the 1780s)*

| Town | *Posad* dwellers (male) 1722[a] | Total population 1780s[b] (000s) | Total population 1811[c] (000s) |
|---|---|---|---|
| St Petersburg | | 297 | 335.6 |
| Moscow | 13,673 | 213 | 270.2 |
| Riga | | 35 | 32.0 |
| Saratov | 2,093 | 31 | 26.7 |
| Kiev | | 30 | 23.3 |
| Astrakhan | 877[d] | 30 | 37.8 |
| Tula | 1,958 | 25 | 52.1 |
| Kazan | 2,279 | 22 | 53.9 |
| Tobolsk | (1–2,000)[e] | 20 | |
| Yaroslavl | 8,484 | 19 | 23.8 |
| Kursk | 2,554 | 19 | 23.5 |
| Kaluga | 6,100 | 17 | 23.1 |
| Orel | 2,773 | 17 | 24.6 |
| Revel | (1–2,000)[e] | 17 | 17.6 |
| Orenburg | | 16 | 5.4 |
| Tver | 2,846 | 16 | 17.5 |
| Smolensk | 2,028 | 15 | 12.4 |
| Archangel | 2,875[f] | 15 | 11.0 |
| Voronezh | 1,532 | 13 | 22.1 |
| Irkutsk | (500–1,000)[e] | 12 | |
| Nezhin | | 12 | |
| Mogilev | | 12 | 5.8 |
| Akhtyrsk | | 11 | |
| Vitebsk | | 11 | 16.9 |
| Tomsk | (500–1,000)[e] | 11 | 8.6 |
| Yelets | 1,605 | 11 | |
| Ostrogozhsk | 328 | 11 | |
| Bolkhov | 3,746 | 11 | |
| Novgorod | 2,570 | 11 | 6.3 |
| Vologda | 2,662 | 10 | 9.6 |
| Simbirsk | 2,097 | 10 | 13.3 |
| Kronshtadt | | 10 | |
| Yeniseisk | (500–1,000)[e] | 10 | |
| Kharkov | | 10 | 10.4 |
| Sumy | | 10 | |
| Minsk | | 10 | |

[a] *Source:* Ya. Ye. Vodarskii, 'Chislennost i razmeshchenie posadskogo naseleniya v Rossii vo vtoroi polovine XVII v.', in *Goroda feodalnoi Rossii* (Moscow, 1966), 271–97.

[b] *Source:* G. Rozman, *Urban Networks in Russia, 1750–1800, and Premodern Periodization* (Princeton, 1976), 159–219.

[c] *Source:* A. G. Rashin, *Naselenie Rossii za 100 let, 1811–1913 gg.: statisticheskie ocherki* (Moscow, 1956), 87–91, 93.

[d] The 1723 Main Magistracy Regulation records Astrakhan as having between 1,000 and 2,000 *posad* households.

[e] Number of households according to the 1723 Main Magistracy Regulation. See Ya. Ye. Vodarskii, 'Spisok gorodov Rossii s ukazaniem primernogo kolichestva posadskikh dvorov (1723 g.)', *Istoricheskii arkhiv* (1961), no. 6, pp. 235–6.

[f] With Kholmogory.

The large and mixed populations of these two cities were naturally much fostered by the presence there of the court and government. In the case of certain other cities provincial reform proved important. In 1708, following a period of widespread unrest with which the old system of local administration was clearly unable to cope, Peter divided his realm into eight large provinces or *gubernii*, to which three more were added in 1713–14. Thenceforward, in the new provincial capitals there resided a governor and a vice-governor (together with numerous other officials), who were responsible for military and civil affairs in their areas. The eleven provincial capitals were Moscow, St Petersburg, Smolensk, Kiev, Azov, Kazan, Archangel, Tobolsk, Riga, Nizhnii Novgorod, and Astrakhan. Table 10.2 reveals that, although all the new provincial capitals were important towns, they were by no means the most significant commercial centres in Russia, and that geographical position was clearly the prime factor in their designation. Even so, the new provinces varied greatly in both size and population.[38]

The enormous size of some of the new provinces soon led to attempts to subdivide them. Finally, in 1719, Russia was redivided into eleven provinces (*gubernii*),[39] these provinces in turn being subdivided into fifty sub-provinces (*provintsii*). The sub-provinces were further divided into districts (*uezdy*). This particular reform was in part a response to new needs associated with the upkeep of the army in peacetime, financial requirements, and the maintenance of law and order. Thenceforward, each province was under the control of a governor (*gubernator*), to whom in turn the military governor (*voevoda*) of each sub-province was made subservient (at least in military and judicial matters). The list of fifty provinces and sub-provincial centres reveals some interesting anomalies.[40] As might be expected, most of the larger towns of this land that was still only slightly urbanized found themselves in this role, but there were exceptions, such as Olonets in the north-west (many of whose registered population were, however, farmers living a great distance from the town), Staraya Rusa (made subservient to Novgorod), or Torzhok (put under Tver). Toropets, with a *posad* population of 2,139 in 1722, was placed under Velikie Luki, with only 884 people in its *posad*. In Voronezh province, Kasimov and Yelatma were placed under Shatsk, a much smaller centre, and in Kiev province neither Bolkhov, Belev, nor Kursk were raised to the status of sub-provincial capitals. In these and similarly anomalous cases, it seems that geographical position, communications, military significance, or even physical fabric had spoken against their elevation and in favour of other centres.[41]

In spite of such seeming anomalies, Russian governments of the first half of the eighteenth century were aware of the importance of towns for the generation of trade, and of the connection between trade and revenue. Of no ruler was this more true than of Peter the Great himself. His attempts to liberalize trade may not always have been to the liking of the urban merchants, but on the other hand Peter strove to encourage a more enterprising attitude among this group by increasing their autonomy from the local military governor. The 1699 Ratusha reform and the founding in 1721 of the urban Magistrat, which entailed entrusting urban government to elected representatives of the newly established guilds, are symptomatic of this policy. However, the degree of real autonomy conferred by this legislation was limited not only by the central government and the activities of its officials, but also by the conservatism of the urban merchants themselves. The towns of Russia were still mainly small and backward, and their populations could hardly meet the aspirations of a tsar like Peter. In 1727, two years after his death, the towns were again made subject to military governors, and the urban Magistrat was greatly reduced in power. Furthermore, in a series of measures which swept away much of the cumbersome and expensive provincial administration bequeathed by Peter, a strict hierarchy of command was once again established, whereby the military governors of ordinary towns were subordinated to the military governors of the sub-provinces, and the latter in turn to the provincial governors.

Apart from the city of St Petersburg, Peter's most lasting contribution to Russian urbanism lay in city planning. Determined that his new capital should symbolize a break with many traditional Russian ideas, Peter employed both foreign and Russian architects to devise ambitious plans for the city. Although many of these proposals were subsequently discarded, the city emerged with the skeleton of a regular plan and broad, impressive streets. Instead of being surrounded by courtyards in the traditional manner, public buildings and houses fronted directly on to the thoroughfares. Some of the latest fashions in European architecture were also applied in the city.[42] Certain of the new ideas were then introduced to Moscow and to various new towns, such as Petrozavodsk, or Yekaterinburg in the Urals. A few fortresses and some provincial capitals also benefited. Even so, overall achievement was modest. City-planning measures were subsequently applied under Peter's successors, notably Anna and Elizabeth, but, with the major exception of St Petersburg, their impact was again limited.

## CATHERINE II AND HER SUCCESSORS (1762–1861)

Down to the middle of the eighteenth century the impact of Russia's rulers upon the development of towns had taken many different forms, but this did not mean that they had been completely successful in moulding urbanism to their will. Their greatest success had been in the way they had utilized the towns as administrative and military centres and, apart from certain interludes, maintained their authority over the territory. They had also managed to bind the urban residents to their places of abode, and thus secured the service and finance they needed. Less effective had been their policies for encouraging trade and commerce, in the system as a whole or in individual parts of it. A dynamic and determined ruler like Peter the Great could ensure the success of his new capital, St Petersburg, but even he could do little to counteract the negative consequences of such policies for other regions of Russia, or to overcome the basic conservatism of the townsmen as a whole.

From the middle of the eighteenth century Russia's towns, though still retaining much of their medieval character, began slowly to change. General population growth and expanding inter-regional commerce inevitably affected many of them and, in spite of the continuing restrictions on their movement,[43] many non-*posad* elements began to take up residence. Some scholars have argued that the actual urban population in the late eighteenth century exceeded 8 per cent of the total, while the *posad* element remained at about 3.5 per cent.[44] In these circumstances, the *posad* population inevitably suffered from commercial competition by their non-*posad* neighbours, as well as by nobles and peasants in the countryside. Towns were also affected by the new fashion for building noble residences. While the old fortifications decayed and disappeared, town planning, rebuilding, and the occasional use of stone gradually altered the traditional appearance of the towns.

The relative growth of towns down to 1811 is shown in Table 10.2, above. Comparisons between the data for the 1780s and those for 1811 should be treated cautiously, since the former are estimates derived from the work of V. M. Kabuzan and G. Rozman and probably include some agricultural suburbs.[45] By the century's end, St Petersburg and Moscow were the premier cities, the former having by now outstripped the latter in size, and both far larger than the third city. Table 10.2 also reveals the rapid growth of some central towns (e.g. Tula), certain ports (Riga, Revel, and Astrakhan), a number of the Volga towns (Kazan, Saratov, and Simbirsk), as well as several of those in the Central Black Earth Region and the Ukraine. The effects of commerce are therefore clearly seen. However, some old provincial and sub-provincial centres now lagged behind (e.g. Novgorod, Pskov, Belgorod, Sevsk, Kostroma, and Vologda).

Eighteenth-century developments meant that some small places which had long served as administrative centres (such as many of the former forts along the Belgorod Line) were by now towns in name only, whereas other trading and commercial centres with larger populations were still officially villages.[46] Some of these anomalies were corrected by Catherine II's provincial reforms in the late 1770s and early 1780s. In timing, at least, these reforms, which were promulgated in 1775, were a response to the catastrophe of the Pugachev rebellion (1773–4), an event which had starkly revealed the inadequacies of provincial administration and control.[47] Under these measures Russia was divided into forty-one provinces (*namestnichestva* or *gubernii*) (fifty by 1796), each with a male population of 300,000–400,000. The provinces in turn were divided into districts (*uezdy*) with 20,000–30,000 male inhabitants each. All the new administrative centres were to be towns in order to ensure that the necessary personnel and facilities would be available. It was also specified that the new administrative centres should be conveniently located, and that they should not be privately owned. The latter condition was laid down to avoid having to compensate private landowners for the loss of their settlements.

All the provincial capitals of 1708 (except Azov), 1719, and 1727 (with the exception of Belgorod, long overshadowed by Kursk) still fulfilled this role. However, the same certainly cannot be said for all the former sub-provincial capitals: many of the smaller ones, having failed to respond to new economic developments, were now relegated to the position of district centres.[48] Population expansion and the acquisition of new territories to the south and the west dictated the choice of a number of entirely new capitals.[49] There were some rather unusual decisions determined more by administrative and policing considerations than by economic realities. The choice of Novgorod, Pskov, Vladimir, and Olonets (soon to be replaced as provincial capital by Petrozavodsk) was evidently dictated by factors connected with geographical location, tradition, and communications rather than inherent economic importance. In the Urals a new town, Perm, was preferred to the less conveniently located industrial centre of Yekaterinburg. Further south the choice fell upon Ufa, which was more centrally located with respect to the unruly Bashkirs than was the economically important town of Orenburg. The pattern of provincial centres is illustrated in Fig. 10.2.

When it came to the designation of district centres the government encountered much more difficulty because of the lack of well-located (and state-owned) urban-type centres in many parts of the country. Occasionally it was necessary to purchase a suitable privately owned village, or to designate a less than imposing state settlement as district centre. At the same time the opportunity was taken to downgrade some of the former towns which had lost any economic vitality they might once have had.[50]

Yu. R. Klokman calculates that Catherine's reforms resulted in the

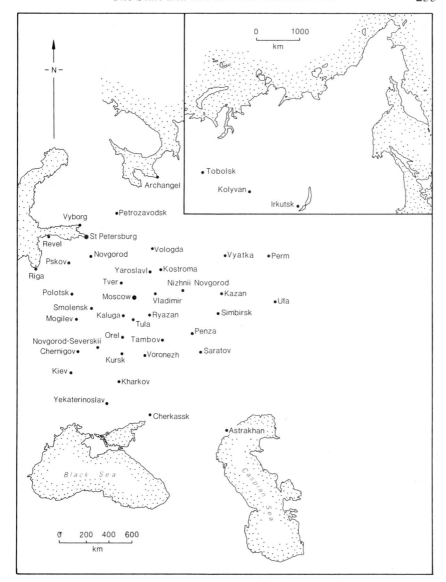

FIG.10.2.  Provincial capitals in Russia after the 1775 Reform

designation of 216 new towns in Russia, taking the total to 499 by 1787.[51] However, perhaps a quarter of the new towns were found to be unsuitable for one reason or another and ceased to be district centres in 1797. Despite this reduction, the net effect of the reform was considerably to increase the number of townsmen in Russia, and, by adopting an essentially territorial definition of the town, to eradicate a number of the traditional distinctions between urban residents. For the first time the town became an actual administrative entity in law.[52]

Robert E. Jones has written that Catherine II's aim was 'to stimulate the development of the provinces by injecting them with elements of a more advanced and more progressive way of life modelled on the example of St. Petersburg'.[53] This was to be done, he believes, by implanting the characteristics of the capital in the towns of provincial Russia. One way of doing this was through the physical redesign and replanning of towns, a process which began in Moscow and St Petersburg with the appointment of a special commission for that purpose in 1762, and then extended to other towns, especially after 1775. Catherine's goals, according to Jones,[54] were threefold: to make the towns healthier and safer for their residents, to mould the behaviour of the citizens in ways considered desirable by the authorities, and to transform the towns into symbols of Imperial order and good government. A further policy, also designed to foster the towns as centres of progress or 'enlightenment', was the granting of a measure of self-government, a goal finally attained in the Charter to the Towns issued in 1785. In these as in other matters success was only partial, counteracted by urban poverty, traditional attitudes, and also faults in the legislation itself.[55] Even so, many scholars have argued that Catherine's measures did serve to encourage urban life and enterprise in a slowly changing Russia.[56]

As the Empire continued to expand and to absorb new territories, so the number of cities in Russia grew. Where these were new foundations, the town-planning policies had the fullest effects, with their emphasis upon ordered streets and outward architectural appearance. Elsewhere—in the older towns, and especially in the administrative centres—the central and wealthier portions particularly benefited. As the nineteenth century wore on, however, and rural to urban migration continued, so these strictly planned central areas, whose symmetry and rationality were so symbolic of enlightened absolutism, became surrounded by poor and unhealthy suburbs.[57] Until the Emancipation of 1861 urbanization still proceeded slowly and only about 10 per cent of the population lived in towns at that point. However, by that stage there were many pointers to a more dynamic future. Between 1811 and 1863, for example, the number of people living in towns and cities grew from 2.8 million to about 6.1 million. By the 1860s, as Table 10.3 shows, both St Petersburg and Moscow had grown into major urban centres. Both were ringed by industrial suburbs, textiles being the most important industry in Moscow, cotton and machinery in St Petersburg. In the Central Region and the Urals many industrial towns and settlements had sprung up.[58] The impressive growth of ports such as Odessa, Nikolaev, and Riga is testimony to an expanding international trade. Elsewhere, such long-standing commercial and administrative centres as Kiev, Kharkov, Nizhnii Novgorod, and Samara were also developing, though not always as a consequence of industrialization. Table 10.3, however, indicates the irregular nature of urbanization in this period. Formerly dynamic and important centres, such as Kazan, Astrakhan, Tula, and Yaroslavl, grew sluggishly by contrast with some of the

TABLE 10.3. *Urban Populations, 1811–1914* (000s)

| Town | 1811 | 1863 | 1897 | 1914 |
|---|---|---|---|---|
| St Petersburg | 335.6 | 539.5 | 1,264.9 | 2,118.5 |
| Moscow | 270.2 | 462.5 | 1,038.6 | 1,762.7 |
| Riga | 32.0 | 77.5 | 282.2 | 558.0 |
| Kiev | 23.3 | 68.4 | 247.7 | 520.5 |
| Odessa | 11.0 | 119.0 | 403.8 | 499.5 |
| Tiflis | 29.9[a] | 60.8 | 159.6 | 307.3 |
| Tashkent | — | — | 155.7 | 271.9 |
| Kharkov | 10.4 | 52.0 | 174.0 | 244.7 |
| Saratov | 26.7 | 84.4 | 137.1 | 235.7 |
| Baku | — | 13.9 | 111.9 | 232.2 |
| Yekaterinoslav | 8.6 | 19.9 | 112.8 | 211.1 |
| Vilno | 56.3 | 69.5 | 154.5 | 203.8 |
| Kazan | 53.9 | 63.1 | 130.0 | 194.2 |
| Rostov-na-Donu | 4.0 | 29.3 | 119.5 | 172.3 |
| Astrakhan | 37.8 | 42.8 | 112.9 | 151.5 |
| Ivanovo-Voznesensk | *c.*5.0 | *c.*11.0 | 54.2 | 147.4 |
| Samara | 4.4 | 34.1 | 90.0 | 143.8 |
| Tula | 52.1 | 56.7 | 114.7 | 139.7 |
| Omsk | 4.6 | 19.5 | 37.3 | 134.8 |
| Kishinev | 42.6[b] | 94.1 | 108.5 | 128.2 |
| Minsk | 11.2 | 30.1 | 90.9 | 116.7 |
| Tomsk | 8.6 | 21.0 | 52.2 | 114.7 |
| Nizhnii Novgorod | 14.4 | 41.5 | 90.1 | 111.2 |
| Yaroslavl | 23.8 | 27.7 | 71.6 | 111.2 |
| Vitebsk | 16.9 | 27.9 | 65.9 | 108.2 |
| Nikolaev | 4.2 | 64.6 | 92.0 | 103.5 |
| Yekaterinodar | 4.3[a] | 9.5 | 65.6 | 102.2 |
| Tsaritsyn | 3.8 | 8.4 | 55.2 | 100.8 |
| Orenburg | 5.4 | 27.6 | 72.4 | 100.1 |

[a] 1825 data
[b] 1840 data

*Source:* A. G. Rashin, *Naselenie Rossii za 100 let, 1811–1913 gg.: statisticheskie ocherki,* (Moscow, 1956), 93, 107, 110.

peripheral cities in western, southern, and south-eastern European Russia. In 1856 less than a quarter of the urban population lived in cities with more than 50,000 people.[59]

The role of the state in such changes was an ambivalent one. As the bureaucracy expanded and the administration became more sophistic-ated, so the enforced 'services' of the townsmen ceased to be so essential. Yet the enhanced need for revenue could only be satisfied by an expanding national economy. This fact had been recognized by Catherine II in her liberalization of trade,[60] though her immediate successors failed to embark on any thorough policy of economic modernization, in spite of their debts.[61] The development of commerce and industry in the countryside meant that the towns were not always as central to the economy as they had once been. The major exceptions were the industrial centres which were now burgeoning, and it was here that the true ambivalence of the government's attitude was felt. Urbanization was to be welcomed if it increased the national prosperity. A number of measures in the early nineteenth century further encouraged economic enterprise among the peasants and made it rather easier for them to move to the towns.[62] On the other hand, urbanization was to be feared if it threatened disorder and depopulated the countryside. Such contradictions and uncertainties in official attitudes found their way into policy-making and militated against any consistent policy either towards industrialization itself or towards the towns.

## THE ERA OF INDUSTRIALIZATION (1861–1917)

Peter the Great and Catherine II bequeathed to their successors the notion that the stimulation of economic activity, within properly defined channels, was an important national goal. Subsequent rulers pursued this goal somewhat inconsistently. However, after the débâcle of the Crimean War, and more particularly by about the 1880s, the rulers of Russia had come to regard industrialization and economic modernization as vital if Russia were to retain her status as a great power. Industrialization was promoted by policies of railway-building, tariff protection, government investment, monetary stability, and the attraction of capital from abroad. Towns and cities began to grow rapidly. In these circumstances the aim of government policy, in so far as any one aim was consistently pursued, was not so much to stimulate urban development as to control it. It also strove to ensure that urbanization threatened none of the regime's goals, among which its own security was perhaps pre-eminent.

The aim of controlling urbanization proved an enormous one, largely because of the unprecedented rate at which towns and cities grew in this period. From 6.1 million people or 10 per cent of the population living in towns and cities in 1863, the total number of urbanites in European Russia grew to 18.6 million or 15.3 per cent of the population by January 1914. If, following V. P. Semenov-Tyan-Shanskii and A. G. Rashin, the populations of non-urban industrial and commercial centres are added, the latter percentage can be raised to 17.3 per cent.[63] These statistics must be treated with caution, especially in view of the importance of seasonal migration between town and country in late Imperial Russia.[64] However, the absolute significance of urban growth cannot be doubted. From a mere 540,000 in 1863, St Petersburg's population had grown to over 2 million by 1914. Hardly less important, with nearly 1.8 million people in 1914, was Moscow, which had grown from 460,000 in 1863. Three other cities—Riga, Kiev, and Odessa—contained half a million people or more by 1914, and there was a total of twenty-nine cities with over 100,000.[65] By comparison, in 1863 only three cities (St Petersburg, Moscow, and Odessa) had had over 100,000 people. The fairly wide geographical scattering of big cities (those with over 100,000 people in 1914) should also be noted, as illustrated in Table 10.3, above, and Fig. 10.3. Thus, although the influence of heavy industry and international trade is apparent in the case of several ports and inland cities, there were also a number of essentially commercial and administrative centres in this category (e.g. Tiflis, Tashkent, Saratov, Kazan, Samara, Omsk, etc.). In other words, the urbanization process cannot be tied too closely to industrialization even by this stage, since other forces were also contributing to it.[66] At the same time, many industrial workers and even those in large-scale industry were locating outside the cities.[67]

The largest demographic contribution to the urbanization process came from rural to urban migration. The Emancipation of the Serfs, enacted in 1861, removed one important barrier to migration, only to erect another by strengthening the power of the rural *mir*. There has been considerable debate in the literature over the extent to which this legislation induced industrial labour shortages and thus hindered economic development.[68] There can be no doubt that there was considerable rural to urban migration encouraged by rural poverty, itself arguably fostered by the terms of the Emancipation Statute. Nevertheless, the need to obtain permission from the *mir* to move to the city, and the general circumstances under which land was held and taxation levied, must also have retarded such movement.[69] Despite the progress of urban growth before 1914, it was totally insufficient to absorb the rural population increases. 'Russia is still a peasant country,' wrote Lenin in 1917, 'one of the most backward in Europe.'[70]

Whatever the nature of the constraints on rural to urban migration in

FIG. 10.3. European Russia: towns with over 100,000 people in 1914

this period, it was still of sufficient magnitude to create major problems for the government. St Petersburg, Moscow, and other industrial cities began to experience the social problems and crises which so often attend the early stages of industrialization: poverty, transience, overcrowding, bad housing, inadequate municipal services, poor health standards, and frequent epidemics.[71] In St Petersburg, for example, the balance between births and deaths was usually a negative one until the middle of the 1880s. The city had the unenviable reputation of being the unhealthiest capital in Europe, suffering regular outbreaks of cholera, typhus, and other diseases.[72] As might be expected in these circumstances, infant mortality was especially severe. In Moscow the situation was hardly better, deaths generally exceeding births until the 1890s.[73] Similarly unhealthy environments characterized such cities as Kostroma, Saratov, Samara, and Nizhnii Novgorod,[74] underlining their dependence on migration from the countryside for the maintenance and increase of their populations. Although a poor urban environment was only one among many factors contributing to the growing revolutionary movement, it can hardly have made the task of preserving social order any easier.[75]

The government's response to the developing urban problems was a wholly inadequate one. The state retained in its hands powers of policing and registering migrants even after the Stolypin Reform of 1906. Nevertheless, it completely failed in the task of catering adequately for the huge numbers who did migrate. As time went on, more and more resources had to be directed into policing and security functions. City planning, which began to atrophy from the mid-nineteenth century, failed to embrace the wider social, economic, and environmental dimensions appropriate to the new age. Government supervision was restricted to narrowly architectural or technical concerns, and even the revived urban-planning movements which appeared at the end of the Imperial era failed to make much impact on official attitudes.[76] At the local level the efforts of municipal authorities to counteract the worst effects of industrial urbanization were frustrated by the suspicious and bureaucratic interference of the regime and its officials. The 1870 municipal reform, which (following the precedent set by St Petersburg in 1846 and by certain other cities rather later) extended a genuine measure of autonomy to many cities in European Russia, certainly gave much greater scope for social improvement, and seems to have produced real benefits in Moscow, Odessa, and elsewhere.[77] Its positive effects, however, were soon curtailed by the repressive measures enacted after Alexander II's assassination in 1881, and by the municipal 'counter-reform' of 1892. A narrow and restrictive franchise, inadequate financing, bureaucratic control, corruption, and simple indolence all militated against a decisive and successful policy towards the cities.

Such failings are not perhaps surprising in a land to which the experience of industrialization was still new, as was altogether apparent in the urban geography of Russia in the late Imperial era. Not only were there relatively few large cities, but those that did exist were as remarkable for their traditional features as for their modern ones. As J. H. Bater has written of St Petersburg:

There was obviously ample evidence of modernization in industry, commerce and urban transport during the decades following the Emancipation. Electric trams, world-wide business linkages, joint-stock banks, international industrial enterprises—all were part of the St. Petersburg scene. While this was happening, there was an underlying and seemingly unchanging commercial system which operated within the framework of traditional, largely peasant values. The itinerant pedlar was no less a part of city life than the shop, nor did bazaars and street peddling necessarily give way wholly to modern commerce.

St Petersburg was, he concludes, 'a city of peasants'.[78]

If such traditional traits characterized the largest cities of the Empire, they were no less apparent in the small ones, of which there were many. Of the 932 cities recorded by T. S. Fedor for 1910, 492 had populations of below 10,000 (accounting for 11 per cent of the urban population), and a further 351 below 50,000 (accounting for another 31 per cent of the urban population).[79] In fifty provinces of European Russia, Rashin lists thirty-one capitals with populations below 100,000, of which eleven had less than 50,000.[80] Many of these cities had grown quite substantially in the latter part of the Imperial era, but were still small by any standards. Interestingly enough, three of the thirty-one (Kostroma, Vologda, and Kaluga) had been among the largest cities in Russia in the seventeenth century.

All this, then, is testimony to the limited nature of urbanization at the end of the Romanov era. Industrialization was proceeding apace, aided and abetted by the government, but in large numbers of cities the new forces of social and economic change had still had only a limited impact, and life went on much as it had done in an earlier epoch.

## CONCLUSION

The dramatic nature of the changes which swept Russia during the three centuries of the Romanov era is reflected in the evolving urban geography of the period. Of the twenty-nine cities with a population of over 100,000 in 1914, only seven (Moscow, Kazan, Astrakhan, Tula, Nizhnii Novgorod, Yaroslavl, and Tsaritsyn) had even existed as cities in Russia in the early seventeenth century. Of the rest, most had been founded by the Russians as they expanded their territories to the south, south-east, and east (or even, as in the case of St Petersburg, to the north-west). Others had been annexed by Russia, especially in the west.

In place of the major cities of the seventeenth century, situated mostly on waterways in an arc around Moscow, Russia's principal cities of the early twentieth century consisted of great seaports, such as St Petersburg, Riga, and Odessa, industrial centres, such as Moscow, Kiev, Kharkov, and Baku, far-flung provincial nodes, such as Tiflis and Tashkent, and some old-established commercial points, such as Kazan. These twentieth-century cities were connected by a network of railways, and linked into a world industrial and commercial economy in a way which would have been inconceivable three centuries before.

That the state had an important influence upon this transformation is a truism which requires no special emphasis here. This chapter has also hinted that the state's influence may not always have been positive. Serfdom, high taxation, and rigid social and economic policies no doubt held back the process of development. Such retarding forces were further strengthened by the effects of conservative attitudes among the townsmen and by adverse environmental circumstances.

Perhaps it is appropriate in conclusion to ponder in what respects exactly the state succeeded in moulding Russian urbanism to its own ends, and in what respects it failed. This chapter has mentioned a number of instances on the positive side. Thus, the state founded and developed many towns in its new territories which went on to become important centres of administration and commerce (many, of course, did not). Not only this but, in the eighteenth century especially, it designated large numbers of settlements in the older parts of Russia as administrative centres and thus points of control for the regime. In general, and especially in the period between the Pugachev rebellion and the late nineteenth century, the policy was successful in promoting tsarist control over the countryside. It may also have served to encourage the spread of trade and commerce. Here, however, the designation of administrative centres was often as much a response to urban growth as a cause of it. The state's influence over which urban centres were to grow was usually a partial and indirect one, St Petersburg and Moscow being the most obvious exceptions.[81]

Also on the positive side must be placed the success with which the state planned and controlled the reconstruction of a large number of towns, especially from the time of Catherine II. Although hampered by a lack of resources (amongst other factors), and from the latter part of the nineteenth century by the strengthened forces of urban growth, the effects of this policy are still evident in numerous Russian cities to this day. A further success for the state lay in the way it was able to harness the services of the various categories of townsmen to its own needs and to the raising of its revenues. Among other things this involved controls over migration and social mobility. Such policies were never completely

successful, and, where they were, may have had deleterious consequences for the economy. But in the early period especially they seemed
essential to the well-being of the state.

This chapter has also mentioned a number of ways in which the state
failed to accommodate urbanism to its purposes. Until late in the tsarist
era, one of the most important was the way in which towns failed to
develop the enterprising and self-confident classes upon which the West
had built its economic success. Both Peter and Catherine felt their lack,
and even in the late nineteenth century the entrepreneurial classes
seemed to some observers to be far too dependent on the government
and far too fearful of competition from abroad. By this stage, of course,
the government, while anxious to promote industry and to protect it
from competition or social unrest, had become fearful for its own
prerogatives and unhappy about policies which might undermine its
authority—hence its suspicion concerning municipal autonomy and its
inability to face up to the problems of industrial urbanism. In the end the
regime's failure to adjust to changing circumstances, in the cities as
elsewhere, proved its own undoing.

The state was therefore far from omnipotent with respect to the cities.
Given its own poverty, the far-flung nature of its territories, and the
fitful character of communications until late in the period, this is hardly
surprising. By the end of the Romanov era, when communications had
improved and the state's command over resources had grown, the forces
of industrialization and urbanization had been unleashed and proved
impossible to cope with. Indeed, the state's policies were unlikely to
succeed given the contradictory character of its goals. On the one hand,
cities were seen as focuses of social order and pillars of state power.
Perhaps because of the particular problems of ensuring social control in
Russian circumstances, the authorities were ever suspicious of spon-
taneity or social fluidity. On the other hand, cities were regarded as
essential to the economic success which the state also needed. Yet that
very association between the city and economic advance encouraged the
economic and social changes which the state feared and strove to
constrain. It was the Romanov dynasty's misfortune that it proved too
brittle for the social and political changes of its day and that it was swept
away in consequence. Had it been more malleable and responsive, its
survival might have been secured. Yet even in those circumstances, its
ability to direct the process of urbanization would have been only partial.
The city would have continued to pose problems, just as it has done in
every society and in every age where the state has striven to control it.

# Conclusions

Discussion of the nature of Russian development on the eve of the 1917 revolution has been a major preoccupation of historians over the past eighty years. For Soviet historians there has been little alternative but to argue that Russia in 1917 was a country in which capitalist development, although assuming specific forms, was well under way, and that a proletarian seizure of power was therefore justified. This is a view with which Western Marxist historians have, with some reservations, been prepared to concur. Mainstream Western historians have put a different interpretation on pre-revolutionary Russian development. They suggest that Russia was undergoing far-reaching economic, social, and political changes in the decades before 1917 that were finally overcoming its centuries-long backwardness, with the result that at last it was assuming its place alongside the industrial countries of Western Europe. A booming industrial sector, improvements in communications, urbanization, increases in total agricultural production, and the introduction of elements of constitutional rule are all cited in support of this thesis. In both these orthodox Marxist and modernization paradigms the existence of local and regional differences in the ways that people made a living from the land, and the social and political relations between people at the local level, are rarely considered in any detail; the focus of attention is on the emergence of a national polity, a national economy, and a class-structured society. Indeed, the forces associated with the rise of industry are usually credited with subordinating geographically specific social and economic processes to the more powerful universalizing force of the capitalist market. Within this theoretical framework geographic space is inevitably treated as a 'stage' upon which momentous events are played out, and regional society is treated as historically transient and intrinsically uninteresting. In recent years some notable publications which examine Russian history at the level of individual regions have been published, including, for example, Robert Edelman's account of the 1905 revolution in the south-west.[1] It is our hope that these signal a move among historians towards considering the active role of regions in the shaping of Russia's history.

The central concern of this book has been to encourage the development of a geographically based history of Russia in which the focus of attention is on local configurations of economy, society, and polity. These configurations, as we have tried to show, were the product of the interaction of macro-level forces with pre-existing and highly varied local societies which had been moulded in specific environmental and historical contexts. In the Introduction we referred to the concept of the 'peasant ecotype', which has been used by anthropologists to describe and analyse the ways in which human societies are adapted to particular sets of resources available in the environment. By selecting places from each of the major natural zones of Russia, we have attempted in the main chapters of this book to draw attention to the environmental element in pre-revolutionary Russia's regional diversity. The natural environment simultaneously set limits upon the way in which local societies could respond to influences from outside, while opening up opportunities for new adaptations. For this reason alone the results of market penetration, state policy, urban culture, and other forces emanating from metropolitan society cannot be expected to have been the same everywhere, quite apart from the many cultural, political, and social factors which also served to give distinctiveness to local societies. The question that we have not attempted to consider is the role that regionally specific societies, in interaction with each other and with the centre, played in major events in Russian history. This is a question which, lying as it does beyond the realm of historical geography as normally defined, demands far more attention from the historians than it has received hitherto.

COLONIZATION AND THE PHYSICAL INCORPORATION OF THE
PERIPHERIES IN ROMANOV RUSSIA

The principal integrative processes considered in this book are human colonization and economic expansion. During the Romanov period these two processes took metropolitan Muscovite and Russian culture to the peripheries of the empire. Their impact found expression in the emergence of areally and temporally different patterns of human occupance. In the first part of the period, approximately until the end of Catherine II's reign, colonization was the more important process. By the time of the Romanovs' assumption of power much of the northern forest lands had already been assimilated into the state of Muscovy, and in the following years attention turned southwards towards the steppe with its valuable agricultural resources and promising trade outlets on the Black Sea. The military conquest and settlement of the steppe became a major preoccupation of the Romanovs. The state used a combination of compulsion and incentive to encourage people to move to the frontier, and from the eighteenth century foreigners also became the target of its settlement policies. The bureaucracy that was required to organize the colonization of the steppe was considerable, and through it the state attempted to control who settled its new lands, and under what terms. These attempts to shape the nature of the society that developed on the peripheries were never wholly successful, however. The problem of distance made absolute control difficult, and in the interstices of state power there was plenty of scope for the emergence of varied local societies and cultures. Furthermore, there was a tension in tsarist policy: on the one hand the state was determined to populate its frontier regions, but, on the other, it was not prepared to dismantle the system of serfdom in the heartland provinces of Russia, thereby releasing people to migrate south. The ambiguous attitude to cossack settlers and runaway serfs in the settling of the Central Black Earth Region, and the invitation to foreigners to settle in the southern and eastern peripheries, have to be seen in the light of this tension in official policy.

The varied societies that developed on the peripheries as a consequence of tsarist colonization policy are described in the early chapters of this book. Voronezh was one of the provinces lying in the central black-earth belt, whose history was bound up with the height of military measures against the nomads. The course of conquest, colonization, and agricultural settlement described in Chapter 3 for Voronezh was paralleled in neighbouring black-earth provinces, such as Kursk, Orel, and Tambov. By the eighteenth century such provinces had become the principal grain-producing regions of Russia and a heartland of serfdom. Even so, their societies long retained vestiges of their former frontier character, such as the *odnodvortsy* discussed in Chapter 2. Their anomalous position in the Russian society of the eighteenth and early nineteenth centuries can be explained only in terms of the special circumstances attending the settlement of the southern frontier in the previous two hundred years. Official attitudes towards these descendants of the erstwhile guardians of the frontier remained ambivalent, denying them the right to join the newly self-confident nobility, but reluctant to infringe their historically grounded privileges. Nineteenth-century observers noted their peasant characteristics, and yet were aware of their unusual history. This consciousness of uniqueness seems to have been shared by the *odnodvortsy* themselves.

From the point of view of the geographer, one of the most interesting points to emerge from an analysis of the settlement of the Central Black Earth Region is the way in which the circumstances attending the use of land in the seventeenth century still made themselves felt two centuries later, under very different economic and social conditions. Thus, the character of service land-tenure, the smallness of initial land allotments, the poverty and dubious social origins of the initial settlers, and the scarcity or even absence of serfs were all important in the long-term development of the *odnodvortsy* settlement. The *odnodvortsy* were, of course, affected by the social and economic changes of the eighteenth and nineteenth centuries, adopting communal landholding and having to respond to land shortages and market developments. Yet even after the Emancipation, scholars found it possible to distinguish between their settlements and those of other Russian peasants in the same region, with their distinctive types of village and customs, and with some settlements resisting communal landholding even down to 1917.

If the frontier experience thus had long-term consequences for the character of certain regions of European Russia, such as the Central Black Earth Region, it is pertinent to enquire whether that experience bears comparison with similar experiences elsewhere in the world. This was done for the seventeenth-century south in Chapter 1 where its settlement history was examined in the light of the frontier thesis of Frederick Jackson Turner. The relevance, and also the limitations, of Turner for understanding the Russian case are apparent. From Chapter 1 it is evident that the frontier was different in different places, and this applies at the international scale as well as the local. Turner was, of course, aware of this, and attempted to develop his concept of the section to explain the long-term consequences of these different frontier experiences. This is a concept which would certainly bear further examination in the Russian context.

The importance of such differences in the frontier experience becomes apparent when the settlement of the Central Black Earth Region is compared with that of the steppe on the southern and eastern margins of European Russia. In the latter case military colonization was, on the whole, much less important than in the former. When Russian forces reached the Black Sea steppe and the Volga, the power of the steppe nomads was much reduced and their resistance relatively easy to overcome. The regions that the state acquired in this final stage of southern expansion in European Russia consisted of continuous stretches of open steppe which, in the lower and middle Volga, were difficult to bring into cultivation. The settlement of these new regions was characterized by many of the same contradictions in state policy as before: migration to the new frontiers had to be encouraged without undermining the labour force of the nobles' estates in the central provinces. The solution to the dilemma was to invite a variety of semi-autonomous groups to settle. These included a range of religious nonconformists from inside and outside Russia, other foreigners, cossacks, and state and court peasants, as well as noble landlords. Each group was settled under its own legal statute, which governed the ownership and inheritance of the land, rights and obligations with respect to its use, and in some cases defined the group's military and tax obligations. Only after the Emancipation in 1861 did the numerous legal differences between the colonies of settlers in the southern and eastern steppe diminish, as privileges previously granted were rescinded.

The Mennonite colonists discussed in Chapter 4 were not exceptional in receiving privileges at the time of their settlement and being able to retain elements of their distinctive culture. Their experience in this respect was, in fact, similar to that of other foreigners and religious nonconformists who came to occupy the steppe. The existence of these groups in the peripheries of European Russia is often overlooked in the literature on Russian agricultural history, yet the dominant feature of these regions was their multi-cultural character. Recent accounts[2] of cossack settlement represent a welcome addition to the sparse literature on the cultural groups inhabiting the southern and eastern steppe, but the role of other groups in the economic development of the peripheries has yet to be described. Where the Mennonites were concerned the contribution made was undoubtedly great, especially in New Russia, where they were at the forefront, first of commercial production of merino sheep, and later of commercial cereal production. By the nineteenth century, when the arc of provinces stretching from Kiev to the middle Volga became the principal cereal-growing region of Russia, the Mennonites were among the most successful capitalist farmers in the south. Yet, neither the Mennonites nor their Russian and non-Russian neighbours had developed systems of farming which could protect them sufficiently against the effects of drought. The Volga was the scene of the worst famines and rural uprisings in the last decades of the nineteenth century.

By confining our discussion to European Russia, we have chosen not to consider the Russian colonization of Asia. Migration first occurred, from the sixteenth century, into the coniferous forests of Siberia and later, in the nineteenth century, into the desert oases of Central Asia. The agricultural settlement of the Asian land mass was hardly undertaken in any systematic way until the latter decades of the nineteenth century. Thus, for most of the period discussed in this book, it was a territory in reserve. When the settlement of Siberia began in earnest it was in response to the overcrowding of land in European Russia. Unlike the earlier phases of settlement, therefore, finding willing migrants was not a problem for the tsarist government. However, there were elements in the story that were familiar since, as Donald Treadgold has shown in his study of migration to Siberia, despite the existence of a complicated bureaucracy established to organize migration, the government was never able fully to control the flow of

people east.[3] Siberia came to be populated by migrants, many of them
illegal, who were accommodated into Siberian villages in an *ad hoc*
manner, rather than in the orderly way that had been envisaged by the
government.


ECONOMIC EXPANSION AND THE REGIONS

The physical colonization of the peripheries of Russia was followed by
their gradual integration into a national market centred on St Petersburg
and Moscow. The most rapid expansion of the economy took place in
the nineteenth century with the development of large-scale industry, but
there had been earlier phases of economic growth, most notably under
Peter the Great and from about the 1750s. As Chapters 3 and 8 make
clear, regional specialization and trade were already in the process of
development before the nineteenth century. Through a network of local
and national markets the products of the forest and steppe were
exchanged: wood products, furs, animal products, and metals moved
south from northern regions and, in exchange, southern cereals were
shipped north. In the pre-railway era these goods were moved by the
system of inland waterways and by overland haulage, though as yet many
peripheral districts and areas far from the waterways remained isolated.
The division between the forested northern provinces and the agriculturally orientated steppe and forest-steppe provinces intensified in
the nineteenth century, and further finer regional economic divisions
emerged. Industrial development led to a proliferation of industrial
regions in the country. Thus, Moscow, St Petersburg, and the Urals,
which had been the scene of pre-nineteenth-century industrial developments, were joined by the southern Ukraine, tsarist Poland, the Baltic
industrial regions, and, in the Transcaucasus, by an industrial region
based on Baku.[4] These regions were centres of the manufacturing
industries and mining. Outside such regions other towns became
important centres of trade. These included the Black Sea and Volga
river ports which, from the end of the century, were linked by rail with
their agricultural hinterlands in the southern and eastern steppe.
  As has already been indicated, the effect of these developments on
existing rural and urban society is one of the most disputed topics in
Russian history. Contemporary commentators could not reach a
consensus and continued to dispute the relative importance of economic
and other factors in the changes which had taken place in nineteenth-
century Russia, up to the time of the 1917 revolution. The controversy
surrounding the definition of agricultural regions in nineteenth-century
Russia was but part of a wider debate concerning the ability of

traditional peasant society to withstand the corroding influence of the market, which brought orthodox Marxist opinion into direct conflict with its populist opposition. The same question about the fate of peasant society informed much contemporary writing about domestic industries in the nineteenth century. The example of changes in peasant manufacturing in Moscow province showed that even places quite distant from Moscow could not expect to escape the effect of the market. By the second half of the nineteenth century peasant manufacturers in the province were firmly tied into a national market; production had become highly specialized, marketing had passed into the hands of middlemen, and differences had emerged between producers in terms of their control over the means of production and labour. But these changes had not had the same impact on peasant society and economy everywhere in the province. In western districts manufacturing was combined with farming in such a way as to create a relatively stable peasant economy characterized by pluri-active, or multi-occupational, households. These contrasted with districts in the east of the province, in which peasant involvement in outwork in the textile industry was associated with high rates of land abandonment. As has been noted several times in the preceding chapters, the fate of the peasantry is central to the debate about the nature of Russian development before the 1917 revolution. The tendency to view any changes in the peasant economy as indicative of the dissolution of the peasantry as a class, and thus of progress towards some higher stage of development, has been a dominant theme among historians of Russia in both East and West. The alternative view that is offered here is of a more adaptable peasantry that, through diversifying the source of income and developing product specializations, was able to a certain degree to resist the corroding force of the market. The possibilities for such adaptations were not evenly distributed, however, depending as they did upon resource availability and market conditions.

A similar story of the ability of peasants to resist threats to their existing order was told in Chapter 7 in the account of peasants' responses to the Stolypin Land Reform. The Stolypin Reform was quintessentially a measure aimed at modernization. It sought to replace the peasant village community with a system of owner-occupied, consolidated farms on the north-west European model. The Reform applied equally to all regions of Russia but, as the case-study of two districts in northern Russia presented in Chapter 7 illustrates, peasants in the different regions apparently saw in the Reform's provisions different possibilities and threats. For some the Reform offered the opportunity to overcome long-standing problems associated with the disposition of land, for others it could be used to secure an allotment of good land, and for yet others it transformed their holding into a commodity for sale or mortgage. Probably only a minority adopted the Reform with a view to introducing modern systems of cultivation. The variability of peasants' motives in responding to the Reform explains the patchy geographical distribution of enclosures that emerged in the years that followed its enactment. Two areas of relatively high adoption did stand out, however: in the north-west and in the southern and eastern steppe. These were regions where the commercialization of farming was most advanced by the beginning of the twentieth century, and changes had already begun to be introduced by communes into their system of resource management before 1906. The paradox of the Stolypin Reform's adoption is that in those regions of Russia where the problems of land-hunger and poverty were most acute and reform most needed, the government's measures made the least headway.

The commune, or *obshchina*, which the government sought to remove from the Russian countryside through the Stolypin Reform, was itself a variable institution. Twentieth-century discussion of the commune has tended to focus on the practice of repartitioning, and the land fragmentation believed to be associated with it, with the result that the full diversity of the economic, cultural, and social dimensions of life in Russian communes has been underestimated. Considering that even after the Stolypin Reform's enactment the majority of Russian peasants remained members of communes, the paucity of detailed investigations of the institution is a particularly large gap in our understanding of the history of the Russian peasantry. The commune's first serious investigators, working in the 1870s, were concerned to reveal the geographic diversity of the commune's forms and to trace its principal features in different parts of the country. The materials they amassed provide a valuable source for reconstructing Russian village life in the years after the Emancipation, and they have been used in Chapter 6 to compare the resource-management practices in communes situated in two geographically contrasted regions of Russia, each with different agrarian and settlement histories. The two accounts reveal certain similarities in the principles of resource use which are only to be expected in community-based management systems, but the differences are more striking, revealing the different priorities of forest-based and steppe-based peasants as they met the challenges posed by their integration into the national economy in the second half of the nineteenth century.

In some respects Chapter 10, which examines the historical geography of towns during the entire period of Romanov Russia, embraces both the integrative processes considered in this book. Towns were, of course, themselves agents of integration, being used by the tsars to cement their control in the various parts of their territories. In the peripheral areas undergoing settlement towns also aided in the economic integration of the new regions. Later on, when Russia was affected by the forces of industrialization, the consequent economic and social changes were inevitably reflected in the towns, many of which began to grow to unprecedented size. The present book has been unable to explore the consequences of such momentous changes for the internal character of towns in any detail, except to note that such changes were more or less significant depending on the nature and location of the town. Here again, the importance of place emerges, for the impact of industrialization was not the same in every town any more than it was in the countryside. The tsars were not totally successful either in controlling urbanization or in moulding it to a pattern created and approved by themselves.

### REGIONAL DIVERSITY IN ROMANOV RUSSIA

In this book it has been possible to examine only a small sample of the diversity of Russia during the Romanov era, and even that was constantly subject to change. The regions studied are by no means necessarily representative of the whole, but have rather been chosen for their inherent interest. Even so, much of what has been said about life at the local level could also be said of other regions. Our desire to emphasize diversity should not be taken as a wish to prove that every region of Russia was totally unique. Rather, the individuality of regions should be examined in the context of the whole, just as the whole must be seen as the sum of diversity.

Early in the twentieth century, the great French geographer Paul Vidal de la Blache argued that 'geography is the science of places, not of men',[5] and became responsible for the foundation of a great school of regional geography which owed much to the study of history. We believe that a geographical approach, one characterized by a concern with place, can greatly advance our understanding of the past, and of the past of Russia in particular. Russia consisted of human beings, but it was also composed of places. How those men and women subsisted in different places, how they responded to their varied environments, and how they were limited by local circumstances are part of the richness of Russian history. Without such an understanding we shall be ill-equipped to grasp the true nature of Romanov Russia. Without such an understanding we should be hard put to grasp what Russia is today.

# Glossary

*ataman* (*atamany*): cossack leader.

*barshchina:* corvée, labour obligation, usually performed for a lord.

*belomestnyi ataman, kazak:* 'white place ataman' or cossack: ataman or cossack freed from certain obligations. 'White places' referred to settlements held by the church or by individuals rather than by the tsar or the state.

*bobyl* (*bobyli*): cottar, labourer, generally landless or with little land.

boyar: a category of nobleman, usually of high rank.

*cherkasy* (pl.): the Muscovite term for Ukrainian cossacks, sometimes used for all cossacks.

*chernososhnye krestyane* (pl.): 'black-ploughing' peasants: a category of state peasant, especially found in the north. 'Black' land was land belonging to the tsar in his capacity as sovereign (i.e. not personally), later regarded as state land, and was subject to normal tax obligations.

*chetvert* (*chetverti*): a land measure, equal to one half of a *desyatina*; also a dry measure, equal to 8 *pudy* of rye in the 17th cent., with many variations.

*dacha:* land grant; in the General Survey a land unit or survey unit.

*desyatina* (*desyatiny*): a land measure equal to 1.092 hectares or 2.7 acres.

*deti boyarskie* (pl.): junior boyars, members of the middle service class or estate of landholders, holding their land on condition of military or other service to the tsar.

*dvor* (*dvory*): household, house, courtyard, court.

*gorod* (*goroda*): town, also fortified core of a town; sometimes used in the earlier period for the town and its surrounding district or *uezd*.

*guberniya* (*gubernii*): province.

*kazak* (*kazaki*): cossack.

*khutor* (*khutora*): outlying landholding or small hamlet, frequently an out-settlement and often newly founded; also an individual farmstead, particularly one associated with the Stolypin Reform.

*koloniya* (*kolonii*): colony, foreign settlement.

*krugovaya poruka:* joint responsibility, collective guarantee.

*meshchanin* (*meshchane*): townsperson, the legal term used from the late 18th cent. for most of the successors to the *posad* inhabitants.

*mir:* one of the peasant names for the commune.

*namestnichestvo* (*namestnichestva*): province, viceroyalty; name given to many provinces in the 1775 reform, governed by governors-general; later reverted to *gubernii*.

*obrok:* quit-rent, payable in cash or kind; sum payable for the use of certain resources.

*obshchina (obshchiny):* 19th cent. term for the commune, usually repartitional.

*obshchinnoe zemlevladenie:* communal landownership.

*odnodvorets (odnodvortsy):* in the 17th cent. referred to members of the middle service class (see above under *deti boyarskie*), mainly living on the frontier, who held land without serfs; from the 18th cent. a legal term denoting descendants of the former frontier servitors, the great majority of whom had no serfs.

*oklad:* land allocation, the official land allocation rate to which a servitor was theoretically entitled.

*otrub (otruba):* a landholding consolidated under the Stolypin Reform but where the farm was physically separated from it.

*perelog:* long fallow.

*podvornoe zemlevladenie:* hereditable landownership.

*pomeste (pomestya):* service estate: in the 17th cent. an estate or landholding held on condition of military or other service to the tsar, not originally a hereditary property.

*posad (posady):* legal term denoting a trading settlement or a town's trading suburb. Unlike a *sloboda*, the *posad* was subject to ordinary tax and service obligations. However, sometimes the term *posad* was used in a loose sense to refer to all of a town's inhabitants, including military servitors who were not normally subject to tax obligations.

*posadskie lyudi* (pl.): the inhabitants of the *posad*, subject to normal tax and service obligations.

*provintsiya (provintsii):* sub-province, subdivision in the 18th cent. of a *guberniya*.

*pud (pudy):* measure of weight equal to 16.38 kilograms or 36.1 pounds.

*pustosh (pustoshi):* land that was not inhabited, though often worked; sometimes an abandoned plot.

*razryad (razryady):* military district.

*sazhen (sazheni):* unit of length equal to 2.13 metres or 7 feet.

*selskoe obshchestvo:* rural society, constituting from the 19th cent. an administrative unit.

*skhod (skhody):* communal assembly.

*sloboda (slobody):* settlement or urban suburb freed from certain impositions.

*sokha:* a forked tillage implement, usually pulled by horses; old measure of tillable land; old measure of tax assessment.

*strelets (streltsy):* musketeer.

*uezd (uezdy):* district, from the 18th cent. an administrative subdivision of the *guberniya*.

*ukhozhai (ukhozhai):* large unsettled territory used, often on a rented or service basis, for hunting, fishing, etc.

*vedro (vedra):* liquid measure equal to 12.30 litres.

*versta (versty):* unit of length equal to 1.067 kilometres or 0.66 miles.

*voevoda (voevody):* military governor.

*volost (volosti):* rural district; in the earlier period, an administrative district inhabited by non-seigneurial peasants; after 1861 a rural administrative district for peasants, a subdivision of the *uezd*.

*votchina* (*votchiny*): hereditary estate.

*zaimka* (*zaimki*): an intake in virgin or uncultivated land.

*zakhvatnoe zemlevladenie:* seizure tenure.

*zalezh:* shifting cultivation, usually in steppe areas, where the term is sometimes used interchangeably with *perelog*.

*zaseki* (pl.): fortified lines, particularly those constructed of felled trees.

*zemstvo* (*zemstva*): elected organ of rural self-government, introduced in 1864.

# Notes

## Notes to Introduction

1. Quoted in P. Dukes, *The Making of Russian Absolutism, 1613–1801* (London, 1982), 4.
2. A. R. H. Baker, 'Historical Geography in Britain', in id. (ed.), *Progress in Historical Geography* (Newton Abbot, 1972), 90–110; W. Norton, *Historical Analysis in Geography* (London, 1984), 30 ff., 62 ff.
3. R. A. French, 'Historical Geography in the USSR', *Soviet Geography: Review and Translation*, 9/5 (1968), 551–61; id., 'Historical Geography in the USSR', in Baker (ed.) (above, n. 2), 111–28; R. A. French and J. H. Bater, 'Approaches to the Historical Geography of Russia', in eid. (ed.), *Studies in Russian Historical Geography* (London, 1983), i 1–10.
4. Bater and French (eds.) (above, n. 3), 6.
5. J. H. Bater, *St Petersburg: Industrialization and Change* (London, 1976); id., *Urban Industrialization in the Provincial Towns of Late Imperial Russia* (Pittsburgh, Pa., 1986); W. L. Blackwell, *The Beginnings of Russian Industrialization, 1800–1860* (Princeton, NJ, 1968); id., *The Industrialization of Russia: An Historical Perspective* (Arlington Heights, I., 1982); J. Bradley, *Muzhik and Muscovite: Urbanization in Late Imperial Russia* (London, 1985); D. R. Brower, *Estate, Class and Community: Urbanization and Revolution in Late Tsarist Russia* (Pittsburgh, Pa., 1983); M. F. Hamm, (ed.), *The City in Russian History* (Lexington, Ky., 1976); id. (ed.), *The City in Late Imperial Russia* (Bloomington, Ind., 1986); J. P. MacKay, *Pioneers for Profit: Foreign Entrepreneurship and Russian Industrialization, 1885–1913* (Chicago, Ill., 1970); R. W. Thurston, *Liberal City, Conservative State: Moscow and Russia's Urban Crisis, 1906–14* (Oxford, 1987); P. Herlihy, *Odessa: A History, 1794–1914* (Cambridge, 1986).
6. A. R. H. Baker, 'Reflections on the Relations of Historical Geography and the Annales School of History', in id. and D. Gregory, (eds.), *Explorations in Historical Geography* (Cambridge, 1984), 1–27, pp. 17–18.
7. See e.g. E. R. Wolfe, *Peasants* (Englewood Cliffs, NJ, 1966); J. Langton, 'Habitat, Economy and Society Revisited: Peasant Ecotypes and Economic Development in Sweden', *Cambria*, 12/3 (1985), 5–24. It is important to note that the ecotype concept argues only that communities adapt themselves to the possibilities offered by local environments; it does not argue, as did the early environmental determinists, that the nature of society was determined by the environment. Langton (above) uses R. Redfield's formulation in *The Little Community* (Chicago, Ill., 1960) that peasant society can only ever be a 'part' society to examine the dual impacts of macro-economic and political forces, on the one hand, and local

environmental potentialities, on the other, on the evolution of peasant societies.

8. See Glossary for *chernososhnye krestyane* and also for 'white places' as described under *belomestnyi ataman, kazak.* The terms 'white' and 'black' were purely legal ones, having no connection with e.g. the 'black' in 'black earths'.

9. The former court peasants, known after 1797 as apanage peasants, retained their separate identity and administration until the time of the Emancipation.

## Notes to Chapter 1

1. F. J. Turner, 'The Significance of the Frontier in American History', *Proceedings of the State Historical Society of Wisconsin*, 41 (1894), 79–112; cited from G. R. Taylor (ed.), *The Turner Thesis: Concerning the Role of the Frontier in American History* (New York, 1966), 1–18, p. 2.

2. F. J. Turner, *The Significance of Sections in American History* (Gloucester, Mass., 1959), 18–19.

3. V. O. Klyuchevskii, *Kurs russkoi istorii* (Moscow, 1937), i. 20–1.

4. P. N. Milyukov, *Russia and its Crisis* (Chicago, Ill., 1905), 5; M. Skiada and A. Shaposhnikov, 'Starye i novye torgovye puti soobshchenii v Voronezh-skom krae', *Voronezhskaya pamyatnaya knizhka za 1870–71* (Voronezh, 1871); N. O. Vtorov, 'O zaselenii Voronezhskoi gubernii', *Voronezhskii vestnik na 1861 g.* (St Petersburg, 1861), 247.

5. Other attempts to examine the Russian frontier in the light of the Turner thesis include: B. H. Sumner, 'The Frontier', ch. 1 of *Survey of Russian History* (London, 1961), 1–46; R. Dow, 'Prostor: A Geopolitical Study of Russia and the United States', *Russian Review*, 1/1 (1941), 6–19; V. T. Bill, 'The Circular Frontier of Muscovy', *Russian Review*, 9/1 (1950), 45–52; D. W. Treadgold, 'Russian Expansion in the Light of Turner on the American Frontier', *Agricultural History*, 26 (1952), 147–52; A. Lobanov-Rostovsky, 'Russian Expansion in the Far East in the Light of the Turner Hypothesis', in W. D. Wyman, and C. B. Kroeber (eds.), *The Frontier in Perspective* (Madison, Wis., 1957), 79–94; J. L. Wieczynski, *The Russian Frontier: The Impact of the Borderlands upon the Course of Early Russian History* (Charlottesville, Va., 1976).

6. See e.g. V. P. Zagorovskii, *Belgorodskaya cherta* (Voronezh, 1969), 59.

7. D. I. Bagalei, *Ocherki iz istorii kolonizatsii i byta stepnoi okrainy Moskovskogo gosudarstva* (Moscow, 1887), 90–108; I. D. Belyaev, *O storozhevoi, stanichnoi i polevoi sluzhbe na polskoi okraine Moskovskogo gosudarstva do tsarya Alekseya Mikhailovicha* (Moscow, 1846).

8. Bagalei (above, n. 7), 20; see also 66–77.

9. D. J. B. Shaw, 'Urbanism and Economic Development in a Pre-industrial Context', *Journal of Historical Geography*, 3/2 (1977), 107–22; id., 'Southern Frontiers of Muscovy, 1550–1700', in J. H. Bater and R. A. French (eds.), *Studies in Russian Historical Geography* (London, 1983), i. 117–42.

10. A. A. Novoselskii, *Borba Moskovskogo gosudarstva s tatarami v pervoi polovine XVII veka* (Moscow and Leningrad, 1948).

11. I. N. Miklashevskii, *K istorii khozyaistvennogo byta Moskovskogo gosudarstva: Zaselenie i selskoe khozyaistvo yuzhnoi okrainy v XVII veke* (Moscow, 1894), 102.

12. e.g. L. B. Veinberg, and A. A. Poltoratskaya, *Materialy dlya istorii Voronezhskoi i sosednikh gubernii*, ii. *Voronezhskie pistsovye knigi* (Voronezh, 1891), 36, 86, 101, etc.

13. Turner (above, n. 1), 2.

14. V. O. Klyuchevskii, *Sochineniya* (Moscow, 1956), i. 279. For reference to xenophobia see P. Dukes, *The Making of Russian Absolutism, 1613–1801* (London, 1982), 23, 55–6.

15. Cited in A. W. Fisher, *The Russian Annexation of the Crimea* (Cambridge, 1970), 21–2.

16. A. A. Golombievskii, 'Vypiska v Razryade o postroenii novykh gorodov', *Izvestiya Tambovskoi uchenoi arkhivnoi komissii*, 33 (1892).

17. *Memuary otnosyashchiesya k istorii yuzhnoi Rusi*, ii (Kiev, 1896), 403.

18. D. Rickey Jr., *Forty Miles a Day on Beans and Hay* (Norman, Okla., 1963), 230–4.

19. B. A. Osipov, *Ocherki po istorii Saratovskogo kraya: Konets XVI i XVII vv* (Saratov, 1976), 63–4; A. N. Zertsalov, *Materialy dlya istorii Sinbirska i yego uezda (Prikhodo-raskhodnaya kniga Sinbirskoi prikaznoi izby 1665–7 gg.)* (Simbirsk, 1896).

20. e.g. M. De-Pule, *Materialy po istorii Voronezhskoi i sosednikh gubernii: Orlovskie akty XVII–XVIII stol.* (Voronezh, 1861), 350–4; see also Sumner (above, n. 5), 31.

21. Turner (above, n. 1), 2.

22. See e.g. *Akty otnosyashchiesya k istorii Voiska Donskogo* (Novocherkassk, 1902); *Akty otnosyashchiesya k istorii Voiska Donskogo izdannym A. A. Lishinym*, i–iii (Novocherkassk, 1891–4); *Akty otnosyashchiesya k istorii yuzhnoi i zapadnoi Rossii* (St Petersburg, 1863– ); *Donskie dela*, i–v (St Petersburg, 1906–17); *Vossoedinenie Ukrainy s Rossiei: Dokumenty i materialy*, 3 vols. (Moscow, 1954).

23. e.g. Veinberg and Poltoratskaya (above, n. 12).

24. Turner (above, n. 1), 2.

25. De-Pule (above, n. 20), 456; G. Germanov, 'Postepennoe raspredelenie odnodvorcheskogo naseleniya v Voronezhskoi gubernii', *Zapiski Imperatorskogo Geograficheskogo obshchestva*, 12 (1857), 185–325, p. 224; see also 'Stateinaya zapis 158 g. goroda Kozlova s posadskimi lyudmi v tyaglo gulyashchikh, torgovykh i remeslovykh', *Izvestiya Tambovskoi uchenoi arkhivnoi komissii*, 33 (1892).

26. Turner (above, n. 1), 5.

27. Ibid. 6.

28. Ibid. 5. The sequence of frontiers proposed by Turner has been subjected to much criticism. As Billington wrote, 'The frontier was actually a broad, westward-moving zone, in which a variety of individuals were applying a number of skills to the conquest of nature, unmindful of any orderly

pattern that later theorists might expect to find in the evolution of society' (R. A. Billington, 'The American Frontier', in P. Bohannan, and F. Plogg, (eds.), *Beyond the Frontier* (New York, 1967), 3–24, p. 9).

29. Veinberg and Poltoratskaya (above, n. 12), 139–41.

30. Vtorov (above, n. 4); Zagorovskii (above, n. 6), 39; id., 'Zemledelcheskoe naselenie v Pridonskikh uezdakh na Belgorodskoi cherte v seredine XVII veka i vozniknovenie pervykh sel "za chertoi" ', in *Yezhegodnik po agrarnoi istorii Vostochnoi Yevropy, 1964* (Kishinev, 1966), 199–207.

31. Veinberg and Poltoratskaya, *Materialy* (above, n. 12), vol. i (Voronezh, 1887), no. 135; D. I. Bagalei, *Materialy dlya istorii kolonizatsii i byta stepnoi okrainy Moskovskogo gosudartstva v XVI–XVII stoletii*, ii (Kharkov, 1890), no. 13, pp. 49–74; Miklashevskii (above, n. 11), 165–178, 260–3.

32. V. M. Vazhinskii, 'Razvitie rynochnykh svyazei v yuzhnykh russkikh uezdakh vo vtoroi polovine XVII veka', *Uchenye zapiski Kemerovskogo pedagogicheskogo instituta*, 5 (1963); id., 'Torgovye svyazi yuzhnykh gorodov Rossi v tretei chetverti XVII veka', in *Goroda feodalnoi Rossii* (Moscow, 1966), 298–307.

33. R. Wade, *The Urban Frontier* (Cambridge, Mass., 1959), 1.

34. Ibid.; D. W. Meinig, *The Great Columbia Plain* (Washington, 1968), 486.

35. *Kniga bolshomu chertezhu* (Moscow and Leningrad, 1950); Belyaev (above, n. 7); Bagalei (above, n. 31) vol. 1, (Kharkov, 1886), no. 1; ibid., vol. ii, nos. 2, 7, 11, 14; *Akty Moskovskogo gosudarstva* i (St Petersburg, 1890), 2–5.

36. L. B. Veinberg, *Materialy po istorii Voronezhskoi i sosednikh gubernii: Drevnie akty XVII stoletiya*, xvi (Voronezh, 1890), no. 778.

37. Bagalei (above, n. 31), vol. i, no. 2, pp. 5–13.

38. Miklashevskii (above, n. 11), 210 ff.; S. I. Tkhorzhevskii, 'Gosudarstvennoe zemledelie na yuzhnoi okraine Moskovskogo gosudarstva v XVII veke', *Arkhiv istorii truda v Rossii*, 18 (Petrograd, 1923).

39. P. N. Chermenskii, 'Donskie votchiny boyar Romanovykh', *Izvestiya Tambovskoi uchenoi arkhivnoi komissii*, 57 (1917), 43–81.

40. I. I. Sokolova, *Sluzhiloe zemlevladenie i khozyaistvennoe sostoyanie priokskikh uezdov russkogo gosudarstva v kontse XVI–pervoi treti XVII vv.* (Moscow, 1975); ead., 'Sluzhiloe zemlevladenie Kashirskogo uezda v kontse XVI–pervoi treti XVII veka (po materialam pistsovykh knig)', *Vestnik Moskovskogo universiteta: Seriya istoriya* (1975), no. 1, pp. 82–92; O. A. Shvatchenko, 'K voprosu o sostoyanii pomestnoi sistemy v Zamoskovskom i Ryazanskom krayakh v kontse 80-kh–90-e gody XVI v.', *Vestnik Moskovskogo universiteta: Seriya istoriya* (1974), no. 2, pp. 56–68.

41. V. M. Vazhinskii, *Zemlevladenie i skladyvanie obshchiny odnodvortsev v XVII veke po materialam yuzhnykh uezdov* (Voronezh, 1974), 61–2.

42. Turner (above, n. 1), 14.

43. Ibid. 15.

44. Ibid. 17.

45. Ibid. 15.

46. J. Keep, 'Light and Shade in the History of the Russian Administration', *Canadian Slavic Studies*, 6 (1972), 1–9.

47. P. Longworth, *The Cossacks* (London, 1971), 23.

48. Veinberg and Poltoratskaya (above, n. 12), vol. i, nos. 60, 101; vol. ii (the 1615 register of inquisition for Voronezh has many references to settlements destroyed by the cossack rebel Ivan Zarutskii); Zagorovskii (above, n. 6), 25; De-Pule (above, n. 20), nos. ccxvii–ccxxiv; Bagalei (above, n. 31), i. 9–11.

49. Zagorovskii (above, n. 6), 253–74; P. Avrich, *The Russian Rebels, 1600–1800* (London, 1973).

50. Avrich (above, n. 49); Zagorovskii (above, n. 6), 253–74; Sumner (above, n. 5), 141 ff.

51. Veinberg and Poltoratskaya (above, n. 12), vol. i, nos. 26, 28, 32, 37, 38, 39, 41, 45, etc.

52. e.g. Zagorovskii (above, n. 6), 249, 255.

53. Avrich (above, n. 49), 24, 89, 104, 162.

54. Tsarev-Alekseev was renamed Novyi Oskol in 1655.

55. Zagorovskii (above, no. 6), 127–8.

56. Veinberg and Poltoratskaya (above, n. 12), vol. i, no. 116; Vtorov (above, n. 4), 256 ff; G. M. Veselovskii and N. V. Voskresenskii, *Goroda Voronezhskoi gubernii* (Voronezh, 1876), Bobrov section.

57. Veinberg and Poltoratskaya (above, n. 12), vol. i, nos. 3, 6; G. N. Anpilogov, *Novye dokumenty o Rossii kontsa XVI–nachala XVII veka* (Moscow, 1967), 325–7.

58. Veinberg (above, n. 36), vol. iii, decree of June 1655; Veinberg and Poltoratskaya (above, n. 12), vol. i, no. 155; De-Pule (above, n. 20), 339, 394, 434; N. Vtorov and K. Aleksandrov-Dolnik, *Drevnie gramoty i drugie pismennye pamyatniki, kasayushchiesya Voronezhskoi gubernii*, ii (Voronezh, 1852), document listed for 1625.

59. Zagorovskii (above, n. 6), 125, 249–51; F. K. Yavorskii, *Voronezhskie petrovskie akty* (Voronezh, 1872), nos. xxvii, xxx, xxii, lvii, lxxxii, xciii, xcv, cvii, cxi.

60. e.g. Veinberg and Poltoratskaya (above, n. 12), vol. i, nos. 91, 92.

61. Ibid., no. 35; Anpilogov (above, n. 57), 310 ff.; Zagorovskii (above, n. 6), 26, 29, 252.

62. Veinberg (above, n. 36), (on Korotoyak); see also: Bagalei (above, n. 31), vol. i, no. 22 (on Ostrogozhsk and Korotoyak); vol. ii, nos. 3 (on Userd) and 6 (on Orlov); Veinberg (above, n. 36), vols. vii (on Voronezh), and xv (on Olshansk, Valuiki); Vtorov and Aleksandrov-Dolnik (above, n. 58), vol. i (Voronezh, 1851), 129; ibid. ii (Voronezh, 1852), 92 (descriptions of Voronezh); for Ostrogozhsk see also *Trudy Voronezhskoi uchenoi arkhivnoi komissii*, 2 (1902), 65–74. For other towns in Russia see e.g. P. P. Smirnov, *Goroda Moskovskogo gosudarstva v pervoi polovine XVII veka*, pts. 1–2 (Kiev, 1917–19); the series *Materialy dlya istorii goroda XVI–XVIII stol.* (Moscow, 1883–4); and such descriptions as A. Ye. Mertsalov, *Ocherki goroda Vologdy po pistsovoi knige 1627 g.* (Vologda, 1885).

63. Zagorovskii (above, n. 6), 40.

64. Veinberg and Poltoratskaya (above, n. 12), vol. i, no. 77; De-Pule (above, n. 20), 394; Vtorov and Aleksandrov-Dolnik (above, n. 58), ii. 1.

65. S. I. Sakovich, 'Kniga zapisnaya melochnykh tovarov Moskovskoi bolshoi

tamozhni', in *Iz istorii torgovli i promyshlennosti Rossii kontsa XVII veka* (Moscow, 1956).
66. Vazhinskii (above, n. 32); see also A. A. Novoselskii, 'Iz istorii Donskoi torgovli', *Istoricheskie zapiski*, 26 (1948), 198–216, p. 205.
67. Turner (above, n. 1), 2.
68. Turner (above, n. 2).
69. See e.g. the critical essays in Taylor (above, n. 1).

Notes to Chapter 2

1. 'As successive terminal moraines result from successive glaciations, so each frontier leaves its traces behind it, and when it becomes a settled area the region still partakes of the frontier characteristics' (F. J. Turner, 'The Significance of the Frontier in American History', in G. R. Taylor (ed.), *The Turner Thesis: Concerning the Role of the Frontier in American History* (New York, 1966), 1–18, p. 2).
2. N. P. Pavlov-Silvanskii, *Gosudarevy sluzhilye lyudi* (St Petersburg, 1909), 89 ff.; J. Blum, *Lord and Peasant in Russia from the Ninth to the Nineteenth Century* (New York, 1967), 169–70.
3. D. J. B. Shaw, 'Southern Frontiers of Muscovy, 1550–1700', in J. H. Bater and R. A. French (eds.), *Studies in Russian Historical Geography* (London, 1983), i. 117–142, pp. 122–3.
4. Pavlov-Silvanskii (above, n. 2), 98.
5. Ibid. 98–9.
6. V. P. Zagorovskii, *Belgorodskaya cherta* (Voronezh, 1969), 29; A. A. Yelfimova, 'K voprosu o roli krestyanstva v formirovanii sluzhilykh lyudei Voronezhskogo uezda v XVII v.', *Izvestiya Voronezhskogo gosudarstvennogo pedagogicheskogo instituta*, 153 (1975), 47–55.
7. Pavlov-Silvanskii (above, n. 2), 119–20; S. V. Rozhdestvenskii, *Sluzhiloe zemlevladenie v Moskovskom gosudarstve XVI veka* (St Petersburg, 1897), 251 ff.
8. R. Hellie, *Enserfment and Military Change in Muscovy* (Chicago, Ill., 1970), 31.
9. Pavlov-Silvanskii (above, n. 2), 119; Rozhdestvenskii (above, n. 7), 270, 282 ff., 347 ff., 361 ff.
10. Zagorovskii (above, n. 6), 30; L. B. Veinberg and A. A. Poltoratskaya, *Materialy dlya istorii Voronezhskoi i sosednikh gubernii*, ii (Voronezh, 1891), 20 etc.
11. Published in Veinberg and Poltoratskaya (above, n. 10), 1–141, 237–61. The 1615 cadastre was officially entitled a *dozornaya kniga*, that for 1629 a *pistsovaya kniga*.
12. Ibid.; also Ye. G. Shulyakovskii (ed.), *Ocherki istorii Voronezhskogo kraya* (Voronezh, 1961), ch. 2.
13. *Chetverti*: see Glossary. The register records *chetverti* 'in three fields' (a practice derived from surveying in a three-field system). Fifteen *chetverti* 'in three fields' is usually equivalent to 45 ordinary *chetverti* or 22.5 *des*.

14. Veinberg and Poltoratskaya (above, n. 10), 26–30.
15. Zagorovskii (above, n. 6), 28. The holdings of the junior boyars and other middle-ranking or high-ranking servitors were known as *pomestya* (service estates), unlike those of the contract servitors.
16. N. A. Blagoveshchenskii, *Chetvertnoe pravo* (Moscow, 1899), 62 ff.
17. Zagorovskii (above, n. 6), 28; but see also V. N. Storozhev, *Voronezhskoe dvoryanstvo po desyatnyam XVII v.* (Voronezh, 1894); also Rozhdestvenskii (above, n. 7), 251–4. The *chetverti* in the renumeration scale were also surveyed 'in three fields': see n. 13, above.
18. e.g. some at Chertovitskoe: see Veinberg and Poltoratskaya (above, n. 10), 56–7.
19. Ibid. 30–2.
20. My calculation from the 1615 cadastre.
21. Calculated from the 1629 cadastre.
22. I. N. Miklashevskii, *K istorii khozyaistvennogo byta Moskovskogo gosudarstva: Zaselenie i selskoe khozyaistvo yuzhnoi okrainy v XVII veke* (Moscow, 1894), 84–5, 109 ff.; also Zagorovskii (above, n. 6), 48.
23. My calculations from the 1615 and 1629 cadastres.
24. A. A. Novoselskii, *Borba Moskovskogo gosudarstva s tatarami v pervoi polovine XVII veka* (Moscow and Leningrad, 1948), 301, 303–4.
25. Miklashevskii (above, n. 22), 109 ff. See also Pavlov-Silvanskii (above, n. 2), 221.
26. D. I. Bagalei, *Materialy dlya istorii kolonizatsii i byta Moskovskogo gosudarstva v XVI i XVII stoletii*, ii (Kharkov, 1890), no. 13, pp. 49–74.
27. V. I. Koshelev, 'Gorodok Orlov i yego voennaya zona v XVII veke', *Izvestiya Voronezhskogo gosudarstvennogo pedagogicheskogo instituta*, 12/1 (1950), 87–144.
28. V. M. Vazhinskii, *Zemlevladenie i skladyvanie obshchiny odnodvortsev v XVII veke* (Voronezh, 1974), 55, 106. The Belgorod and Sevsk military districts (*razryady*) embraced most of the southern and south-western frontier regions in the latter half of the seventeenth century. See ch. 10 below for details.
29. A. A. Novoselskii, 'Rasprostranenie krepostnicheskogo zemlevladeniya v yuzhnykh uezdakh Moskovskogo gosudarstva v XVII v.', *Istoricheskie zapiski*, 4 (1938), 21–40.
30. Zagorovskii (above, n. 6), 114–19.
31. D. I. Bagalei, *Ocherki iz istorii kolonizatsii i byta stepnoi okrainy Moskovskogo gosudarstva* (Moscow, 1887), 113–14; Zagorovskii (above, n. 6), 29.
32. Zagorovskii (above, n. 6), 111, 126–9, 149.
33. Ibid. 129 ff.
34. Pavlov-Silvanskii (above, n. 2), 198; Vazhinskii (above, n. 28), 118, 145.
35. Pavlov-Silvanskii (above, n. 2), 221–2; see also Veinberg and Poltoratskaya (above, n. 10), i (Voronezh, 1887), petition dated 1625.
36. Hellie (above, n. 8), 192, 211–25.
37. Pavlov-Silvanskii (above, n. 2), 212–4; Zagorovskii (above, n. 6), 110; V. M. Vazhinskii, 'Usilenie soldatskoi povinnosti v Rossii v XVII v.', *Izvestiya Voronezhskogo gosudarstvennogo pedagogicheskogo instituta*, 157 (1976), 52–68, pp. 60–1.

38. Zagorovskii (above, n. 6), 141.
39. Bagalei (above, n. 26), ii. 49–74.
40. Vazhinskii (above, n. 28), 120, 122–5.
41. V. P. Zagorovskii, 'Soldatskie sela i soldatskoe zemlevladenie v Voronezh-skom krae XVII veka', *Iz istorii Voronezhskogo kraya*, 4 (Voronezh, 1972), 90–7, pp. 93–4.
42. V. P. Zagorovskii, 'Zemledelcheskoe naselenie v Pridonskikh uezdakh na Belgorodskoi cherte v seredine XVII veke i vozniknovenie pervykh sel "za chertoi" ', in *Yezhegodnik po agrarnoi istorii Vostochnoi Yevropy, 1964* (Kishinev, 1966), 199–207, p. 201.
43. 'Opisanie goroda Tambova i Verkhotsenskikh volostei, uchinennoe knyazem Vasilem Vas. Kropotkinym v 7186 (1678) godu', in *Letopis zanyatii Arkheograficheskoi komissii, 1865–6* (St Petersburg, 1868), 30–55.
44. 'Perepisnaya kniga goroda Verkhososenska 1709 g.', in L. B. Veinberg, *Materialy po istorii Voronezhskoi i sosednikh gubernii: Drevnie akty XVII veka* (Voronezh, 1885–90), vol. 13.
45. Vazhinskii (above, n. 28), ch. 2; P. Ivanov, 'Syabry-pomeshchiki', *Zhurnal Ministerstva Narodnogo Prosveshcheniya*, 350 (Dec. 1903), 406–42; Ye. I. Samgina, 'Sluzhiloe zemlevladenie i zemlepolzovanie v Chernskom uezde v pervoi polovine XVII veka', in *Novoe o proshlom nashei strany* (Moscow, 1967), 264–76.
46. Vazhinskii (above, n. 28), 62 ff.
47. Ibid. 119 ff.; Zagorovskii (above, n. 6), 244 ff.
48. Zagorovskii (above, n. 6), 246–7.
49. Polnoe Sobranie Zakonov Rossiiskoi Imperii (St Petersburg, 1830), vol. v 3287. V. Semevskii, *Krestyane v tsarstvovanie Yekateriny II* (St Petersburg, 1901), ii. 726.
50. Semevskii (above, n. 49), n. 2.
51. N. K. Tkacheva, 'Iz istorii odnodvortsev v XVIII v.', in *Yezhegodnik po agrarnoi istorii Vostochnoi Yevropy, 1968* (Leningrad, 1972), 133–41, p. 134; Ye. Shchepkina, *Tulskii uezd v XVII veke* (Moscow, 1892), pp. xxx–xxxi.
52. 'Chetvertnye krestyane', in F. A. Brokgauz and I. A. Efron (eds.), *Entsiklopedicheskii slovar* (St Petersburg, 1903), xxxviii, 726–36, p. 729.
53. M. T. Belyavskii, *Odnodvortsy Chernozemya* (Moscow, 1984).
54. Semevskii (above, n. 49), 762.
55. Belyavskii (above, n. 53), 35–7, 50, 77–8, 84, 98, 102, 140.
56. 'Chetvertnye krestyane' (above, n. 52), 729.
57. Tkacheva (above, n. 51), 140; see also V. Levshin, 'Topograficheskoe, istoricheskoe, statisticheskoe i kameralnoe opisanie Tulskoi gubernii, po nachertaniyu IVEO', TsGIA, f. 91, d. 285, ll. 69ob. ff., 95ob., 126.
58. Bagalei (above, n. 31), 298 ff.; Semevskii (above, n. 49), 726, 728–9; Belyavskii (above, n. 53), 51–2 etc. The *odnodvortsy* were forbidden in 1785 to migrate between provinces without permission (Semevskii (above, n. 49), 768).
59. 'Chetvertnoe krestyane' (above, n. 52), 734. The sample was incomplete.
60. Blagoveshchenskii (above, n. 16), 234–6 etc.
61. F. A. Shcherbina, 'Chetvertnoe zemlevladenie v Voronezhskoi gubernii', *Voronezhskii yubileinyi sbornik*, i (Voronezh, 1886), 467–80, pp. 472, 476; P.

N. Pershin, *Zemelnoe ustroistvo dorevolyutsionnoi derevni*, i (Moscow, 1928).

62. V. M. Protorchina (ed.), *Voronezhskii krai v XVIII v.* (Voronezh, 1980), 42.
63. TsGVIA, f. VUA, yed. khr. 18668, l. 13.
64. e.g. TsGADA, f. 1355, yed. khr. 263, l. 12, nos. 93–101; ibid., yed. khr. 294, l. 12, no. 47; ibid., yed. khr. 240, l. 15, no. 72.
65. Veinberg and Poltoratskaya (above, n. 10), pp. 85, 96, 110, 244; also TsGADA, f. 1355, yed. khr. 51, 55.
66. Yelfimova (above, n. 6); Shcherbina (above, n. 61), 473–6.
67. TsGVIA, f. VUA, yed. khr. 18799; cf. TsGADA, f. 1355, op. 1, yed. khr. 37.
68. Belyavskii (above, n. 53), 97–8.
69. 'Opisanie' (above, n. 43); cf. TsGADA, f. 1355, op. 7, d. 1650.
70. Semevskii (above, n. 49), 744.
71. TsGADA, f. 1355, op. 1, d. 97.
72. V. I. Nedosekin, *Nakazy v Zakonodatelnuyu Komissiyu 1767 g. ot Voronezhskoi gubernii kak istoricheskii istochnik* (Voronezh, 1953), 13.
73. TsGADA, f. 1355, yed. khr. 228, nos. 66, 67.
74. Comparisons for Voronezh district based on Veinberg and Poltoratskaya (above, n. 10); TsGADA, f. 1355, yed. khr. 51, 249. For Tambov see above, n. 69.
75. TsGVIA, f. VUA, yed. khr. 18668, l. 13.
76. Belyavskii (above, n. 53), 129.
77. G. Germanov, 'Postepennoe raspredelenie odnodvorcheskogo naseleniya v Voronezhskoi gubernii', *Zapiski Imperatorskogo Geograficheskogo obshchestva*, 12 (1857), 185–325, p. 224.
78. Levshin (above, n. 57), ll. 179 ff.
79. Baron A. von Haxthausen, *The Russian Empire: Its People, Institutions and Resources*, i trans. R. Faire (London, 1968), 371.
80. See D. Thorner, B. Kerblay, and R. E. F. Smith (eds.), *A. V. Chayanov on the Theory of Peasant Economy* (Homewood, Ill., 1966). The broader concept of the peasantry as a distinctive social formation is, however, a controversial one.
81. Semevskii (above, n. 49), 743.
82. Ibid. 768.

## Notes to Chapter 3

1. J. M. Letiche (ed.), *A History of Russian Economic Thought: From the Ninth through the Nineteenth Centuries* (Berkeley, Calif., 1955), 209 ff.: N. P. Nikitin, 'Zarozhdenie ekonomicheskoi geografii v Rossii', *Voprosy geografii*, 17 (1950), 43–104; N. L. Rubinshtein, 'Topograficheskie opisaniya namestnichestv i gubernii XVIII v.: Pamyatniki geograficheskogo i ekonomicheskogo izucheniya Rossii', *Voprosy geografii*, 31 (1953), 39–89; D. M. Lebedev, *Ocherki po istorii geografii v Rossii 1725–1800 gg.* (Moscow, 1957); L. A. Goldenberg, *Russian Maps and Atlases as Historical Sources* (Toronto, Ont., 1971).
2. e.g. R. H. Brown, *Mirror for Americans* (New York, 1943); H. C. Darby, *The*

*Domesday Geography of Eastern England* (Cambridge, 1952); see also J. O. M. Broek, *The Santa Clara Valley, California: A Study in Landscape Change* (Utrecht, 1932).

3. See E. R. Wolf, *Peasants* (Englewood Cliffs, NJ, 1966).
4. L. V. Milov, *Issledovanie po Ekonomicheskikh primechaniyakh k Generalnomu mezhevaniyu* (Moscow, 1965), 213; the short economic notes to the General Survey were consulted in the following places: TsGADA, f. 1355, yed. khr. 51 (Voronezh district), 202 (Biryuch), 213 (Bobrov), 228 (Boguchar), 240 (Valuiki), 249 (Zadonsk), 263 (Zemlyansk), 282 (Kalitva), 287 (Korotoyak), 294 (Livensk), 297 (Nizhnedevitsk), 314 (Ostrogozhsk), 325 (Pavlovsk).
5. Milov (above, n. 4).
6. TsGIVA, f. VUA, yed. khr. 18668, published as *Opisanie Voronezhskoi namestnichestva 1785 g.* (Voronezh, 1982).
7. B. D. Grekov, 'Opyt obsledovaniya khozyaistvennykh anket XVIII veka', republished in his *Izbrannye trudy*, iii (Moscow, 1960), 225–80.
8. *Trudy Volnogo Ekonomicheskogo obshchestva*, 8 (1768), 160–86; L. B. Veinberg, *Materialy po istorii Voronezhskoi i sosednikh gubernii*, xv–xvi (Voronezh, 1889–90).
9. H. L. C. Bacmeister, *Topograficheskie izvestiya, sluzhashchie dlya polnogo geograficheskogo opisaniya Rossiiskoi imperii*, i–iv (St Petersburg, 1771–4); A. Shchekatov, *Slovar geograficheskii Rossiiskogo gosudarstva*, 7 vols. (Moscow, 1801–8); Ye. F. Zyablovskii, *Zemleopisanie Rossiiskoi imperii* (St Petersburg, 1810).
10. Ye. Bolkhovitinov, *Istoricheskoe, geograficheskoe i ekonomicheskoe opisanie Voronezhskoi gubernii* (Voronezh, 1800).
11. S. G. Gmelin, *Reise durch Russland zur Untersuchung der drey Natur-Reiche . . . in den Jahren 1768–74*, i–iv (St Petersburg, 1770–84).
12. Answers in Veinberg (above, n. 8).
13. Veinberg (above, n. 8), vol. xvi, no. 783, answer 10.
14. Gmelin (above, n. 11), 102.
15. S. V. Kirikov, *Chelovek i priroda Vostochnoevropeiskoi lesostepi v X–nachale XIX vv.* (Moscow, 1979), 110.
16. TsGADA, f. 1355, yed. khr. 325, no. 14.
17. Gmelin (above, n. 11), 89; Zyablovskii (above, n. 8), 178–201.
18. Veinberg (above, n. 8), vol. xvi, no. 774, answer 28.
19. Ibid., no. 781, answer 28.
20. Gmelin (above, n. 11), 44.
21. Veinberg (above, n. 8), vol. xv, no. 774, answer 11.
22. Veinberg (above, n. 8), vol. xvi, no. 779, answer 11.
23. Veinberg (above, n. 8), vol. xvi, nos. 783, 784, answer 11; for other aspects of environmental change see Kirikov (above, n. 15), 131–64; id., *Izmeneniya zhivotnogo mira v prirodnykh zonakh SSSR (XVIII–XIX vv.): Stepnaya zona i lesostep* (Moscow, 1959).
24. Kirikov (above, n. 15), 100–2.
25. Ibid. 102.
26. V. M. Protorchina (ed.), *Voronezhskii krai v XVIII v.* (Voronezh, 1980), 76–

7. The full data in Table 3.1 are derived from the 1785 topographical description (above, n. 6).

27. The *cherkasy* of the Ostrogozhsk Regiment constituted one of the cossack regiments permitted to settle in Slobodskaya Ukraina in the 17th and early 18th cents. Much of the territory of this regiment was eventually included in Voronezh province.

28. F. A. Shcherbina, 'Zaselenie Ostrogozhskogo kraya', in *Sbornik statisticheskikh svedenii po Voronezhskoi gubernii*, ii/2 (Voronezh, 1887), 17–33, p. 19; A. G. Slyusarskii, *Sotsialno-ekonomicheskoe razvitie Slobozhanshchiny (XVII–XVIII vv.)* (Kharkov, 1964), 114.

29. V. Semevskii, *Krestyane v tsarstvovanie Yekateriny II* i (St Petersburg, 1903), 30, ii (St Petersburg, 1901), 744, 852. *Barshchina* was labour service undertaken for the lord, *obrok* was quitrent in cash or kind payable to the lord in lieu of labour service: some privately owned peasants were subject to both obligations.

30. Ibid. i. 30.

31. Based on the economic notes to the General Survey (above, n. 4).

32. TsGADA, f. 1355, yed. khr. 249, no. 2.

33. Ibid., yed. khr. 228, no. 36.

34. e.g. ibid., yed. khr. 263, nos. 26, 28, 36, 39, 43 etc.; ibid., yed. khr. 249, nos. 1, 4, 7, 11, 13 etc.; ibid., yed. khr. 297, nos. 2, 3, 6, 13, 14, 16 etc.

35. Ibid., yed. khr. 51, nos. 58, 70.

36. Ibid., yed. khr. 263, nos. 30, 52, 88.

37. Ibid., yed. khr. 228, no. 53.

38. Ibid., d. 120, ll. 3–5.

39. Data calculated from material in Bolkhovitinov (above, n. 10).

40. Semevskii (above, n. 29), i. 585. But the 1785 topographical description suggests 16.5 *desyatiny* (TsGVIA, f. VUA, yed. khr. 18668, ll. 13–17).

41. See Semevskii (above, n. 29), i. 585.

42. Milov (above, n. 4), 170. See Glossary for meanings of terms *perelog* and *zalezh*. Our glossary definitions accord with those adopted by Smith (see R. E. F. Smith, *The Origins of Farming in Russia* (Paris, 1959), 50), although the terms are frequently used interchangeably in the literature—e.g. Milov (above, n. 3, 165) suggests that their meanings in the 18th-cent. steppe may have been virtually identical, the choice of term possibly depending on regional tradition.

43. Semevskii (above, n. 29), i. 30.

44. Milov (above, n. 4), 168 ff.

45. TsGVIA, f. VUA, yed. khr. 18668, l. 88.

46. Milov (above, n. 4), 172–5; see also his article 'O roli perelozhnykh zemel v russkom zemledelii vtoroi poloviny XVIII v.', in *Yezhegodnik po agrarnoi istorii Vostochnoi Yevropy, 1961 g.* (Riga, 1963), 279–88.

47. TsGVIA, f. VUA, yed. khr. 18668, l. 88.

48. Veinberg (above, n. 8), vol. xvi, no. 783, answer 18.

49. Bolkhovitinov (above, n. 10), 22 ff.

50. N. L. Rubinshtein, *Selskoe khozyaistvo Rossii vo vtoroi polovine XVIII v.* (Moscow, 1957), 448–9.

51. TsGVIA, f. VUA, yed. khr. 18668, l. 9ob.
52. Ibid., l. 88.
53. Veinberg (above, n. 8), vol. xvi, no. 784, answer 18.
54. A. Shafonskii, *Chernigovskogo namestnichestva topograficheskoe opisanie s kratkim geograficheskim i istoricheskim opisaniem Malyya Rossii* (Kiev, 1851), 167–8. Shafonskii's book was written in 1786.
55. TsGVIA, f. VUA, yed. khr. 18668, l. 88.
56. Veinberg, vol. xvi, no. 780, answer 18.
57. Ibid., (above, n. 8) no. 783, answer 18.
58. Grekov (above, n. 7), 244.
59. Veinberg (above, n. 8), vol. xv, 1781 returns for Boguchar, answer 18.
60. TsGVIA, f. VUA, yed. khr. 18668, l. 88.
61. Veinberg (above, n. 8), vol. xv, 1781 returns for Biryuch and Boguchar, answer 18.
62. Veinberg (above, n. 8), vol. xvi, no. 783, answers 2, 4.
63. An examination of farming practices among German Mennonite colonists living in Samara Province in the late 19th cent. will be found in ch. 4.
64. Milov (above, n. 4), 171.
65. TsGADA, f. 1355, yed. khr. 213, no. 142.
66. Rubinshtein (above, n. 50), 124.
67. Veinberg (above, n. 8), vol. xv, Bobrov district, answer 4.
68. e.g. Veinberg (above, n. 8), vol. xvi, no. 783, answer 4.
69. TsGADA, f. 1355, yed. khr. 240, l. 1; ibid., yed. khr. 263, l. 1.
70. Bolkhovitinov (above, n. 10), 22 ff.
71. Milov (above, n. 4), 213 ff.
72. Grekov (above, n. 7), 243.
73. Ibid. 254.
74. TsGADA, f. 1355, yed. khr. 203, l. 11, no. 41.
75. Grekov (above, n. 7), 254.
76. I. K. Kirilov, *Tsvetushchee sostoyanie Vserossiiskogo gosudarstva* (Moscow, 1977), 187–8.
77. TsGADA, f. 1355, yed. khr. 249, l. 1.
78. TsGVIA, f. VUA, yed. khr. 18668, l. 49.
79. TsGADA, f. 1355, yed. khr. 325, l. 4.
80. Ibid., yed. khr. 213, l. 17. There is a discrepancy with TsGADA, f. 16, d. 654, l. 7, no. 14, as regards the social composition of this village.
81. See above, nn. 3, 9.
82. TsGADA, f. 16, d. 654, ll. 53–4. Data also in Bolkhovitinov (above, n. 10), the 1785 topographical description, and the economic notes to the General Survey.
83. Calculated from the 1785 topographical description and Bolkhovitinov (above, n. 10).
84. Bolkhovitinov (above, n. 10), 62.
85. Yu. R. Klokman, *Sotsialno-ekonomicheskaya istoriya russkogo goroda: Vtoraya polovina XVIII v.* (Moscow, 1967), 253.
86. Ibid. 255; Bolkhovitinov (above, n. 10), 29, 76.
87. Gmelin (above, n. 11), 90.

88. Veinberg (above, n. 8), vol. xvi, no. 783, answer 5.
89. Bolkhovitinov (above, n. 10), 22 ff.


Notes to Chapter 4

1. Quoted in Count de Rochechouart, *Memoirs of the Count de Rochechouart 1788–1822*, trans. Frances Jackson (London, 1920), 78–9.
2. W. M. Kollmorgen, 'A Reconnaissance of Some Cultural-Agricultural Islands in the South', *Economic Geography*, 17 (1941), 409–30; 'Agricultural-Culture Islands in the South: Part II', *Economic Geography*, 19 (1943), 109–17.
3. Baron A. von Haxthausen, *The Russian Empire: Its People, Institutions, and Resources*, i, trans. R. Faire (London, 1968), 420–1.
4. V. P. Semenov-Tyan-Shanskii and V. L. Lamanskii, *Rossiya—polnoe geograficheskoe opisanie nashego otechestva: Srednee i nizhnee Povolzhe i Zavolzhe*, vi (St Petersburg, 1901).
5. P. S. Pallas, *Travels through the Southern Provinces of the Russian Empire, in the years 1793 and 1794. Translated from the German of P. S. Pallas, Counsellor of State to his Imperial Majesty of all the Russias*, 2 vols. (London, 1802–3), i. 101–2, 65.
6. M. Holderness, *Journey from Riga to the Crimea with some Account of the Manners and Customs of the Colonists of New Russia*, 2nd edn. (London, 1827), 160–1.
7. Ibid. 159.
8. Official *zemstvo* and other investigations in the late 19th cent. were to show that foreigners farmed under a variety of environmental conditions and had negotiated differing agreements with the state about the conditions of settlement which could affect their prosperity. By that time, however, culturalist explanations for the relative economic success of the various groups had become the norm. This was true even during the 1880s and 1890s when journals such as *Russkii vestnik* published anti-German articles; the same characteristics that in earlier years had been admired in the German settlers were now held against them. How Mennonites fared during this period is discussed in H. L. Dyck, 'Russian Mennonitism and the Challenge of Russian Nationalism', *Mennonite Quarterly Review*, 554 (1889), 307–41.
9. A. Klaus, *Nashi kolonii: Opyty i materialy po istorii i statistike inostrannoi kolonizatsii v Rossii* (St Petersburg, 1869), 152.
10. X. Hommaire de Hell, *Travels in the Steppes of the Caspian Sea, the Crimea, the Caucasus etc.* (London, 1839), 77.
11. Baron A. von Haxthausen, (above, n. 3), 430.
12. E. Henderson, 'Biblical Researches and Travels in Russia including a Tour in the Crimea and the Passage of the Caucasus with Observations on the State of the Rabbinical and Kharaite Jews, and the Mohammedan and Pagan Tribes Inhabiting the Southern Provinces of the Russian Empire', *Reports of the British and Foreign Bible Society*, 7 (1822), 24.

13. For an account of the development of policy see R. P. Bartlett, *Human Capital: The Settlement of Foreigners in Russia 1762–1804* (Cambridge, 1979).
14. The conditions for the Germans have been described by several authors. Having taken a year to make the journey from Germany, the migrants would arrive in the empty, treeless steppe to live for the first years in earthen dugouts while they waited for the delivery of building materials from the north. See e.g. F. C. Koch, *The Volga Germans in Russia and the Americas* (Pittsburgh, Pa., 1977).
15. The history of the Mennonites can be found in a number of sources but the main English-language source is *The Mennonite Encyclopaedia: A Comprehensive Reference Work of the Anabaptist-Mennonite Movement* (4 vols. Scottdale, Pa., 1959). The history of the Mennonites who migrated to Russia is to be found in D. G. Rempel, 'The Mennonite Commonwealth in Russia: A Sketch of its Founding and Endurance 1789–1919', *Mennonite Quarterly Review*, 48 (1974), Part 1, pp. 5–54; Part 2, ibid. 259–308; J. B. Teows, 'The Mennonites and the Siberian Frontier 1907–1930: Some Observations', *Mennonite Quarterly Review*, 47/2 (1973), 83–101; J. Urry, 'The Open and the Closed', D.Phil. thesis (Oxford, 1978); and, in Russian, Klaus (above, n. 9), 101–228.
16. Urry (above, n. 15), 759.
17. Ibid. 109–10.
18. The terms of the 'twenty requests' negotiated by the first Mennonites to migrate to Russia are given in Rempel (above, n. 15), 283–6.
19. Urry (above, n. 15), 145.
20. Ibid. 144.
21. This was because, with the outbreak of hostilities at Berislav, Count Potemkin, then Viceroy of the new territories, diverted the Mennonites to settle on his own estate.
22. Urry (above, n. 15), 156.
23. Ibid. 437. Among the large landholders were Wilhelm Martin with 75,000 *des.* and D. Schroeder with 21,000 *des.*
24. Klaus (above, n. 9), 180–92, gives a detailed account of the troubles in the colonies over the land question.
25. Urry (above, n. 15), 715.
26. Rempel (above, n. 15), 32.
27. The reason for this was the combined effect of increasing population pressure in the central Russian provinces and, particularly after 1861, the increase of peasant mobility. The European peripheries and later Siberia became destinations for land-hungry peasants from the centre of European Russia.
28. The terms of the settlement included a clause to the effect that the Mennonites had to set a good example to the surrounding population. Settlers were given 65 *des.* of land for families of two and more workers and 32 *des.* for those of under two. In the case of Aleksandrtal settlers, this had to be paid for over 20 years and with a 5% interest rate. Military service

was waived for the new immigrants, but after 20 years an exemption fee would have to be paid in lieu, and state and local taxes were waived for just three years after arrival.

29. A. M. Yegorev, 'Opisanie khozyaistva Menonitov', in *Trudy Imperatorskogo moskovskogo obshchestva selskogo khozyaistva: Prilozhenie k V vypusku* (Moscow, 1881), 87–142; Samarskoe gubernskoe zemstvo, *Sbornik statisticheskikh svedenii po Samarskoi gubernii: Otdel khozyaistvennoi statistiki*, i. *Samarskii uezd*, vyp. 1 (Samara, 1883); Samarskoe gubernskoe zemstvo, *Podvornoe i khutorskoe khozyaistvo v Samarskoi gubernii: Opyt agronomicheskogo obsledovaniya*, i (Samara, 1909); Otsenochno-statisticheskii otdel Samarskogo gubernskogo zemstva, *Podvornaya perepis krestyanskikh khozyaistv Samarskoi gubernii: Samarskii uezd* (Samara, 1913).
30. *The Mennonite Encyclopaedia*, iv. 46.
31. Yegorev (above, n. 29), 19.
32. *The Mennonite Encyclopaedia*, i. 144, 176.
33. Borma's fragmented pattern of individual strips was characteristic of communes in the province, where, according to a 1913 government survey, more than 50% of households held their land in 11 or more strips and 98.7% had their most distant strip more than 5 *versty* from the homestead. After 1906 a majority of peasants in Borma enclosed their land into *khutora* or *otruba*. The village was held up by contemporary advocates of enclosure as a model, and reference was made to it in several publications. Interestingly, no mention is made in the literature of the proximity of Mennonite colonies, which might well have influenced the decision in Borma to consolidate.
34. *Podvornaya perepis krestyanskikh khozyaistv* (above, n. 29).
35. Yegorev (above, n. 29), 92.
36. J. Pallot, 'The Development of Peasant Land Holding from Emancipation to Revolution', in J. Bater and R. A. F. French (eds.), *Studies in Russian Historical Geography* (London, 1983), i. 83–107.
37. Klaus (above, n. 9), 123.
38. Ibid. 131.
39. *Podvornoe i khutorskoe khozyaistvo* (above, n. 29).
40. Klaus (above, n. 9), 157–61.
41. *Sbornik statisticheskikh svedenii* (above, n. 29), 93.
42. Calculated from *Podvornaya perepis krestyanskikh khozyaistv* (above, n. 29).
43. F. A. Sev, *K voprosu o merakh uluchsheniya krestyanskogo khozyaistva* (Samarskoe gubernskoe zemstvo; Samara, 1912).
44. *Trudy mestnykh komitetov o nuzhdakh selskokhozyaistvennoi promyshlennosti*, xxxv (St Petersburg, 1903), 226.
45. Sev (above, n. 43), 4 *et passim*.
46. *Uchastkovoe agronomicheskaya organizatsiya v Samarskom uezde* (Samara, 1909); *Uchastkovaya agronomicheskaya organizatsiya Samarskogo uezdnogo zemstva, 1910* (Samara, 1912).
47. *Sbornik statisticheskikh svedenii* (above, n. 29), 96.
48. Yegorev (above, n. 29), 103.

49. *Sbornik statisticheskikh svedenii* (above, n. 29), 91.
50. Details of the farming system in Am Trakt are in *Podvornoe i khutorskoe khozyaistvo* (above, n. 29).
51. *Sbornik statisticheskikh svedenii* (above, n. 29), 92.
52. Yegorev (above, n. 29), 107–9.
53. *Sbornik statisticheskikh svedenii* (above, n. 29), 92.
54. Yegorev (above, n. 29), 118.
55. Ibid. 117.
56. Ibid. 123–4.
57. Ibid. 102.
58. *Sbornik statisticheskikh svedenii* (above, n. 29), 92–3.
59. Yegorev (above, n. 29), 126.
60. *Sbornik statisticheskikh svedenii* (above, n. 29), 92–3.
61. Yegorev (above, n. 29), 136.
62. Ibid. 131.
63. *Sbornik statisticheskikh svedenii* (above, n. 29), 96.
64. Ibid. 91.
65. Urry (above, n. 15), 736.

## Notes to Chapter 5

1. The fifth edition is used here: A. S. Yermolov, *Organizatsiya polevogo khozyaistva: Sistemy zemledeliya i sevooboroty*, 5th edn. (St Petersburg, 1914). In an article pre-dating by one year the first edition of the book Yermolov examined the geography of farming systems: 'Kulturnye raiony v Rossii i russkie sevooboroty', *Selskoe khozyaistvo i lesovodstvo*, 128/2 (1878).
2. Yermolov, *Organizatsiya* (above, n. 1), 71.
3. Ibid. 72.
4. Ibid. 73.
5. Yermolov, 'Kulturnye raiony' (above, n. 1).
6. *Andreevskii entsiklopedicheskii slovar*, xxii (St Petersburg, 1897), 906.
7. Apart from the two historically most common types of short fallow, *chernyi par* (black fallow) and *zelenyi par* (green fallow), a third type, *zanyatyi par* or 'occupied fallow', had made an appearance by the end of the 19th cent. This was fallow in which intertillage crops, such as potatoes, sugar-beet, and turnips, or annual grasses were grown. The introduction of intertillage crops in the fallow generally marked the beginning of the transition to multiple-field rotations and must be distinguished from the cultivation in the fallow of cereals or flax, which was often a sign of the degeneration of the system into *pestropole*, the unregulated, continuous cropping of land. While forage crops could help to maintain the fertility of the land, cereals merely took goodness out of it and caused declining yields.
8. For a discussion in English of the origins of the three-field system see R. A. French, 'The Introduction of the Three-Field System', in J. H. Bater and R. A. French (eds.), *Studies in Russian Historical Geography* (London, 1983),

i. 65–79; R. E. F. Smith, *The Origins of Farming in Russia* (Paris, 1959); id., *Peasant Farming in Muscovy* (Cambridge, 1977).

9. In 1798, during Paul I's reign, a School of Agriculture was founded near St Petersburg to demonstrate and promote the system of the seven-field rotation but in 1803 the School was closed and the model farms it had set up were disbanded. Apart from government-inspired schemes, individual landowners tried to promote multiple-field rotations: S. D. Stroganov, for example, founded a school of agriculture with the purpose of spreading intensive farming systems, but this also failed in its immediate objective.

10. *Andreevskii entsiklopedicheskii slovar*, xxxiii (St Petersburg, 1901), 796.

11. P. A. Vikhlaev, *Vliyanie travoseyaniya na otdelnye storony krestyanskogo khozyaistva*, vyp. 7. *Travoseyanie i arenda zemel* (Moscow, 1913).

12. Yermolov 1914 (above, n. 1), 80.

13. Yermolov, 1914 (above, n. 1), 160–6.

14. The only English-language source to give a comprehensive account of the pre-revolutionary geography of agricultural production is G. Pavlovsky, *Agricultural Russia on the Eve of the Revolution* (New York, 1968).

15. A. I. Skvortsov, *Khozyaistvennye raiony yevropeiskoi Rossii* (St Petersburg, 1910). Skvortsov divided Russia into 34 economic regions. He wrote that the identification of such regions 'above all else requires the analysis of their physical-geographical conditions because these conditions determine their character' (quoted in N. Nikitin, 'Khozyaistvennye raiony yevropeiskoi Rossii', *Trudy Vysshego seminariya selsko-khozyaistvennoi ekonomiki i politiki*, 1 (Moscow, 1921) ).

16. The 'father' of regional geography who used physiographic regions as the basis for economic regionalization was K. I. Arsenev, author of two major works on Russia's regions, *Nachertanie statistiki Rossiiskogo gosudarstva*, i (St Petersburg, 1818), and *Statisticheskii ocherk Rossii* (St Petersburg, 1848). Others who wrote in the same vein were D. I. Rikhter, V. P. Semenov-Tyan-Shanskii, and V. V. Viner. For a history of attempts at regionalization see Yu. G. Saushkin, *Ekonomicheskaya geografiya: Istoriya, teoriya, metody, praktika* (Moscow, 1977).

17. There is now a large literature in English on the work of Organization and Production scholars, e.g. T. Shanin, *Peasants and Peasant Society* (Harmondsworth, 1979); S. Soloman, *The Soviet Agrarian Debate* (Boulder, Colo., 1977).

18. A. N. Chelintsev, *Opyt izucheniya organizatsii krestyanskogo selskogo khozyaistva na primere Tambovskoi gubernii* (Kharkov, 1919). Chelintsev continued to advocate the labour–consumption theory of the peasant farm until 1928, when he was forced under political pressure to denounce the theory as 'reactionary in substance'; see N. Jasny, *Soviet Economists of the Twenties: Names to be Remembered* (Cambridge, 1972), 203–4.

19. A. N. Chelintsev, 'Selskokhozyaistvennye raiony yevropeiskoi Rossii, kak stadii selskokhozyaistvennoi evolyutsii i kulturnyi uroven selskogo khozyaistva v nikh', *Ocherki po selskokhozyaistvennoi ekonomii*, 3 (1910); id., 'Raiony yevropeiskoi Rossii, ustanavlivaemye po tipam organizatsii selskogo

khozyaistva', *Trudy Imperatorskogo volnogo ekonomicheskogo obshchestva*, 3–4 (1912); id., *Selsko-khozyaistvennaya geografiya Rossii* (Berlin, 1923).

20. Chelintsev 'Selskokhozyaistvennye raiony' (above, n. 19), 7.
21. E. Boserup, *The Conditions of Agricultural Growth: The Economics of Agrarian Change under Population Pressure* (New York, 1965); C. Geertz, *Agricultural Involution: The Processes of Ecological Change in Indonesia* (Berkeley, Calif., 1963); D. Grigg, *The Agricultural Systems of the World* (Cambridge, 1974).
22. Chelintsev, 'Selskokhozyaistvennye raiony' (above, n. 19), 3.
23. Chelintsev, 'Raiony' (above, n. 19), 51.
24. Ibid. 51–2.
25. Ibid. 53. A corollary of Chelintsev's evolutionary view of agriculture was that systems of production associated with particular methods of soil conservation could only exist temporarily: all systems were perpetually in a state of becoming something else. For this reason Chelintsev preferred to use the term 'agricultural formation' rather than 'agricultural system' to describe the complex of land use and crop and livestock production associated with different soil conservation practices.
26. Reproduced from 'Raiony' (above, n. 19), a paper he read to the Imperial Free Economic Society in St Petersburg. The data he used were taken from a variety of population and economic censuses enumerated between 1887 and 1906. The figures related to both peasant farms and landowner estates. In Chelintsev's view changes in the landowner economy followed the lead given by peasants and he devoted a part of his paper to an analysis of 1,570 noble estates in an attempt to prove the point. Inclusion of landowner data did not, he argued, subvert his demographic argument.
27. In his papers on regionalization Chelintsev did not explain how the threshold values for the various components of agriculture he plotted were arrived at, but he seems to have drawn some from his earlier empirical work on peasant farm organization. He had, for example, examined the economics and labour needs of various types of livestock husbandry on peasant farms in 'Khozyaistvennaya forma skotovodstva', *Zhivotnovodstvo*, 5 (1911).
28. Chelintsev, 'Raiony' (above, n. 19), 59.
29. Ibid. 70.
30. Ibid. 90–2.
31. Chelintsev, *Selsko-Khozyaistvennaya geografiya* (above, n. 19), 317–20.
32. The first Russian translation of *The Isolated State* was made in 1857: J. H. von Thunen, *Yedinenoe gosudarstvo v otnoshenii k obshchestvennoi ekonomii*, trans. Mathieu Wolkoff (Karlsruhe, 1857).
33. A. V. Chayanov, 'Opyty izucheniya izolirovannogo gosudarstva', *Trudy Vysshego seminariya selskokhozyaistvennoi ekonomiki i politiki*, 1 (Moscow, 1921), 5–36.
34. G. I. Baskin, 'Raionirovanie territorii, kak neobkhodimaya predposylka i tverdaya osnova ekonomicheskogo stroitelstva', *Vestnik Samarskogo gubernskogo statisticheskogo byuro*, 1 (1920); id., *Kriticheskaya otsenka materialov Vserossiiskoi selskokhozyaistvennoi perepisi i sistema yee razrabotki: Samarskaya*

*guberniya* (Otsenochno-statisticheskoe otdelenie Samarskogo gubernskogo zemstva: Samara, 1921).

35. Baskin, *Kriticheskaya otsenka* (above, n. 34), 67–8.
36. See e.g. E. S. Karnaukhova, *Razmeshchenie selskogo khozyaistva Rossii v periode kapitalizma* (Moscow, 1951). On p. 46 Karnaukhova writes '. . . by giving exclusive attention to "natural" factors in resolving questions of an economic kind they [Chelintsev etc.] revealed the bourgeois and petty-bourgeois influences on them'.
37. V. K. Yatsunskii, 'Voprosy ekonomicheskogo raionirovaniya v trudakh V. I. Lenina', *Voprosy geografii*, 31 (1953), 304.
38. Ibid.
39. Lenin used local authority household censuses enumerated in the 1880s and 1890s for his research. The Organization and Production scholars later employed dynamic censuses, which recorded data for a sample of households over a period of years, to examine the same problem. These showed that the tendency for households to polarize into groups of rich and poor was counteracted by opposite levelling tendencies. Over time households seemed to rise and fall on the socio-economic scale in a pattern of cyclical mobility. An analysis of this research is in T. Shanin, *The Awkward Class. Political Sociology of the Peasantry in a Developing Society, Russia 1910–1925* (London, 1972).
40. The best account of agrarian-Marxist research is to be found in T. Cox, *Peasants, Class, and Capitalism: Rural Research of L. N. Kritsman and His School* (Oxford, 1986).

## Notes to Chapter 6

1. J. Thirsk, *The Rural Economy of England: Collected Essays* (London, 1984), 35–6.
2. For a summary of the debate in English see J. Blum, *Lord and Peasant in Russia from the Ninth to the Nineteenth Century* (New York, 1967), 504–35, and in Russian V. A. Aleksandrov, *Obychnoe pravo krepostnoi derevni: XVIII–nachalo XIX v.* (Moscow, 1984), 71–109.
3. The argument was that, with the imposition of the poll tax by Peter I, villages developed a mechanism for allocating and reallocating land between households according to the number of souls they had, in order to ensure that all were able to raise the income to pay their tax burden.
4. There are several scholars, most of them populists, who can be classed as subscribing to the evolutionary view of the repartitional commune, e.g. K. R. Kacharovskii, A. A. Kaufman, N. Oganovskii, N. P. Pavlov-Silvanskii, F. Shcherbina, V. I. Semevskii, A. Vasilchikov.
5. A recent counter to the view that repartitioning was of late origin has been made by the Soviet scholar Shapiro, who argues that partible inheritance practised in Russia from early times was a forerunner of the equalizing repartition. See A. L. Shapiro, 'Problemy genezisa i kharaktera russkoi

obshchiny v svete novykh izyskanii sovetskikh istorikov', in *Yezhegodnik po agrarnoi istorii: Problemy istorii russkoi obshchiny,* 6 (Vologda, 1976), 36–46.

6. S. A. Grant, '*Obshchina* and *Mir*', *Slavic Review*, 35 (1976), 636–51.
7. K. R. Kachorovskii, *Russkaya obshchina: Vozmozhno li, zhelatelno li yeya sokhranenie i razvitie?* 2nd edn. (Moscow, 1906).
8. A. A. Kaufman, *Russkaya obshchina v protsesse yeya zarozhdeniya i rosta* (Moscow, 1908).
9. Shapiro (above, n. 5), 46.
10. Aleksandrov (above, n. 2), 102.
11. V. A. Aleksandrov, *Selskaya obshchina v Rossii (XVIII–nachalo XIX v.)* (Moscow, 1976), 42.
12. J. Pallot, 'The Northern Commune: Archangel Province in the Late Nineteenth Century', paper presented to the Conference on the Commune and Communal Forms, School of Slavonic and East European Studies, London, 1986.
13. Arkhangelskii gubernskii statisticheskii komitet, *Arkhangelskaya guberniya po statisticheskomu opisaniyu 1785 goda (Itogi podvornoi perepisi)* (Archangel, 1916), 30.
14. Ibid. 35.
15. See e.g. O. Lofgren, 'Peasant Ecotypes: Problems in the Comparative Study of Ecological Adaptation', *Archaeologia Scandinavia* (1976), 100–15; E. R. Wolfe, *Peasants* (Englewood Cliffs, NJ, 1966).
16. *Sbornik materialov dlya izucheniya selskoi pozemelnoi obshchiny* (St Petersburg, 1880).
17. *Dokumenty po istorii krestyanskoi obshchiny*, i (Moscow, 1983).
18. L. I. Kuchumova, 'Iz istorii obsledovaniya selskoi pozemelnoi obshchiny v 1877–1880 godakh', *Istoriya SSSR* (1978), no. 2, pp. 115–27.
19. Communal assemblies could exist at a variety of levels. Where a land commune was made up of several settlements there could be two types of assembly, a general assembly (*obshchii skhod*) and a village assembly (*poselennyi* or *malyi poderevenskii skhod*). Above village level there was the rural society assembly (*selskii skhod*) and the rural district assembly (*volostnoi skhod*). Where a commune and rural society were coextensive there was generally a single assembly. Assemblies had different rules for attendance and for what matters were to be discussed at which level. See L. I. Kuchumova, 'Selskaya pozemelnaya obshchina yevropeiskoi Rossii v 60–70-e gody XIX v.', *Istoricheskie zapiski*, 116 (Moscow, 1981), 323–47.
20. According to one historian there were no general repartitions in Archangel province before 1830. This would mean that before the Emancipation most communes must have had no more than two repartitions. See P. A. Kolesnikov, 'Osnovnye etapy razvitiya severnoi obshchiny', in *Yezhegodnik po agrarnoi istorii*, 6 (above, n. 5), 3–35, p. 23.
21. *Dokumenty* (above, n. 17), 196.
22. A survey carried out by the Archangel *zemstvo* over a series of years in the 1880s reveals that this practice was widespread, with the result that households must have had a core collection of strips for long periods of

time. See Arkhangelskaya gubernskaya zemskaya uprava, *Selskaya pozemel-naya obshchina v Arkhangelskoi gubernii*, i–iv (Archangel, 1882–9).

23. I. V. Vlasova, *Traditsii krestyanskogo zemlepolzovaniya v Pomore i zapadnoi Sibiri v XVII–XVIII vv.* (Moscow, 1984), 90.
24. Ibid. 84–7.
25. Ibid. 88.
26. *Dokumenty* (above, n. 17), 105.
27. Ibid. 206–7.
28. Ibid. 138.
29. For a good discussion of this system see M. M. Gromyko, *Traditsionnye normy povedeniya i formy obshcheniya russkikh krestyan XIX v.* (Moscow, 1986).
30. *Sbornik materialov* (above, n. 16), 205–23.
31. There was some discussion in the 19th cent. about the way to classify repartitions. The Imperial Free Economic Society used a threefold division in its survey: the radical repartition, or *korennoi peredel*; repartitioning by lot, *zherebevka*; and partial repartition, *pereverstka*. These referred respectively to repartitions in which the number of strips in the arable was changed; repartitions in which the number of strips in the arable remained the same and the need to add or subtract strips from individual households was made by exchanges with immediate neighbours (with the result that there was a type of 'shuffling' of strips among all households in a commune regardless of whether their total entitlement to land had changed); and repartitions in which the number of strips in the arable remained the same and exchanges were made between only some households in the commune. See *Dokumenty* (above, n. 17), 40. Later, *zemstva* statisticians recording repartitions usually employed a different classification, which differentiated between qualitative and quantitative repartitions, the former changing the position of households' strips, and the latter the size and number of strips held. Quantitative repartitions were further subdivided into 'general' (*obshchie*), which involved all households in a commune, and 'partial' (*chastnye*), which involved just a number of households. For a discussion of the various classifications see O. A. Khauke, *Krestyanskoe zemelnoe pravo* (Moscow, 1913), 1–10.
32. *Sbornik materialov* (above, n. 16), 212.

Notes to Chapter 7

1. J. Pallot, 'The Geography of Enclosure in Pre-Revolutionary Russia—Tver, Tula and Samara Compared', Ph.D. thesis (London, 1977), 142–56.
2. The Witte Commission, set up in 1902, played a particularly important role in turning opinion against the commune. The results of this nation-wide survey of peasant farming, which was printed in over 80 volumes, constituted an indictment of many aspects of farming in the commune, but, as David Macey has shown, there was some exaggeration of the supposed

defects of the commune and of the support for individualization in the summary volumes. See D. Macey, 'The Russian Bureaucracy and the "Peasant Problem": The Pre-History', Ph.D. thesis (Columbia University, 1976).

3. There is a large English-language literature on the history and politics of the Stolypin Land Reform. The most recent studies are: R. Hennessy, 'The Agrarian Question in Russia 1905–1907: The Inception of the Stolypin Land Reform', in *Marburger Abhandlungen zur Geschichte und Kultur Osteuropas*, 16 (Giessen, 1977) Macey (above, n. 2); G. Yaney, *The Urge to Mobilize: Agrarian Reform in Russia 1861–1930* (Urbana, Ill., 1982). A class analysis of the Reform is to be found in the classic Soviet texts: S. M. Dubrovskii, *Stolypinskaya zemelnaya reforma* (Moscow, 1963); S. M. Sidelnikov, *Agrarnaya politika samoderzhaviya v periode imperializma* (Moscow, 1980). A more recent Soviet interpretation which sees the reform as a 'wager on the middle peasant' is to be found in V. S. Dyakin, *Krizis samoderzhaviya v Rossii 1895–1917* (Leningrad, 1984).

4. For details of the organizational changes see G. Yaney, *The Systematization of Russian Government* (Urbana, Ill., 1982); id., *The Urge to Mobilize* (above, n. 3).

5. Kofod wrote numerous books on consolidated farms. In his first, a study of *khutora* formed before 1905 in the western provinces, he discussed the pros and cons of different ways of arranging land in *khutora*. See A. A. Kofod, *Krestyanskie khutora na nadelnoi zemle*, 2 vols. (St Petersburg, 1905).

6. J. Pallot, '*Khutora* and *Otruba* in Stolypin's Program of Farm Individualization', *Slavic Review*, 42/2 (1984), 242–5.

7. Such a view was expressed, for example, by Lykoshin, head of the land section of the Ministry of Internal Affairs, in his tour of inspection of the provinces in 1909. See TsGIA, f. 408, 1909, op. 1, d. 113, 'S materialom po poezdke tov. min. MVD Lykoshin dlya osmotra rabot zemleustroitelnykh komissii po razbivke na khutora krestyanskikh zemel'. In this he wrote, 'allotting pasture separately cannot but have a dangerous psychological effect, reinforcing the belief that it is impossible to carry on farming without having a parcel of land left outside . . . for the pasture of livestock'. A similar view was expressed by N. Chaplin, chief of the land section of the Ministry of Justice: see TsGIA, f. 1291, 1909–10, op. 120, yed. khr. 19, 'Po otchetu upravlyayushchego mezhevoyu chastyu Chaplina o zemleustroitelnykh rabotakh v 11 guberniyakh'. These and similar reports are discussed in Pallot (above, n. 6).

8. G. T. Robinson, *Rural Russia under the Old Regime* (Berkeley, Calif., 1969), 72–7.

9. J. Pallot, 'The Development of Peasant Land Holding from Emancipation to Revolution', in J. H. Bater and R. A. French, (eds.), *Studies in Russian Historical Geography*, (London, 1983), i. 83–108.

10. O. A. Khauke, *Krestyanskoe zemelnoe pravo* (Moscow, 1914).

11. TsGIA, f. 408, 1909, op. 1, d. 116, 'Po uchastii chlenov zemleustroitelnykh kommissii v sezde nepremennykh chlenov gubernskikh prisutstvii po vyrabotke pravila o zemleustroistve'.

12. The ability of peasants to withdraw from projects at the last minute was noted, and lamented, by Chaplin (above, n. 7), l. 16.

13. The most readily available data on the Reform's implementation record the number of petitions for various types of land settlement received by local Committees, the passage of these petitions through the planning stages, the final approval of projects, and the numbers of complaints and withdrawals. See e.g. Ministerstvo Zemledeliya, *Otchetnye svedeniya o deyatelnosti zemleustroitelnykh kommissii na 1 yanvarya 1916 goda* (Petrograd, 1916).

14. TsGIA, f. 408, 1914, op. 1, yed. khr. 894, 'Spiski selenii Vitebskoi i dr. gubernii so svedeniyami o zakonchennykh zemleustroitelnykh rabotakh na 1 yan. 1914'.

15. Novgorodskaya gubernskaya zemskaya uprava, Otsenochno-statisticheskoe otdelenie, *Lichnoe vladenie nadelnoyu zemleyu Cherepotskogo, Ustyuzhenskogo i Kirillovskogo uezdov* (Novgorod, 1913), 26.

16. *Spisok naselennykh mest Novgorodskoi gubernii: Ustyuzhenskii uezd*, viii (Novgorod, 1911).

17. *Lichnoe vladenie* (above, n. 15), 63.

18. According to the 1914 survey, the number of *khutora* formed by 1 Jan. of that year was between 25 and 47. This is likely to be more reliable than the earlier *zemstvo* estimate of 61. Often *khutora* were included in records before their owners had moved on to them. This may account for the discrepancy in the figures.

19. *Lichnoe vladenie* (above, n. 15), 70. Apart from consolidations, some households took out title to their land without consolidating, while large numbers in the district were automatically converted to hereditary tenure under the 14 June law. The result was that four-fifths of the total number of households in the district held their land in hereditary tenure by 1911.

20. Khauke (above, n. 10), 285.

21. *Lichnoe vladenie* (above, n. 15), 68.

22. TsGIA, f. 408, 1909, op. 1, d. 113, 'S materialom' (above, n. 7).

23. TsGIA, f. 408, 1914, op. 3, d. 14, 'Otchety i svedeniya o deyatelnosti zemleustroitelnykh komissii v Yaroslavskoi gubernii'.

24. I. V. Mozzhukhin, *Zemleustroistvo v Bogoroditskom uezde* (Moscow, 1917).

25. Ibid. 232.

26. Khauke (above, n. 10), 282.

27. A. A. Kofod, *Russkoe zemleustroistvo* (St Petersburg, 1914), 51.

28. V. P. Danilov, 'Ob istoricheskikh sudbakh krestyanskoi obshchiny v Rossii', in *Yezhegodnik po agrarnoi istorii: Problemy istorii russkoi obshchiny*, 6 (Vologda, 1976), 103–34, pp. 103–6. Danilov in particular takes issue with the numerical results of the Reform calculated by Pershin, the first major post-revolutionary historian of land settlement: P. N. Pershin, *Uchastkovoe zemlepolzovanie v Rossii: Khutora i otruba, ikh rasprostranenie za desyatiletie 1907–1916 gg. i sudby vo vremia revolyutsii (1917–1920 gg.)* (Moscow, 1922).

29. Danilov (above, n. 28), 106.

30. Khauke (above, n. 10), 147.

31. Ibid.

32. According to the law, only communes which obtained certification of their tenure change were officially said to have changed to hereditary tenure. Such certification could be given to a whole commune even if the request came from only one of its members. By 1916 some 470,000–500,000 households were in 'certified' communes. The first generation of writers on the Stolypin Reform tended to include in their calculations all the 3.5m. households that were eligible under the 14 June law rather than the smaller number (0.5m.) that received certificates of the change in tenure. For a full discussion of this issue see D. Atkinson, 'The Statistics on the Russian Land Commune', *Slavic Review*, 32 (1973), 713–87.
33. Where individual households withdrew from the commune but continued to make use of common resources the legal status of these resources remained unchanged. The problem of legal definition arose when whole villages consolidated but left some land in common use. See Khauke (above, n. 10), 177.
34. Ibid. 182.
35. There is a good account of the 'mobilization' of land in the aftermath of the Stolypin Reform in Dubrovskii (above, n. 3).
36. V. Chernov, 'V khaose sovremennoi derevni', *Sovremennik*, 6 (1911), 186.
37. See n. 32.
38. The most comprehensive account of the commune in the 1920s is in V. P. Danilov, *Sovetskaya dokolkhoznaya derevnya*, 2 vols. (Moscow, 1979).
39. R. W. Davies, *The Industrialisation of Soviet Russia*, ii. *The Soviet Collective Farm 1929–1930* (London, 1980), 56–63.

Notes to Chapter 8

1. V. I. Lenin, *Sochineniya* (Moscow, 1941), i. 137–8.
2. B. N. Mironov, *Vnutrennii rynok Rossii vo vtoroi polovine XVIII–pervoi polovine XIX v.* (Leningrad, 1981), 9–10, 20–2, and refs.; S. H. Baron, 'The Transition from Feudalism to Capitalism in Russia: A Major Soviet Historical Controversy', *American Historical Review*, 77 (1973), 715–29.
3. e.g. T. Shanin, *The Awkward Class: Political Sociology of the Peasantry in a Developing Society, Russia 1910–1925* (London, 1972); id., *Russia as a Developing Society*, i. *The Roots of Otherness: Russia's Turn of Century* (Basingstoke, 1985); A. Gerschenkron, *Economic Backwardness in Historical Perspective* (Cambridge, Mass., 1962); P. Gatrell, *The Tsarist Economy, 1850–1917* (London, 1986).
4. R. Hellie, 'The Foundations of Russian Capitalism', *Slavic Review*, 26 (1967), 148–54; G. Rozman, *Urban Networks in Russia, 1750–1800, and Premodern Periodization* (Princeton, 1976).
5. Yu. A. Tikhonov, 'Problema formirovaniya vserossiiskogo rynka v sovremennoi sovetskoi istoriografii', in *Aktualnye problemy istorii Rossii epokhi feodalizma* (Moscow, 1970), 200–23.
6. J. Langton, 'The Industrial Revolution and the Regional Geography of England', *Institute of British Geographers: Transactions*, NS 9 (1984), 145–67;

id. and G. Hoppe, 'Town and Country in the Development of Early Modern Western Europe', *Historical Geography Research Series*, 11 (Nov. 1983); A. Everitt, 'Country, County and Town: Patterns of Regional Evolution in England', *Transactions of the Royal Historical Society*, 5th ser., 29 (1979), 79–108; J. Thirsk, *England's Agricultural Regions and Agrarian History* (Basingstoke, 1987); J. D. Marshall, 'Why Study Regions?', *Journal of Regional and Local Studies*, 5 (1985), 6 (1986); F. F. Mendels, 'Agriculture and Peasant Industry in Eighteenth Century Flanders' in W. N. Parker and E. L. Jones (eds.), *European Peasants and their Markets* (Princeton, NJ, 1975), 179–204.

7. A number of the salient issues are discussed by Langton and Hoppe (above, n. 6) and by R. A. Dodgshon, 'The Modern World System by Immanuel Wallerstein: A Spatial Perspective', *Peasant Studies*, 6 (Jan. 1977), 8–19. For the purposes of the present chapter, the term 'all-Russian market' is taken to imply a degree of regional market interdependence involving the interchange of bulk goods as well as of high-value goods. The broader questions concerning the nature of the transition from 'feudal' to 'capitalist' or from 'traditional' to 'modern' lie beyond the scope of this chapter.

8. For some examples of regional studies see E. G. Istomina, *Novgorodskaya guberniya vo vtoroi polovine XVIII v. (opyt istoriko-geograficheskogo issledovaniya)* (Moscow, 1969); R. E. Jones, *Provincial Development in Russia: Catherine II and Jakob Sievers* (New Brunswick, NJ, 1984); for examples of published topographical descriptions of the late 18th cent. see e.g. *Opisanie Tobolskogo namestnichestva* (Novosibirsk, 1982); *Opisanie Voronezhskogo namestnichestva 1785 g.* (Voronezh, 1982).

9. P. I. Lyashchenko, *Istoriya narodnogo khozyaistva SSSR* (Leningrad, 1950), 279 ff.; N. D. Chechulin, *Goroda Moskovskogo gosudarstva v XVI veke* (St Petersburg, 1889); P. P. Smirnov, *Goroda Moskovskogo gosudarstva v pervoi polovine XVII v.*, 2 vols. (Kiev, 1917–19); H. L. Eaton, 'Decline and Recovery of Russian Cities, 1500–1700', *Canadian–American Slavic Studies*, 11 (1977), 220–52.

10. V. A. Chernyshev, *Sukhoputnye sredstva soobshcheniya v Rossii XVI–XVII vv.* (Leningrad, 1980).

11. M. E. Falkus, *The Industrialisation of Russia, 1700–1914* (London, 1972), 17. Eaton (above, n. 9), 223, quoting Vodarskii, gives a figure of 15.5m. for 1719.

12. J. Blum, *Lord and Peasant in Russia* (New York, 1968), 164–5.

13. P. I. Lyashchenko, *History of the National Economy of Russia to the 1917 Revolution* (New York, 1949), 220.

14. Lyashchenko (above, n. 9), 275 ff.; *Ocherki istorii SSSR: Period feodalizma: XVII v.* (Moscow, 1955), 113 ff.

15. Falkus (above, n. 11), 26 ff.; A. Kahan, 'Continuity in Economic Activity and Policy during the Post-Petrine Period in Russia', *Journal of Economic History*, 25 (Mar. 1965), 61–85.

16. Falkus (above, n. 11), 23, 27; J. H. Bater, *St Petersburg: Industrialization and Change* (London, 1976), 17–27.

17. Falkus (above, n. 11), 17.
18. P. Dukes, *The Making of Russian Absolutism 1613–1801* (London, 1982), 167.
19. Falkus (above, n. 11), 29. This figure applies to 'industrial enterprises' and not to peasant handicrafts etc. The statistics for this period are unreliable, partly because of problems of definition.
20. Mironov (above, n. 2), 57.
21. *Ocherki istorii SSSR: Period feodalizma: Rossiya vo vtoroi polovine XVIII v.* (Moscow, 1956), 123; S. G. Strumilin, 'O vnutrennem rynke Rossii XVI–XVIII vv. (po povodu knigi Kafengrauza)', *Istoriya SSSR* (1959), no. 4, pp. 75–87.
22. Mironov (above, n. 2), 112; W. L. Blackwell, *The Beginnings of Russian Industrialization, 1800–1860* (Princeton, 1968), 26.
23. Mironov (above, n. 2), 110.
24. See the discussions at Catherine II's Legislative Commission of 1767 reported by P. Dukes, *Catherine the Great and the Russian Nobility* (London, 1967), 137–9.
25. Mironov (above, n. 2), 57, 62.
26. N. L. Rubinshtein, *Selskoe khozyaistvo Rossii vo vtoroi polovine XVIII v.* (Moscow, 1957), 20–1. Rubinshtein's regionalization consisted of the following: (1) the Central Industrial Region (Moscow, Vladimir, Yaroslavl, Kostroma, and Kaluga provinces); (2) the North-West Trade and Industrial Region (Tver, Smolensk, Novgorod, Pskov, St Petersburg); (3) the Northern Industrial Region (Archangel, Olonets, Vologda); (4) the Central Black Earth Trade and Agricultural Region (Tula, Ryazan, Orel, Kursk, Tambov, Voronezh, Penza); and peripheral areas still being settled and developed: (5) the Middle Volga (Nizhnii Novgorod, Kazan, Simbirsk, part of Saratov); (6) the Urals (Vyatka, Perm, Orenburg); (7) the South-West (Kharkov and Left Bank Ukraine). He excluded the very peripheral areas, including the Baltic lands, Belorussia, New Russia, North Caucasus, and Astrakhan province, which had their own economic peculiarities. Most unfortunately, he did not publish a detailed justification of his very interesting scheme. Mironov (above, n. 2), 90–5, attempts to regionalize Russia for the late 18th and early 19th cents. by constructing coefficients of industrial and agricultural specialization for groups of provinces. According to his findings, there was a greater degree of differentiation between regions in industrial than in agricultural indicators, although the agricultural differences between the black-earth and the non-black-earth zones as a whole were more noticeable than the industrial differences.
27. V. M. Kabuzan, *Izmeneniya v razvitii naseleniya Rossii v XVIII–pervoi polovine XIX v.* (Moscow, 1971), app. 2. Kabuzan's calculations are based on provincial boundaries for the early 19th cent., although in the case of Vologda province there was no change after the 1780s.
28. S. A. Kovalev, *Selskoe rasselenie* (Moscow, 1963), 156.
29. TsGVIA, f. VUA, yed. khr. 18643. Rozman (above, n. 4), 190, suggests a total urban population of 30,000 out of a provincial population of 572,000 for 1782, but does not detail his sources. Kabuzan (above, n. 27) gives

8,925 urban estate males out of a total male population of 278,081 for the same date. However, he also gives 7,337 non-taxable males for the province, a portion of whom would have lived in towns. The above-mentioned archival source gives the number of non-estate persons (*raznochintsy*) living in the towns of the province as constituting a quarter of the recorded urban population. This confirms the impression that the urban population was in fact greater than the numbers of merchants and townspeople would imply.

30. *Ocherki* (above, n. 21), 122; Rozman (above, n. 4), 191; H. Kellenbenz, 'The Economic Significance of the Archangel Route', *Journal of European Economic History*, 2 (Winter 1973), 541–81; U. M. Polyakova, 'Gorodovaya obyvatelskaya kniga Arkhangelska 1786–88 gg. kak istochnik dlya izucheniya sotsialnogo stroya severnogo goroda', in *Materialy po istorii Yevropeiskogo Severa SSSR*, i (Vologda, 1970), 121–51.

31. E. G. Istomina, *Vodnye puti Rossii vo vtoroi polovine XVIII–nachale XIX veka* (Moscow, 1982), 167–80.

32. Rozman (above, n. 4), 190, implies an urbanized population of 5.2% in 1782 in the context of a national urbanized population of 8.4–8.8% (see id., 'Comparative Approaches to Urbanization: Russia, 1750–1800' in M. F. Hamm (ed.), *The City in Russian History* (Lexington, Ky., 1976), 69–85, p. 78), although he fails to give adequate supporting information. For other estimates see above, n. 29.

33. TsGVIA, f. VUA, yed. khr. 18642, l. 1; ibid., yed. khr. 18646, ll. 19, 42; Rubinshtein (above, n. 26), 30–4.

34. TsGVIA, f. VUA, yed. khr. 18646, ll. 24 ff.; ibid., yed. khr. 18643.

35. Yu. R. Klokman, *Sotsialno-ekonomicheskaya istoriya russkogo goroda: Vtoraya polovina XVIII veka* (Moscow, 1967), 307. TsGVIA, f. VUA, yed. khr. 18643 gives a total population of 7,500, while Rozman (above, n. 4), 183, suggests 10,000. The distinction between merchants (*kuptsy*) and townspeople (*meshchane*) dates from a manifesto of 1775, when the former *posad* dwellers were categorized into these two groups, depending upon their wealth.

36. TsGVIA, f. VUA, yed. khr. 18646, l. 15. The date is Rubinshtein's.

37. Ibid., l. 48; also ibid., yed. khr. 18643; Klokman (above, n. 35), 307.

38. e.g. J. Patten, *English Towns, 1500–1700* (Folkestone, 1978); A. D. Dyer, *The City of Worcester in the Sixteenth Century* (Leicester, 1973); Rozman (above, n. 4).

39. Klokman (above, n. 35). According to Mironov (above, n. 2), 31, Chulkov's statistics derive largely from the 1760s. On Vologda, see also TsGVIA, f. VUA, yed. khr. 18646, l. 15.

40. TsGVIA, f. VUA, yed. khr. 18646, ll. 15, 56; Klokman (above n. 35), 306–7.

41. TsGVIA, f. VUA, yed. khr. 18646, l. 19. Istomina (above, n. 31), 176–7, notes that the value of goods reaching Archangel via the Sukhona–Dvina waterway equalled 10.9m. roubles in 1812.

42. Klokman (above, n. 35), 306–7.

43. Mironov (above, n. 2), 78–9, 122.

44. Rubinshtein (above, n. 26), 210, 230–1. Rubinshtein's data from the General Survey (ibid. 323) suggest that 94% of the province was forest, 2% arable (giving 2.5 *desyatiny* per male), and 1.2% hayland (1.3 *desyatiny* per male).
45. TsGVIA, f. VUA, yed. khr. 18646, 1. 19.
46. Ibid., ll. 19, 47, 77. Also Rubinshtein (above, n. 26), 244–5; Istomina (above, n. 31), 177; L. V. Milov, *Issledovanie ob 'Ekonomicheskikh primechaniyakh' k Generalnomu mezhevaniyu* (Moscow, 1965), 251–5, describes evidence which he believes indicates the development of commercial activity on the estates of Vologda district.
47. TsGVIA, f. VUA, yed. khr. 18646, 1. 97. Rubinshtein's data (above, n. 26), 379, indicate that Vologda province was a slightly grain-deficient region. Much grain was imported, but much was also exported, especially northwards.
48. Rubinshtein (above, n. 26), 323.
49. TsGVIA, f. VUA, yed. khr. 18646, 1. 27.
50. Ibid., ll. 83, 97.
51. Ibid., 1. 90.
52. Ibid., ll. 35, 42.
53. Ibid., 1. 86. See also ll. 56, 77, 90.
54. Ibid., 1. 20.
55. Ibid., 1. 86.
56. Ibid., 1. 23.
57. Ibid., 1. 32.
58. Ibid., ll. 23, 27, 47, 62, 86, 90, 97.
59. V. Levshin, 'Topograficheskoe, istoricheskoe, statisticheskoe i kameralnoe opisanie Tulskoi gubernii, po nachertaniyu IVEO', TsGIA, f. 91, d. 285 (1811).
60. Ibid., 1. 179. Also l. 192.
61. Ibid., ll. 7 ff.
62. Ibid., 1. 10.
63. V. I. Semevskii, *Krestyane v tsarstvovanie Yekateriny II* (St Petersburg, 1901), i. 27–8. But for a critique of this view see I. de Madariaga, *Russia in the Reign of Catherine the Great* (London, 1981), 97 and refs.
64. TsGIA, f. 91, d. 285, 1. 231ob.
65. Ibid., 1. 75ob. There were also others from outside the province.
66. Ibid., ll. 42–3. There were 35 distilleries, mainly small and using serf labour; outside Tula there were two paperworks (one owned by a lord, the other by a Tula merchant with 100 assigned peasants), and 2 sail-making plants, one with almost 1,000 assigned workers. Other plants were tiny, with the exception of a silk factory in Venev district, with 70 assigned and 50 hired workers, and a noble-owned ironworks in Odoev district, using 18 serfs and 200 hired hands.
67. Semevskii (above, n. 63), 28 n.; W. R. Augustine, 'Notes towards a Portrait of the Eighteenth Century Russian Nobility', *Canadian Slavic Studies*, 4 (Fall, 1970), 373–425, pp. 403, 404, 406.
68. Rozman (above, n. 4), 178–9; according to the topographical description of

1775 (TsGADA, f. 16, yed. khr. 979, l. 2ob.), Tula had 3,407 male merchants and townsmen and 4,469 weapon-makers and smiths.

69. TsGIA, f. 91, d. 285, l. 53.
70. Ye. V. Myshkovskii, *Ocherki po istorii russkoi manufaktury XVII–XVIII vv. (po materialam Tulskogo oruzheinogo zavoda)* (Leningrad, 1975); Rozman (above, n. 4), 179, suggests that in 1782 about 45% of the registered male population was employed in weapons production.
71. Levshin lists 1,134 male merchants, 5,063 townsmen, and 5,345 male weapon-makers at the beginning of the 19th cent. The weapon-makers were a legal rather than a strictly occupational category.
72. An inventory of 1813–14 lists 18 factories employing over 15 people in the towns of the province, mainly in Tula itself (P. G. Ryndzyunskii, *Gorodskoe grazhdanstvo doreformennoi Rossii* (Moscow, 1958), 39). In addition to the activities listed, an inventory of 1775 mentions tailors, boot- and shoe-makers, a coach-builder, and a mint (TsGADA, f. 16, d. 979, ll. 2–3). Leather-working was obviously an important activity in Tula.
73. i.e. with Siberia, China, Central Asia, Western Europe, etc. See TsGIA, f. 91, d. 285, ll. 64–5.
74. A much larger one existed in the district.
75. There were 6 distilleries.
76. TsGIA, f. 91, d. 285, l. 109ob.
77. Ibid., l. 124.
78. Ibid., l. 124ob.
79. See the discussion in Milov (above, n. 46), 179 ff.
80. 'Opisanie Tambovskogo namestnichestva', in *Sobranie sochinenii vybrannykh iz Mesyatseslovov na raznye gody*, vi (St Petersburg, 1790), 375–459, pp. 387, 420, 431, etc.; also TsGADA, f. 1355, op. 1, d. 1650, ll. 2 etc.
81. S. Larionov, *Opisanie Kurskogo namestnichestva* (Moscow, 1786); 'Topograficheskoe Kurskoi gubernii opisanie', TsGVIA, f. VUA, d. 18799.
82. Mironov (above, n. 2), 97, notes that the Kursk Korennaya Fair was third in the country in trade turnover, with a turnover of 3m. silver roubles in 1791.
83. *Topograficheskoe opisanie Kharkovskoi gubernii* (Moscow, 1788); also TsGADA, f. 1355, op. 1, yed. khr. 2009.
84. TsGVIA, f. VUA, yed. khr. 18728, 18732, 18727; TsGADA, f. 1355, op. 1, yed. khr. 18/377; A. P. Pronshtein, *Zemlya Donskaya v XVIII v.* (Rostov, 1961).
85. Rubinshtein (above, n. 26), 399.
86. Ibid. 273–5.
87. Patten (above, n. 38); F. F. Mendels, 'Proto-Industrialization: The First Phase of the Industrialization Process', *Journal of Economic History*, 32 (1972), 241–61. Mironov (above, n. 2), 71–2, points out that traditional patterns of exchange, largely superseded in Western Europe by the late 18th and 19th cents., continued to dominate in Russia.
88. Mendels (above, n. 87); L. A. Clarkson, *Proto-Industrialization: The First Phase of Industrialization?* (Studies in Economic and Social History; Basingstoke, 1985).

89. Rozman (above, n. 4) attempts to use stages in the evolution of urban networks as a basis for historical periodization and for cross-national comparisons of societal development.
90. W. Christaller, *Central Places in Southern Germany* (Englewood Cliffs, NJ, 1966); B. J. L. Berry, *Geography of Market Centers and Retail Distribution* (Englewood Cliffs, NJ, 1967).
91. I. D. Kovalchenko, 'O tovarnosti zemledeliya v Rossii v pervoi polovine XIX v.', in *Yezhegodnik agrarnoi istorii Vostochnoi Yevropy, 1963 g.* (Vilnyus, 1964), 469–86.
92. Regional markets for rye and oats in the 18th and 19th cents. have been analysed by Kovalchenko and Milov from a study of price fluctuations. See I. D. Kovalchenko and L. V. Milov, *Vserossiiskii agrarnyi rynok XVIII–nachalo XX v.: Opyt kolichestvennogo analiza* (Moscow, 1974).

Notes to Chapter 9

1. The main investigators of domestic industries in the 19th cent. were the pro-peasant neo-populist theorists. An example of their analysis is in V. Vorontsov, *Ocherki kustarnoi promyshlennosti v Rossii* (St Petersburg, 1886). Vorontsov and his colleagues were defenders of domestic industry, which they saw as part of the natural economy of the peasantry and not a precursor of factory production. The issues in the 19th-cent. Russian debate about 'proto-industrialization' anticipated today's discussions about its role in European industrialization. For a summary of this debate see L. A. Clarkson, *Proto-Industrialization: The First Phase of Industrialization?* (Studies in Economic and Social History; Basingstoke, 1985). The standard Soviet work on peasant manufacturing is P. G. Ryndzyunskii, *Krestyanskaya promyshlennost v poreformennoi Rossii* (Moscow, 1966).
2. V. I. Lenin, *Collected Works*, iii. *The Development of Capitalism in Russia* (London, 1977), 451.
3. Ibid. 376–7.
4. Ibid. 377–8.
5. In 1871 Ivanovo was elevated to the status of a town (*gorod*) and renamed Ivanovo-Voznesensk. The most comprehensive work on the distribution of factory villages (*fabrichnye sela*) in 19th- and early 20th-cent. Russia is Ya. Ye. Vodarskii, *Promyshlennye seleniya tsentralnoi Rossii v periode genezisa i razvitiya kapitalizma* (Moscow, 1972).
6. The Commission was made up of well-known economists and statisticians under the chairmanship of Ye. N. Andreev. The findings of the Commission were published in 16 volumes in St Petersburg from 1879 under the title *Trudy Osoboi komissii po issledovaniyu kustarnoi promyshlennosti v Rossii*.
7. A. A. Isaev, *Promysly Moskovskoi gubernii*, 2 vols. (Moscow, 1876–7); *Sbornik statisticheskikh svedenii po Moskovskoi gubernii: Otdel khozyaistvennoi statistiki*, vi/1, vi/2, vii/1, vii/2, vii/3 (Moscow, 1879–82). These latter volumes were also published separately as, *Promysly Moskovskoi gubernii*, i,

ii, iii, v (Moscow, 1879–82), and *Zhenskie promysly Moskovskoi gubernii*, iv (Moscow, 1882).

8. P. A. Vikhlaev, *Moskovskaya guberniya po mestnomu obsledovaniyu 1898–1900*, iv/2 (Moscow, 1908).

9. The reliance upon non-agricultural sources of income was not an entirely new phenomenon in the province. A. V. Pogozhev noted in one *zemstvo* report that factory labour and domestic industries had been used to supplement farm income at the turn of the century but that conclusions about the precise numbers of households involved was difficult because of the unsystematic nature of the sources: see *Sbornik statisticheskikh svedenii po Moskovskoi gubernii: Otdel sanitarnoi statistiki: Obshchaya svodka po sanitarnym issledovaniyam fabrichnykh zavedenii Moskovskoi gubernii*, iv/1 (Moscow, 1890), 7–8.

10. N. Matisen, *Atlas manufakturnoi promyshlennosti Moskovskoi gubernii* (Moscow, 1872), p. v.

11. *Sbornik statisticheskikh svedenii* (above, n. 9), 1.

12. M. Tugan-Baranovskii, *Russkaya fabrika v proshlom i nastoyashchem*, i. *Razvitie russkoi fabriki v XIX veke*, 7th edn. (Moscow, 1898), 177.

13. The maps have been compiled using data contained in the two censuses of handicraft industries in Moscow province, Fig. 9.1 from *Moskovskaya guberniya po mestnomu obsledovaniyu* (above, n. 8) and Figs. 9.2–9.5 from Isaev and *Promysly Moskovskoi gubernii* (above, n. 7). The latter lists every village and, for some industries, household in which particular industries were encountered.

14. P. Belov, 'Kartina kustarnogo proizvodstva v sele Cherkizova Moskovskogo uezda', *Mir Bozhii*, 6 (1900), 30–5, p. 31.

15. *Zhenskie promysly Moskovskoi gubernii*, iv (above, n. 7), 143.

16. *Promysly Moskovskoi gubernii*, i (above, n. 7), 3.

17. Ibid. iv (above, n. 7), 64.

18. R. E. Johnson, *Peasant and Proletarian* (New Brunswick, NJ, 1979). Johnson shows that strong regional and local networks existed among textile workers in Moscow factories.

19. *Zhenskie promysly Moskovskoi gubernii*, iv (above, n. 7), 35–7.

20. *Promysly Moskovskoi gubernii*, i (above, n. 7), 171–2.

21. Ibid. iv (above, n. 7), 53–8.

22. *Zhenskie promysly Moskovskoi gubernii*, iv (above, n. 7), 27.

23. Ibid. 9–10.

24. Ibid. 145–6.

25. *Promysly Moskovskoi gubernii*, ii (above, n. 7), 144.

26. For an account of the organization of the markets of Moscow and other towns in the 19th cent. see R. Gohstand, 'The Geography of Trade in Nineteenth-Century Russia', in J. H. Bater and R. A. French (eds.), *Studies in Russian Historical Geography* ii (London, 1983), 229–378.

27. The figures for the landless must be interpreted with special care since landlessness had many explanations. Some of the landless were *bobyli*, some were newly formed households to which the commune did not allocate land, some had surrendered land or been deprived of it because of

age, infirmity, death of a spouse, or insufficient labour, and some had left the commune to go elsewhere in Russia. Among the landless there were households which voluntarily gave up their entitlement to land in order to pursue industrial work either in the village or elsewhere, and also those which were forced against their will to give it up because of poor husbandry, which might have resulted from spending too much time in other, possibly industrial, labours. Unfortunately, the *zemstvo* census did not distinguish between these various reasons for landlessness.

28. Vikhlaev (above, n. 8), 624–36.
29. *Promysly Moskovskoi gubernii*, v (above, n. 7), 60–1.
30. Isaev (above, n. 7), 3–18.
31. See ch. 5 n. 40.
32. *Promysly Moskovskoi gubernii*, ii (above, n. 7), 7.
33. Vikhlaev (above, n. 8), 11.
34. Vorontsov (above, n. 1), 209–10.
35. *Promysly Moskovskoi gubernii*, i (above, n. 7), 259.
36. Ibid. 260.
37. Ibid. iii (above, n. 7), 170.
38. *Zhenskie promysly Moskovskoi gubernii*, iv (above, n. 7), 148.

Notes to Chapter 10

1. Population estimates for the 17th cent. are extremely dubious. Estimates of the urban *posad* populations are based on enumerations of 1646–7, 1649–52, and 1678. Using Vodarskii's estimates of the *posad* and total populations, it appears that the *posad* component of the Russian population approximated to 2.4% in 1646 and 2.8% in 1678. However, these figures exclude the many ancillary elements in cities and the urban-based servitors. Incorporating the latter, the proportions are 7% and 6.9% respectively. It is assumed here that the proportions would have been lower at the beginning of the century. See Ya. Ye. Vodarskii, 'Chislennost i razmeshchenie posadskogo naseleniya v Rossii vo vtoroi polovine XVII v.', in *Goroda feodalnoi Rossii* (Moscow, 1966), 271–97; id., *Naselenie Rossii v kontse XVII–nachale XVIII veka* (Moscow, 1977), 129–34, 192–3; H. L. Eaton, 'Decline and Recovery of the Russian Cities from 1500 to 1700', *Canadian–American Slavic Studies*, 11/2 (1977), 220–52, pp. 222–3, 226.
2. J. M. Hittle, *The Service City: State and Townsmen in Russia, 1600–1800* (Cambridge, Mass., 1979), 13.
3. Ibid., esp. 5–7.
4. Most notably in the West by Max Weber in his *The City* (Glencoe, Ill., 1958), esp. 91–6, 200; see also S. H. Baron, 'The Town in "Feudal" Russia', *Slavic Review*, 28/1 (Mar. 1969), 116–22; id., 'The Weber Thesis and the Failure of Capitalist Development in "Early Modern" Russia', *Jahrbucher fur Geschichte Osteuropas*, 18 (1970), 320–36. This view is, of course, resolutely opposed by most Soviet historians and in the West by G. Rozman in his *Urban Networks in Russia, 1750–1800, and Premodern*

*Periodization* (Princeton, 1976), and his 'Comparative Approaches to Urbanization: Russia, 1750–1800', in M. F. Hamm (ed.), *The City in Russian History* (Lexington, Ky., 1976), 69–85.

5. J. H. Bater, *St Petersburg: Industrialization and Change* (London, 1976), esp. ch. 5.
6. W. Christaller, *Central Places in Southern Germany* (Englewood Cliffs, NJ, 1966); J. E. Vance Jr., *The Merchant's World: The Geography of Wholesaling* (Englewood Cliffs, NJ, 1970).
7. The literature on the historical evolution of urban networks is reviewed in H. Carter, *An Introduction to Urban Historical Geography* (London, 1983), ch. 5.
8. Vodarskii, 'Chislennost' (above, n. 1). The 'white places' were settlements or suburbs free from certain impositions or obligations levied on the *posad* dwellers (see Glossary for definition of *posad*). The *posad* was designated 'black', being under the control of the tsar and subject to ordinary tax and service obligations.
9. Eaton (above, n. 1), 220–4, 230 ff.
10. Vodarskii's calculations show some 108,000 male *posad* dwellers in 1652, and about 134,000 in 1678. Eaton argues that the latter figure is not really comparable with the former, since it includes Siberia and excessive totals for Moscow and certain other cities. Accepting Vodarskii's adjusted figure of 45,000 for the number of *posad* households in 1678, and his average of 2.8 males per household at that time, would give a total of 126,000 *posad* dwellers in 1678, an increase of 17% over 1652. Eaton points out that the average size of the *posad* had fallen over the period. However, the 1678 statistics include a larger number of *posady* than those for 1652, and there are many small *posady* in the 1678 figures which do not appear in the data for the earlier year. In mid-century the number of *posad* dwellers in Siberia was probably minimal.
11. See Vodarskii, 'Chislennost' (above, n. 1), 282–90.
12. Ibid. 279.
13. Ibid. 282–90. In Tyumen in Siberia 177 houses out of 345 belonged to servitors in 1624. See V. N. Kurilov, 'Uchastie sluzhilykh lyudei v stanovlenii g. Tyumeni kak torgovo-promyshlennogo tsentra v XVII v.', in O. N. Vilkov (ed.), *Goroda Sibiri* (Novosibirsk, 1974), 76–86, p. 77.
14. Hittle (above, n. 2), 23–5; D. H. Miller, 'State and City in Seventeenth Century Muscovy', in Hamm (above, n. 4), 34–52, pp. 35–6.
15. Calculated from L. B. Veinberg and A. A. Poltoratskaya, *Materialy dlya istorii Voronezhskoi i sosednikh gubernii*, ii (Voronezh, 1891), 2–4, and Ye. U. Chistyakova, 'Remeslo i torgovlya na Voronezhskom posade v seredine XVII v.', *Trudy Voronezhskogo gosudarstvennogo universiteta*, 25 (1954), 46–63, p. 61.
16. Kurilov (above, n. 13); see also A. A. Lyutsidarskaya, 'Promyshlennoe razvitie g. Tomska vo vtoroi polovine XVII v.', in Vilkov (above, n. 13), 60–75.
17. In the seventeenth century the word *gorod* referred to the fortified core of an urban settlement (with a specific type of fortification) containing the

office of the *voevoda*, the urban settlement as a whole, or the settlement together with its administrative district (*uezd*). Legally, a *gorod* usually implied some kind of administrative centre, often fortified. However, some private towns were also referred to as *goroda*. See Vodarskii, *Naselenie* (above, n. 1), 117.

18. Hittle (above, n. 2), 36–7; P. P. Smirnov, *Goroda Moskovskogo gosudarstva v pervoi polovine XVII veka*, i/1 (Kiev, 1917), 110; i/2 (Kiev, 1919), 191 ff.

19. R. A. French, 'The Early and Mediaeval Russian Town', in J. H. Bater and R. A. French (eds.), *Studies in Russian Historical Geography* ii (London, 1983), 249–77, pp. 267–8; R. A. French, 'The Urban Network of Later Mediaeval Russia', in G. J. Demko and R. J. Fuchs, *Geographical Studies on the Soviet Union* (Chicago, Ill., 1984), 29–51, pp. 48–9.

20. Kurilov (above, n. 13); F. G. Safronov, 'Osnovanie goroda Yakutska', in Vilkov (above, n. 13), 37–47; V. P. Shunkov, *Ocherki po istorii kolonizatsii Sibiri XVII–nachale XVIII v.* (Moscow and Leningrad, 1946); T. Armstrong, *Russian Settlement in the North* (Cambridge, 1965), 16–21.

21. Its centre migrated between Yablonov, Belgorod, and even Kursk.

22. L. M. Tverskoi, *Russkoe gradostroitelstvo do kontsa XVII veka: Planirovka i zastroika russkikh gorodov* (Moscow and Leningrad, 1953), 50, 61 ff.; V. A. Shkvarikov, *Ocherki istorii planirovki i zastroiki russkikh gorodov* (Moscow, 1954), 50 ff.

23. French, 'The Early and Mediaeval Russian Town' (above, n. 19), 268–74.

24. Hittle (above, n. 2), 41.

25. See above, n. 17.

26. Hittle (above, n. 2), 31.

27. I. I. Dityatin, *Ustroistvo i upravlenie gorodov Rossii*, i (St Petersburg, 1875), 108.

28. J. Keep, 'Shade and Light in the History of Russian Administration', *Canadian Slavic Studies*, 6 (1972), 1–9, p. 8.

29. French, 'The Urban Network' (above, n. 19), 49.

30. V. M. Kabuzan, *Narodonaselenie Rossii v XVIII–pervoi polovine XIX v.* (Moscow, 1963), 180–92. French gives 356 towns at the end of Peter's reign, based on Kirilov's survey. Laws of 1708 and 1719 list 339 and 280 towns respectively, while the official Main Magistracy register of 1723 lists 211 towns or proposed towns (see Ya. Ye. Vodarskii, 'Spisok gorodov Rossii s ukazaniem primernogo kolichestva posadskikh dvorov (1723 g.)', *Istoricheskii arkhiv* (1961), no. 6, pp. 235–6). A. A. Kizevetter gives 269 towns, based on a source for 1738 (see his *Posadskaya obshchina v Rossii XVIII st.* (Moscow, 1903), 83–5). Yu. R. Klokman (see his *Sotsialno-ekonomicheskaya istoriya russkogo goroda: Vtoraya polovina XVIII veka* (Moscow, 1967), 33) estimates 350 towns before the 1775 provincial reforms.

31. Vodarskii, *Naselenie* (above, n. 1), 133, gives 189 *posady* for 1722. *Ocherki istorii SSSR: Period feodalizma: Rossiya vo vtoroi chetverti XVIII v.* (Moscow, 1957), 179, gives 202 *posady* for the 1740s.

32. *Ocherki istorii SSSR: Period feodalizma: Rossiya vo vtoroi polovine XVIII v.* (Moscow, 1956), 151. Vodarskii, *Naselenie* (above, n. 1), 133–4, gives

193,000 for 1722, and estimates a total of 231,000 urban traders and associated persons for 1719. In 'Chislennost' (above, n. 1), 290, he gives 196,000 *posad* dwellers for 1722. These figures do not include peasants and numerous other elements living in the towns; all figures refer to males.

33. Klokman (above, n. 30), 31.

34. I. K. Kirilov, *Tsvetushchee sostoyanie Vserossiiskogo gosudarstva* (Moscow, 1977), 214. Other towns also had significant military populations, e.g. Astrakhan with 6,500 people, many of whom engaged in trade. See ibid. 229; N. B. Golikova, *Ocherki po istorii gorodov Rossii kontsa XVII–nachala XVIII v.* (Moscow, 1982).

35. M. S. Anderson, *Peter the Great* (London, 1978), 120 ff.

36. *Ocherki* (above, n. 31), 180.

37. Ibid. 187.

38. *Ocherki istorii SSSR: Period feodalizma: Rossiya v pervoi chetverti XVIII v., preobrazovaniya Petra I* (Moscow, 1954), 322. Azov was re-ceded to the Turks in 1711 and Voronezh, with a strategic location but not the largest southern town, became provincial capital. The province was renamed in 1725.

39. The former Smolensk province was annexed to Riga, and a new province was designated around Revel.

40. For the list see *Ocherki* (above, n. 38), 326.

41. Further administrative changes were put in train soon after Peter's death. In 1726, for example, Novgorod was made the centre of a new province, independent of that of St Petersburg, and Smolensk was once again given its own province, thus being separated from Riga. In 1727 a new Belgorod province was formed out of territory formerly belonging to Kiev province.

42. Bater (above, n. 5), 17 ff. and refs.

43. The *posad* dwellers were once again permitted to move between *posady*, with official permission, from 1744.

44. Rozman, 'Comparative Approaches' (above, n. 4); V. K. Yatsunskii, 'Nekotorye voprosy metodiki izucheniya istorii feodalnogo goroda v Rossii', in *Goroda feodalnoi Rossii* (Moscow, 1966),83–9, p. 87.

45. Ibid. 87–8.

46. According to economic questionnaires distributed in the 1760s, agriculture was the only occupation of urban residents in 4% of towns, the main occupation in 67%, and important in a further 12% (B. N. Mironov, *Vnutrennii rynok Rossii vo vtoroi polovine XVIII–pervoi polovine XIX v.* (Leningrad, 1981), 202).

47. R. E. Jones, 'Catherine II and the Provincial Reform of 1775: A Question of Motivation', *Canadian Slavic Studies*, 4/3 (1970), 497–512.

48. Nineteen of the fifty provincial (*guberniya*) and sub-provincial (*provintsiya*) centres of 1719–27 did not retain their status with Catherine's reforms.

49. e.g. Vitebsk, Yekaterinoslav, Mogilev, Saratov.

50. The disestablished towns appear in the sources as 'former towns' and their inhabitants, who were usually peasants, continued in their former role. It is unclear what happened where these former towns had had merchants in their populations—the merchants may have been attached to other towns.

51. Klokman (above, n. 30), 319.
52. R. E. Jones, *Provincial Development in Russia: Catherine II and Jakob Sievers* (New Brunswick, NJ, 1984), 104; Hittle (above, n. 2), 205–6, 220.
53. Jones (above, n. 52), 94.
54. Ibid. 98.
55. J. Hartley, 'Town Government in St. Petersburg Guberniya after the Charter to the Towns of 1785', *Slavonic and East European Review*, 62/1 (1984), 61–84.
56. Klokman (above, n. 30), 207 ff.; I. de Madariaga, *Russia in the Age of Catherine the Great* (London, 1981), 622 n. 48; R. E. Jones, 'Urban Planning and the Development of Provincial Towns during the Reign of Catherine II', in J. G. Garrard (ed.), *The Eighteenth Century in Russia* (Oxford, 1973), 321–44.
57. Shkvarikov (above, n. 22), 200; Bater (above, n. 5), esp. ch. 6.
58. Yatsunskii (above, n. 44), 86–7.
59. T. S. Fedor, *Patterns of Urban Growth in the Russian Empire* (Chicago, Ill., 1975), 124.
60. De Madariaga (above, n. 56), 464–70.
61. W. L. Blackwell, *The Beginnings of Russian Industrialization, 1800–1860* (Princeton, NJ, 1968), 121–2 and chs. 5–7.
62. Fedor (above, n. 59), 57, 68; P. G. Ryndzyunskii, *Gorodskoe grazhdanstvo doreformennoi Rossii* (Moscow, 1958).
63. A. G. Rashin, *Naselenie Rossii za 100 let, 1811–1913 gg.: Statisticheskie ocherki* (Moscow, 1956), 99.
64. e.g. Bater (above, n. 5), 255, 313–14.
65. Rashin (above, n. 63), table 54, p. 93. The table does not include Poland or Finland.
66. For debate on this issue see R. H. Rowland, 'Urban In-migration in Late Nineteenth Century Russia', in Hamm, (above, n. 4), 115–24; R. L. Thiede, 'Industry and Urbanization in New Russia from 1860 to 1910', ibid. 125–38; R. A. Lewis and R. H. Rowland, 'Urbanization in Russia and the USSR, 1897–1966', *Annals of the Association of American Geographers*, 49 (1969), 776–96; R. L. Thiede, 'Urbanization and Industrialization in Pre-Revolutionary Russia', *Professional Geographer*, 25 (1973), 16–21; R. A. Lewis and R. H. Rowland, 'A Further Investigation of Urbanization and Industrialization in Pre-Revolutionary Russia', *Professional Geographer*, 26 (1974), 177–82.
67. Fedor (above, n. 59), 139 ff.
68. Ibid. 163 ff.; A. Gerschenkron, 'Agrarian Policies and Industrialization in Russia, 1861–1917', in *The Cambridge Economic History*, iv/2 (Cambridge, 1966), 706–800; M. E. Falkus, *The Industrialization of Russia, 1700–1914* (London, 1972), 47–50.
69. Bater (above, n. 5), 254; W. L. Blackwell, 'Introduction to Part Four', in Bater and French (above, n. 19), ii. 377–86, p. 384. The restrictions on the granting of passports to peasants were eased in 1894, and especially by the Stolypin Reform of 1906.

70. Quoted in H. Rogger, *Russia in the Age of Modernisation and Revolution, 1881–1917* (London, 1983), 127.
71. J. H. Bater, 'Modernization and the Municipality: Moscow and St. Petersburg on the Eve of the Great War', in Bater and French (above, n. 19), ii. 305–27; M. F. Hamm, 'The Breakdown of Urban Modernization: A Prelude to the Revolutions of 1917', in Hamm (above, n. 4), 182–199, esp. pp. 196–7.
72. Rashin (above, n. 63), 233–4; Bater (above, n. 5), 342–53.
73. Rashin (above, n. 63), 239; Bater (above, n. 71), 307.
74. Rashin (above, n. 63), 241–4.
75. Hamm (above, n. 71), 197–8; Bater (above, n. 5), 369, 380–2; L. Haimson, 'The Problem of Social Stability in Urban Russia, 1905–17', *Slavic Review*, 23 (1964), 619–42 and ibid. 24 (1965), 1–23.
76. S. F. Starr, 'The Revival and Schism of Urban Planning in Twentieth-Century Russia', in Hamm (above, n. 4), 222–42, p. 224.
77. Bater (above, n. 71); id. (above, n. 5), 353 ff.; W. Hanchett, 'Tsarist Statutory Regulation of Municipal Government in the Nineteenth Century', in Hamm (above, n. 4), 91–114; F. W. Skinner, 'Trends in Planning Practices: The Building of Odessa, 1794–1917', ibid., 139–59; see also the other relevant essays in Hamm's book. See also M. F. Hamm (ed.), *The City in Late Imperial Russia* (Bloomington, Ind., 1986).
78. Bater (above, n. 5), 391.
79. Fedor (above, n. 59), 124.
80. Rashin (above, n. 63), 77–9, table 45.
81. Another exception is Odessa, whose official designation as a free port in 1819 had 'an inestimable effect on the growth of the city' (Skinner (above, n. 77), 144).

## Notes to Conclusion

1. R. Edelman, *Proletarian Peasants: The Revolution of 1905 in Russia's Southwest* (Ithaca, NY, 1987).
2. P. Longworth, *The Cossacks: Five Centuries of Turbulent Life on the Russian Steppe* (New York, 1969); R. H. McNeal, *Tsar and Cossack, 1855–1914* (London, 1987).
3. D. W. Treadgold, *The Great Siberian Migration* (Princeton, NJ, 1957).
4. For a full description of the industrial geography of Imperial Russia in the 19th cent. see W. L. Blackwell, The Historical Geography of Industry in Tsarist Russia in J. H. Bater and R. A. French (eds.), *Studies in Russian Historical Geography* ii (London, 1983), 387–422.
5. P. Vidal de la Blache, 'Les characteres distinctifs de la geographie', *Annales de geographie*, 22 (1913), 298.

# Index

Academy of Sciences 55, 58, 60, 71, 141
Administrative Reform (1775) 75, 256, 257, 311
agriculture 2, 4, 6, 20, 25, 37, 45, 55, 60, 62, 69–75, 76, 80, 88, 90–111, 112–35 *passim*, 196, 198, 199, 201, 203, 208, 209, 220, 235, 238, 245, 265, 266, 269, 270, 292–3, 294, 311
agricultural regions 119–35, 293
Aleksandrtal 86, 87, 90–110, 290
Alexander II, Emperor 221, 263
Am Trakt 86, 87, 102, 104, 105, 107
arable farming 6, 7, 8, 21, 69–73, 75, 76, 80, 92, 95–6, 97–108, 110, 112–35 *passim*, 201, 202, 203, 204, 268, 269, 304
Archangel 199, 201, 203, 212, 214, 243, 249, 250, 251, 303
  province 114, 140 *passim*, 142–57, 160, 198, 296, 302
artisans 85, 92
Astrakhan 208, 245, 250, 251, 253, 257, 262
  province 114, 131, 302

Baku 259, 263, 270
Baltic provinces 1, 122, 130, 172, 270, 302
*barshchina* 66, 74, 139, 197, 206, 274, 287
Belgorod 42, 44, 245, 246, 253, 254, 283, 311
Belgorod Line 17, 18, 22, 27, 28, 42, 43, 44, 49, 63, 64, 65, 75, 82, 253
Belorussia 129, 302
Bessarabia 131
black earths (chernozems) 6, 8, 35, 37, 47, 48, 60, 61, 62, 72, 74, 90, 97, 104, 115, 116, 117, 130, 131, 140, 204, 208, 278, 302
Black Sea 1, 6, 60, 74, 115, 207, 215, 266, 268, 270
*bobyli, see* cottars
Boguchar 53, 61, 63, 64, 67, 68, 69, 71, 74, 248, 286, 288
Bolkhovitinov Yevgenii 58, 69, 71, 74, 76, 77, 287, 288

cadastres (*pistovye knigi*) 18, 38, 39, 40, 41, 51, 282
Catherine II 11, 47, 48, 52, 55, 58, 64, 75, 76, 81–2, 165, 197, 252, 254, 256, 258, 263, 264, 266, 311
Central Asia 1, 86, 269, 305

Central Black Earth Agricultural Region 12–32, 33, 35, 66, 74, 119, 203, 207, 208, 209, 211, 212, 214, 220, 249, 253, 267, 268
Central Industrial Region 126, 203, 204, 207, 208, 211, 212, 213, 214, 216, 219, 220, 256, 302
Chayanov, A. V. 54, 119, 132
Chelintsev, A. N. 112, 119, 120–31, 132, 135, 138
Chernigov province 72, 82, 130
cities, *see* towns
climate 5–8, 60, 68, 86, 90–1, 96–7, 98, 99, 103, 105, 108, 115, 195, 213
colonies (*kolonii*) 73, 81, 83–111 *passim*, 269, 274, 290
colonization 2, 3, 7, 13, 23, 25, 45, 66, 67, 68, 78, 82, 122, 196, 209, 266, 267, 268, 269, 270
common pasture 93, 108, 109, 136, 153, 154, 160, 165, 190
communal assembly (*skhod*) 147, 148, 152, 156, 162–3, 166, 170, 171, 275, 296
Cornies, Johann 88, 110
cossacks 8, 17, 20–32 *passim*, 36, 37, 38, 40, 41, 42, 44, 48, 53, 64, 69, 73, 74, 77, 82, 199, 212, 245, 267, 268, 269, 274, 280, 287
cottars 38, 40, 41, 42, 44, 148, 274, 307
crop rotations 71, 95, 97, 98, 99, 100, 101, 102, 111, 115, 116, 117, 118, 130

Dokuchaev, V. V. 5
Don cossacks, land (province) of 131, 208, 209
Don r. 17, 22, 23, 27, 28, 30, 40, 58, 60, 67, 68, 74, 114
Dnepr r. 17
drought 96, 97, 103, 104, 269

Elizabeth, Empress 197, 252
Emancipation Statute 11, 137, 157, 167, 183, 256, 259, 262, 268, 269, 272, 278

fairs 56, 62, 74, 75, 76, 77, 197, 198, 201, 202, 205, 208, 209, 213, 214
fallow 71, 73, 97, 98, 99, 101, 102, 103, 104, 106, 108, 111, 113–19, 121, 122, 125, 126, 129, 160, 275, 292
famine 22, 28, 30, 221, 269
feudalism 29, 30, 45, 133